T0396146

High Plasticity Magnesium Alloys

High Plasticity Magnesium Alloys

FUSHENG PAN

Chinese Academy of Engineering, Beijing, P.R. China

National Engineering Research Center for Magnesium alloys (CCMg), Chongqing University, Chongqing, P.R. China

BIN JIANG

National Engineering Research Center for Magnesium alloys (CCMg), Chongqing University, Chongqing, P.R. China

JINGFENG WANG

National Engineering Research Center for Magnesium alloys (CCMg), Chongqing University, Chongqing, P.R. China

YAOBO HU

National Engineering Research Center for Magnesium alloys (CCMg), Chongqing University, Chongqing, P.R. China

SUQIN LUO

School of Materials Science and Engineering, Chongqing Jiaotong University, Chongqing, P.R. China

ELSEVIER

Elsevier
Radarweg 29, PO Box 211, 1000 AE Amsterdam, Netherlands
The Boulevard, Langford Lane, Kidlington, Oxford OX5 1GB, United Kingdom
50 Hampshire Street, 5th Floor, Cambridge, MA 02139, United States

Notices
Knowledge and best practice in this field are constantly changing. As new research and
experience broaden our understanding, changes in research methods, professional practices, or
medical treatment may become necessary.

Practitioners and researchers must always rely on their own experience and knowledge in
evaluating and using any information, methods, compounds, or experiments described herein. In
using such information or methods they should be mindful of their own safety and the safety of
others, including parties for whom they have a professional responsibility.

To the fullest extent of the law, neither the Publisher nor the authors, contributors, or editors,
assume any liability for any injury and/or damage to persons or property as a matter of products
liability, negligence or otherwise, or from any use or operation of any methods, products,
instructions, or ideas contained in the material herein.

British Library Cataloguing-in-Publication Data
A catalogue record for this book is available from the British Library

Library of Congress Cataloging-in-Publication Data
A catalog record for this book is available from the Library of Congress

ISBN: 978-0-12-820110-7

For Information on all Elsevier publications
visit our website at https://www.elsevier.com/books-and-journals

Publisher: Matthew Deans
Acquisitions Editor: Glyn Jones
Editorial Project Manager: Naomi Robertson
Production Project Manager: Prem Kumar Kaliamoorthi
Cover Designer: Christian Bilbow

Typeset by MPS Limited, Chennai, India

Contents

Preface

Magnesium (Mg) alloys have attracted significant attention for their abundant resource, lightweight potential, and functional characteristics. There are abundant Mg resources in the world with more than 30 billion tons of minerals, including dolomite, magnesite, and magnesium chloride in salt lakes and sea water, which have been estimated to be exploitable for more than a thousand years. The density of Mg is $1.74\,\mathrm{g\,cm^{-3}}$, which is about two-thirds of Al and one-fourth of Fe. Thus Mg alloys show great lightweight potential for use in structural parts. In addition, Mg and its alloys exhibit good damping properties, electromagnetic shielding properties, biocompatibility, large hydrogen storage capacity, and a high theoretical specific capacity for batteries. Thus Mg alloys are considered to have great application prospects in many fields, such as aerospace, transportation, 3C, construction, biomedical, and energy fields.

However, Mg alloys have a hexagonal close-packed crystal structure with only three feasible slip systems, which can be activated at room temperature, resulting in poor room temperature plasticity. Besides, basal texture is easily formed in Mg alloys after plastic deformation, further resulting in low plasticity. The low plasticity extremely limits the wider application of Mg alloys; thus, much effort has been made to improve the plasticity of Mg alloys. In the past decade, many new Mg alloys with high plasticity and many new technologies to improve plasticity with enough strength have been developed. The high-plasticity magnesium alloys in this book are referred to as the as-cast magnesium alloys with a plasticity higher than 10% and the wrought magnesium alloys with a plasticity higher than 15%.

In the past two decades a new alloying theory of Mg alloys called the solid solution strengthening and ductilizing (SSSD) theory, which simultaneously improves plasticity and strength, was proposed. Many high-plasticity Mg alloys have also been developed successfully. The purpose of this book is to introduce the new theories for designing alloys with high plasticity, and the development of new alloys based mainly on the theory. The book comprises six chapters. The first chapter gives a brief overview of high-plasticity Mg alloys, some factors that influence the plasticity, and testing and characterization of plasticity for magnesium alloys. The second chapter describes the SSSD theory, which provides the fundamentals of alloy designing for Mg alloys with high plasticity. The following four chapters display the alloy designing and processing technologies of four kinds of Mg alloys with high plasticity, which are Mg−Gd−Zr alloys with ultra-high plasticity, Mg−Mn and Mg−Sn based alloys with medium strength and high plasticity, and Mg−Gd−Y−Zn−Mn alloys with ultra-high strength and high plasticity, respectively.

I would like to take this opportunity to sincerely appreciate all the chapter contributors for sharing their work and knowledge. They are Prof. Bin Jiang, Prof. Jingfeng Wang, Prof. Yaobo Hu, Prof. Xianhua Chen, Prof. Dingfei Zhang, Prof. Aitao Tang, Prof. Jian Peng, Dr. Suqin Luo, Dr. Jia She, Dr. Hanwu Dong, Dr. Ying Zeng, Dr. Tingting Liu, Dr. Zhengwen Yu, Dr. Peng Peng, Dr. Song Huang, Dr. Shijie Liu, Dr. Xiuhong Huang, Dr. Wenhui Yao, Mr. Tianshuo Zhao, Mr. Juan Deng, Mr. Tianxu Zhen, Mr. Chao Zhang, Mr. Shida Ma, etc. I would like to give my special thanks to Prof. Jiangfeng Song who reviewed and proofread the whole book. I also extend my thanks to the publishing staff from Elsevier and Chongqing University Press. Without their great efforts devoted to this book, the completion of this book would not have been possible.

Based on the principle of the SSSD of alloys, a new method of strengthening and ductilizing by the addition of dissimilar metals is being developed by my group for magnesium-based composites. This related work will be discussed in another book.

Fusheng Pan
Chongqing University
Chongqing, China

CHAPTER 1

Overview

1.1 High-plasticity magnesium alloys and their processing technologies

Among metal structural materials, plastic-processed products account for more than 70%. In addition to the steel and aluminum (Al) alloys, magnesium (Mg) alloys, especially wrought Mg alloys are important structural engineering materials. Various lightweight and functional parts of magnesium alloys are produced by plastic forming technologies, which will provide the characteristics of high material utilization, good product appearance and internal quality. The forming products of magnesium alloys introduce impressive weight-reducing effects in the automobile, railway transportation, 3C, defense, aerospace, and general machinery areas. However, the ductility of Mg alloys is low at room temperatures. First, Mg alloy has a hexagonal close-packed (HCP) crystal structure. Only three slip systems on the (0001) basal plane and along with the directions of $[11\bar{2}0]$, $[2\bar{1}\,\bar{1}0]$, and $[\bar{1}2\bar{1}0]$ can be activated at room temperature (R. T.). As the slip in $[2\bar{1}\bar{1}0]$ direction can be divided into those along with the other two directions, Mg alloy has two independent slip systems at R. T. The number of independent slip systems in Mg alloys can not satisfy the von-Mises criterion for polycrystalline material that requires five independent slip systems during the homogeneous plastic deformation. Second, two independent slip systems are on the same plane, and the critical resolved shear stress (CRSS) of basal slip is lower than that of prismatic and pyramidal slip. The basal texture is easily formed in Mg alloys after plastic deformation, resulting in low ductility. Therefore Mg alloy products need to be heated, processed, and annealed many times to obtain high mechanical properties, leading to long procedures, low yield, and high overall cost, which greatly hinder the large-scale application of magnesium alloys. Recent research of magnesium alloys focuses on developing new alloys and new processes to improve ductilities with enough strengths.

High-plasticity magnesium alloys refer to the as-cast magnesium alloys with an elongation higher than 10% and the wrought magnesium alloys with that higher than 15%. The typical high-plasticity magnesium alloys include Mg−Al, Mg−Zn, Mg−Mn, Mg−RE, Mg−Li, and Mg−Sn alloys, among which Mg−Al and Mg−Zn alloys are the most common ones. The compositions, states, phases, and mechanical properties of typical high-plasticity magnesium alloys reported in the literature are listed in Table 1.1.

High Plasticity Magnesium Alloys
DOI: https://doi.org/10.1016/B978-0-12-820110-7.00001-X

Table 1.1 Phases and mechanical properties (R.T.) of Mg alloys with high plasticities.

Alloy	State	Phase	Rm/ xxMPa	Rp0.2/ MPa	A/%
AZ31	Extruded+T4, thickness 0.95 mm	α	254.3	157.6	16.68
	Grain size 17.3 μm	α	275	152	22.0
AZ61+1.0Ce	Annealed	$\alpha+\beta$	—	166.88	16.5
AZ61	Grain size 9.9 μm	$\alpha+\beta$	320	175	19.8
AZ81A	T4	α	275	83	15
AZ91D	T4	α	275	90	15
	T6	$\alpha+\beta$	275	145	6
AZ91	Grain size 9.9 μm	$\alpha+\beta$	395	225	18.2
AM20	Die cast	$\alpha+\beta+MnAl_4/MnAl_6$	210	90	20
AM50A	Die cast	$\alpha+\beta+MnAl4/MnAl_6$	230	125	15
AM60A, AM60B	Die cast	$\alpha+\beta+MnAl4/MnAl_6$	240	130	13
AE41	Die cast	$\alpha+\beta+Mg_{12}Ce$	234	103	15
AE42	Die cast	$\alpha+\beta+Mg_{12}Ce$	244	110	17
AS21	Die cast	$\alpha+\beta+Mg_2Si$	220	120	13
AS41A, AS41B	Die cast	$\alpha+\beta+Mg_2Si$	240	140	15
ZK60	Ex. ratio 16	$\alpha+MgZn$	351		17.2
ZK60+0.94Y	Ex. ratio 16	$\alpha+MgZn$	386.5		15
ZC62	Die cast	$\alpha+MgZn$	227	119	11.0
ZC62	T5	$\alpha+MgZn$	237	138	9.5
WE43A	T5	$\alpha+\beta+Mg_9Nd$	195	270	15
	T6	$\alpha+\beta+Mg_9Nd$	160	260	15
WE43A+0.42Zr	T6	$\alpha+\beta+Mg_9Nd$	253	—	25
Mg−Nd−Zr	Extruded	$\alpha+Mg_9Nd$	204	132	27.0
	Extruded+T5	$\alpha+Mg_9Nd$	238	142	24.5
	Extruded+T6	$\alpha+Mg_9Nd$	249	102	20.6
K1A	F	α	180	55	19
MB8	Annealed	$\alpha+\beta$ (Mn)	250	170	18
Mg−4Li−1Al-	As-cast	α	157		17
Mg−8Li−1Al-	As-cast	$\alpha+\beta$	131		35
Mg−8.7Li	As-cast, 350°C Ex.	$\alpha+\beta$	132	93	52
Mg−10.6Li−1.57Al	Plate, T6	β	117	100	40
Mg−10.8Li−3.44Al−4.96Zn	Rolled	β	267	200	29
	Rolling+annealing	β	273	207	15
LA141	Stripe, T7	β	145	125	23
	As-cast	β	122	85	17

Alloying can improve the mechanical properties of magnesium alloys by the solid solution effects and second phases. Second-phase compounds formed in magnesium alloys, except those in Mg−Li alloys, are brittle and hard, which are harmful to plasticity. The adverse effects of these phases are related to their properties, crystalline structure, shape, size, amount and distribution. The more the amount and the larger the particles' size, the more harmful to the ductility. Smooth spheres are the least harmless. The sharper their surfaces, the more harmful, since cracks incline to originate from here. Particles and strips are less harmful than plates and pieces. The most damaging effects to the plasticities are often introduced by net-like intergranular structures. The second phase should be decreased in amount and size and distributed dispersedly to improve magnesium alloys' ductility. Especially, net-like structures of the second phases are not appreciated. Plastic second phases and alloying metal particles are beneficial to the elongations of magnesium alloys.

1.2 Mg—Al alloys

Mg—Al alloys are the most widely used magnesium alloys. They belong to the wrought Mg alloys with moderate strengths, relatively high plasticities, reasonably high corrosion resistance, and low price. In Mg—Al alloys, some of AZ, AM, AE alloys are of high plasticity (Table 1.1). For example, in AZ series magnesium alloys, AZ31 and AZ61 alloys have high plasticities, high strength, and appropriate corrosion resistance. AZ31 and AZ61 alloys have elongations higher than 19%.

The content of the Al element is less than 10 wt.% in Mg—Al alloys. The as-cast magnesium alloy cooled with a typical rate consisting of α(Mg)+β(Mg$_{17}$Al$_{12}$) at R. T. due to nonequilibrium solidification. The amount of β(Mg$_{17}$Al$_{12}$) phase increases with increasing the content of Al in Mg—Al alloys. Microalloying elements, such as Ca, Ti, Bi, Sb, Sn, Sr and rare earth (RE) elements, can change the morphology of β-Mg$_{17}$Al$_{12}$ phase, such as the amount, size, distribution, orientation. Microalloying also leads to the refinement of α-Mg grains and results in the improvement of mechanical properties. During the solidification, microalloying elements concentrate at the solid—liquid interfaces and hinder the enlargement of the grains and β-Mg$_{17}$Al$_{12}$ phase, and thus the grains of α-Mg matrix are refined. On the other hand, some alloying elements combine with Mg to form second phases. This competition weakens the combination of Mg with Al element, which decreases the amount of β-Mg$_{17}$Al$_{12}$ and improves the distribution of this phase. The second phases with low contents in the alloy, such as Al$_2$Ca, Mg$_3$Bi$_2$, Mg$_3$Sb$_2$, A1-RE, present to be needle-like or particles. These second phases have high thermostabilities and improve the properties of the alloys at elevated temperatures. Microalloying elements also can dissolve in the β-Mg$_{17}$Al$_{12}$ phase and enhance its thermostability. However, many microalloying elements will bring abundant second phases and form a net-like structure, which decreases the mechanical properties. It is better to add some elements with a small amount of each element, improving Mg—Al alloys' strength and ductility simultaneously. For Mg—Al alloys with Al content lower than 10 wt.%, the β-Mg$_{17}$Al$_{12}$ phase can dissolve in the α-Mg matrix after the solution treatment, resulting in an improvement in the plasticity of the alloy. For example, after solution treatment, the elongation of AZ91D-0.41Sm alloys increased to 14.5%, which is 150% of the as-cast state. Mg—6Al—Zn—0.9Y—1.8Gd alloy reached an elongation of 22.3% with the strength of 254.8 MPa.

Mg—Al—Mn (AM) alloys with high plasticity are often used to produce the structural part that bears the high loadings. AM series alloys, such as AM20, AM5, and AM60 alloys, have elongations of 20%, 15%, and 15%, respectively. At room temperatures, the phases in Mg—Al—Mn alloys are $\alpha(Mg)+\beta(Mg_{17}Al_{12})+MnAl$ when the Mn content is below 1 wt.%. The brittle β-Mn phase will appear as the Mn content is enhanced to more than 1 wt.%, which often decreases the ductility. Unlike the situations of AZ alloys, the plasticity of Mg—Al—Mn alloys do not be enhanced by solution treatment, and while they will be improved by aging treatment after solution. Alloying elements, such as Sr, Nd, Ce, etc., with proper contents, can refine the grain size of AM magnesium alloys and improve the microstructure and mechanical properties. With the addition of 0.5 wt.% Sr, AM50 alloy presents an ultimate tensile strength (UTS) of 233 MPa and a relatively high elongation of 16.3%, and a high elongation of 20% will be achieved with 1.5 wt.% Ce. The ductility of AM60 alloy increases to 18% with the addition of 4 wt.% Nd element, while adding Ti and Sc elements show no enhancement. Compared with the addition of single alloying elements, complex alloying with several elements can decrease the consumption of alloying metal and introduce better properties. AM60—1.6RE-0.15B alloy presents an elongation of 18%, while alloying with Sr or Ce shows little enhancement to Nd-containing AM60 alloy's plasticity. After the addition of 0.05 wt.% carbon fiber, the UTS of AM60 can reach 242.4 MPa, with the elongation maintaining 13.2%.

Mg—Al—RE (AE) alloys have good creep and heat resistant properties. Some of these alloys show favorable plasticities. Mischmetal (MM) can remarkably improve the creep resistant ability of Mg—Al alloys, especially when the Al content is lower than 4 wt.%. However, more MM will decrease the fluidity of alloy melts significantly and make the alloy hard to be cast. Thus at the beginning stage of developing AZ magnesium alloys, Al and RE contents are 2%—4% and 1%—2%, respectively. The typical alloys are AE41, AE42, and AE21. AE42 alloy with an elongation of 17% belongs to the alloys of high ductility and high comprehensive mechanical properties. When the Al content in AE alloy is higher than 4 wt.%, alloying elements such as Ca, Sm, Sb are usually used to improve the mechanical properties. The strengthening effects in Mg—Al—RE alloys originate from two aspects. One is Al and RE elements' reaction, leading to Al—RE binary phases, such as $Al_{11}RE_3$, with high melting points, resulting in decreased $Mg_{17}Al_{12}$ of low melting point and the rearrangement of the second phases. The other is the dissolving of RE atoms in the Mg matrix, which hinders the dislocation motions and the diffusion of atoms, leading to the pinning of grain boundaries and dislocations, resulting in the enhancement of strength. Since there are many Al—RE second phases with high thermostability, AE series magnesium alloys have good mechanical properties after heat treatment. For example, after solution treatment following with aging, AE51—0.5Sb and AE51—0.8Ca alloys show UTS of 241 and 232 MPa, elongations 11%—12%. AE51—1.0Sm alloy shows improved properties of 244 MPa in UTS and 15.6% in elongation.

1.3 Mg–Zn alloys

Among the wrought Mg–Zn alloys, ZK60 alloy has the best mechanical properties with high strength and ductility. The extruded ZK60 alloy has an elongation of more than 17%. Therefore ZK60 alloy is widely used as structural materials.

In Mg–Zn alloys, the Zn element introduces reliable solution strengthening effects and improves plasticity as well. The addition of RE and Zr elements in Mg–Zn alloys is also beneficial to grain refinement and ductility enhancement. The Zr addition of more than 0.5 wt.% in Mg–Zn alloys can act as nucleating cores during solidification to improve the nucleation rate. Thus the grain refinement and the strength improvement are introduced to the as-cast alloys, and the plasticity can be enhanced evidently, seen in Table 1.2.

The RE elements added to Mg–Zn alloys are usually Y, Nd, Gd, Ce, La, etc. These elements enhance the mechanical properties at room temperatures by purifying the alloy melts and improving the microstructures. The as-cast Mg–Zn alloys show 7%–12% elongations and good strengths higher than Mg–Al alloys. After adding RE elements and plastic deformation, Mg–Zn alloys present elongations of more than 20%.

In Mg–Zn–Zr alloys, RE elements introduce divorced eutectic Mg–Zn–RE compounds, locating at grain boundaries. The enhancement of RE contents in a specific range increases the amount of the eutectic, and thus the mechanical properties of the as-cast alloys are primarily improved. For Mg–Zn–Zr alloys, the addition of 0.8

Table 1.2 Mechanical properties of rare earth-containing Mg–Zn alloys with relatively high ductilities.

Alloy	State	$\sigma_{0.2}$/MPa	σ_b/MPa	δ (%)
Mg–4.5Zn–2Gd	Aged	215	121	6.43
Mg–4.5Zn–2Nd	Annealed	228	79	11.8
Mg–3.5Zn–1.0Gd	Ac-cast	171	81	6.7
Mg–4.3Zn–0.7Y	Hot-rolled	370	220	19.7
Mg–4.3Zn–0.7Y–0.2Zr	Hot-rolled	325	180	23.5
Mg–5Zn–2Nd–0.5Y–0.6Zr	Ac-cast	210	100	9
Mg–5Zn–2Nd–1Y–0.6Zr	Ac-cast	220	105	12
Mg–1.73Zn–1.54Y	Hot-rolled	268	214	27
Mg–4Zn–0.6Y	Hot-rolled	300	–	25
Mg–3.9Zn–0.7Zr-1.7Ce	Ac-cast	230	–	5
Mg–3.9Zn–0.7Zr–1.7Ce	Extruded	375	–	8
Mg–5Zn–3Gd–0.6Zr	Ac-cast	200	100	5.8
Mg–5Zn–3Gd–0.6Zr	T4	230	109	10.5
Mg–5Zn–0.6Zr–Nd	Ac-cast	195	100	7
Mg–5Zn–0.6Zr–2Nd	Ac-cast	135	95	3
Mg–4.5Zn–0.9La	Ac-cast	154	–	8.3

wt.% Nd largely improves the ductility, with a little enhancement of strength. Mg−Zn−Zr alloys with 1 wt.% Nd have the highest strength and elongation, with the value of 11.8%. When the Nd content is increased to more than 1 wt.%, the amount of Mg−Zn−Nd phase increase to form continuous nets, and the ductility decreases.

For RE elements, Y is the most often used elements that added to Mg−Zn binary alloys. There are three phases in Mg−Zn−Y ternary alloys: long period stacking ordered (LPSO) X-$Mg_{12}YZn$ phase, cubic W-$Mg_3Zn_3Y_2$ phase, and pentagonal dodecahedron quasi-crystal I-Mg_3Zn_6Y phase. The second phases formed in Mg−Zn alloys are related to molar ratios of Y/Zn. With the decrease of Y/Zn molar ratio, the phases in Mg−Zn alloys are changed from X-phase to W-phase and finally to I-phase. The molar ratio or its range related to the phases are like these: the phases are (α-Mg +I) at the Y/Zn molar ratio of 0.164, (α-Mg+I+W) phase presents at the ratio range of 0.164−0.33, (α-Mg+W) phases exist when the ratio increases to 0.33, (α-Mg+W +X) phases come into being at the ratio range of 0.33−1.32, and the phases are (α-Mg + X) at the ratio of 1.32. A small amount of I-phase is beneficial to the dynamic recrystallization, leading to grain refinement and plasticity improvement. The LPSO X-phase can accommodate strain and coordination plastic deformation, which benefits to the enhancement of the ductility. A large amount of W-phase is harmful to the strength and ductility and introduces disadvantages to Mg−Zn alloys. For example, X-phase can markedly increase the UTS and YS of the extruded ZK20 alloy without sacrificing the ductility. With the increase of X-phase, the extruded ZK40−11.67Y alloy shows a high UTS of 408 MPa and a high YS of 300 MPa, which are relatively higher than those of the extruded ZK20+3.67Y alloy. Although the elongation of ZK40−11.67Y alloy decreased to 7.5%, it is still beyond 5%.

In Mg−Zn alloys, with the increase of Gd content, the second phase changes from (I-phase+Mg_7Zn_3 phase) to I-phase, finally to (I-phase+W-phase). The grain sizes are decreased with an evident decrease of second dendritic distances. The shape of the intergranular structure changes from particle-like and thread-like to close-net-like. At the Zn/Gd ratio of 5.8, the ratio of the second phases does not change, and the amounts of the phase and those of intergranular phases increase with the Gd content. In Mg−Zn−Gd alloys with a certain Zn content, the UTS and elongation increase with the Gd content, and while there is a peak value for the YS. On the other hand, in Mg−Zn−Gd alloys with a certain Zn/Gd ratio, the strength increases with decreased ductility when increasing the Zn and Gd contents. $Mg_{95.5}Zn_{3.5}Gd_{1.0}$ alloy possesses the best comprehensive mechanical properties, with a maximum elongation of 6.7%. With a Gd addition of 2 wt.%, the strength reaches the highest, with a UTS as high as 215 MPa and an elongation of 6.43%. As the Gd content increases to more than 2 wt.%, the increase of $Mg_3Gd_2Zn_3$ phase and the coarsening of the grains leads to the decrease of both strength and ductility.

In Mg—Zn alloys, La element exists as the $Mg_{12}La$ phase, a strengthening phase with a high melting point. $Mg_{12}La$, with a high thermo-stability, mainly distributes in grain boundaries and inhabits the grain growth and grain slides. With the increase of La content (0.3, 0.6, and 0.9 wt.%), the grains of the as-cast Mg-4.5Zn alloy are refined, and the strength and plasticity are both improved. When the La content is at 0.9%, the alloy elongation reaches 8.3%.

1.4 Mg—Mn alloys

The Mg—Mn alloy has good corrosion resistance, weldability, and plasticity, but its strength is low. It can be used to manufacture parts with little bearing capacity but with high corrosion resistance and good weldability. Mg—Mn alloys have a peritectic reaction, and no compound can be formed between manganese and magnesium. Adding a small amount of RE to the Mg—Mn alloy can refine the grains, purify the grain boundaries, and further improve the plasticity. Typical Mg-(1.5—2.5)Mn-0.4Ce belongs to wrought magnesium alloy. After proper heat treatment, the strength can reach 250 MPa, and the elongation is higher than 20%. It has an extensive application prospect, and can be used to produce plates, bars, profiles, forgings and so on.

The researches of Mg—Mn alloys can be originated from 1960s. Researchers found that after the aging treatment, short-rod-like α-Mn phases occur in Mg—Mn alloys, with the phase relation of $\{111\}_{Mn}//(0002)_{Mg}$, $\{1\bar{1}0\}_{Mn}//\{10\bar{1}0\}_{Mg}$, $\{111\}_{Mn}//\{2\bar{1}\,\bar{1}0\}_{Mg}$ and $\{110\}_{Mn}//\{0001\}_{Mg}$. These precipitates are mainly parallel or vertical to the basal plane of the Mg matrix. Casually, they are polygon α-Mn particles, whose relationships are random to the Mg matrix, and neither parallel nor vertical, without preferences. Since the second phases parallel to the matrix show no quite evident effects in strengthening, Mg—Mn alloys are not good at acquiring strengthening effects from heat treatment.

With the addition of Mn element, the corrosion resistance of Mg alloys is markedly enhanced, especially at the Fe/Mn ratio is about 0.02. In AZ series magnesium alloys, 0.3—0.5 wt.% Mn is usually applied to improve the corrosion resistant ability. Mn addition introduces many Mn particles in Mg alloys, which effectively hinders the dislocation slips and improve the creep resistant ability, leading to an outstanding creep resistant ability to Mg alloys at elevated temperatures.

Usually, the effects of Mn element on the grain refinement of as-cast Mg alloys are not appreciated, and the mechanical properties at room temperatures are not quite enhanced. Mg—Mn binary alloys show relatively coarsen grain after deformation processes and low yield strength and plasticity at room temperatures. Table 1.3 lists the room temperature mechanical properties of Mg—Mn alloys after deformed at different processes.

Table 1.3 Room temperature mechanical properties of Mg−Mn alloys after different deformation processes.

Alloy	Ratio	Speed (m/min)	Temperature (K)	YS (MPa)	UTS (MPa)	Elo. (%)	Average grain size (µm)
Mg−0.99Mn	30	10	573	189	261	5.3	70
Mg−0.99Mn	30	1	573	183	243	6.1	8
Mg−0.90Mn	7	−	623	142	215	2.9	85
Mg−1.62Mn	30	−	648	∼210	∼270	7.0	23

The additions of some alloying elements, such as Al, Ca, Ce, Gd, Nd, etc., benefit the yield strengths and plasticity of Mg−Mn alloys. Al element introduces effects markedly to the microstructures and mechanical properties of Mg−1Mn alloy. With the increase of Al content, the extruded alloys show complete recrystallized microstructures and refined grains. The basal texture is largely weakened. The mechanical properties at room temperatures are obviously improved. After adding the Ca element, Mg−1.3Mn alloy shows to be in a complete recrystallized state, and the size of the recrystallized grains decreases rapidly with the increase of the Ca content. When the Ca content in Mg-1.3Mn alloy is higher than 0.5 wt.%, the basal texture is markedly weakened. The orientations incline to be the $\langle 11\bar{2}1 \rangle$ direction of the grains which are parallel to the extrusion direction, and the dislocations of basal planes are easy to be activated by forces, which makes the room temperature strength and ductilities to be not appreciated. The additions of a small amount of Ce and La elements in Mg−1.62Mn (ME10) alloy can refine the grains markedly, and the basal texture is obviously weakened. As the situation of Ca element, the alloys with Ce and La additions also show $\langle 11\bar{2}1 \rangle$ directions of the grains parallel to the extrusion direction. The ductility at room temperature is markedly improved to about 20%, with a low yield strength of about 130 MPa. Summarily, the additions of alloying elements, such as Ca, Ce, La, etc., to Mg−Mn alloys can weaken the basal texture and make the $\langle 11\bar{2}1 \rangle$ orientations of the grains to be parallel to the extrusion direction, usually called "rare earth texture," and beneficial to the starting of the dislocations of the basal planes, which improves the plasticity at room temperature and the yield strength as well.

In Mg−1Mn alloy, the $Mg_{17}Sr_2$ phase, inherited from the additions of Sr element, often acts as the cores of the heterogeneous nucleating for the recrystallization grains during the deformation processes, which is beneficial for complete recrystallization. On the other hand, as the increase of the Sr content, there is an increase in the amount of $Mg_{17}Sr_2$ particles and also in the recrystallization level. The recrystallization grains are markedly refined, and the ductility at room temperature is obviously decreased. The yield strength, UTS, and elongation are ∼210, ∼250 MPa, ∼4%, respectively. Moreover, the speed and temperature of the extrusions show large effects on the texture of Mg−1Mn−1.3Sr and Mg−1Mn−2.1Sr alloys. After the extrusion of $1 \, \text{m} \, \text{min}^{-1}$ at 300°C, the orientations of the grains present a strong basal texture, and the alloys are not completely recrystallized. With the increase of the extrusion speed and temperature, the basal texture is largely weakened, with randomly distributed grain orientations, and the recrystallization is at a high level, which is good for the improvement of mechanical properties at room temperatures.

After the addition of 1.0 wt.% Nd and the following extrusion at 300°C, the grains of Mg−1Mn alloy are markedly refined, and the grain size increases rapidly with the increase of the extrusion speed. Since the $Mg_{41}Nd_5$ phase introduces particle stimulating nucleation (PSN) effect, the basal texture in the alloys is significantly weakened, and the grain orientations are quite random, which is good for the improvement of the ductility at room temperatures. Therefore after extruding, Mg−1Mn−1Nd alloy possesses an extraordinary room temperature plasticity, and with the increase of Nd content, the basal texture is largely weakened, leading to a markedly improved elongation.

The studies of Mg−2Mn−1Ce alloy manifest that with the increase of the extrusion temperature, the recrystallization level increases with an obviously weakened basal texture. The grain orientations incline to be $\langle 11\bar{2}2 \rangle$ and $\langle 20\bar{2}1 \rangle$ directions. The basal slips are easy to be activated, which leads to the enhancement of the ductility at room temperatures. The recrystallization behaviors are mainly caused by the PSN effects of $Mg_{12}Ce$ precipitates during the deformation process. Since the massive Mn particles precipitated in the matrix are quite small in size, they can not act as the heterogeneous nucleating cores to lead to the nucleating and the enlargement of recrystallization grains.

Researchers from Chongqing University and Yanshan University have carried out some studies on Mg−Mn alloys. Zhang et al. from Chongqing University have studied the recrystallization behaviors of Mn particles in Mg−Mn alloys during the deformation processes at elevated temperatures. In Mg−Mn alloys, the Mn element introduces little solid solution strength effects. However, during the deformation at elevated temperatures, size of the precipitated Mn particles are high enough to act as the heterogeneous nucleating cores of recrystallization. These particles are beneficial for the nucleating and enlargement of the recrystallization grains and weaken the basal texture, leading to the improvement of ductility at room temperatures.

Peng et al. from Yanshan University has studied the effects of Sn and Ce elements on the microstructure and mechanical properties of Mg−Mn alloys. It is found that with the increase of Sn content, an increasing amount of Mg_2Sn phase, precipitated in the alloy to refine the grain and improve the strength. The Ce addition shows to weaken the basal texture and decrease the yield strength, while the plasticity is improved.

The damping properties of Mg−Mn alloys are examined by Wang et al. from Chongqing University. In Mg−0.44Mn alloy, the massive Mn particles hinder the dislocation effectively and enhance the damping properties. Nevertheless, the increase of Mn content enhances the size of Mn particles and is harmful to the improvement of the damping properties.

Recently, Fusheng Pan's team from Chongqing University put great attention on Mg−Mn alloys. They have proposed "solid solution strengthening and ductilizing (SSSD)" theory. By controlling recrystallization via Mn particles, they developed a series of high plastic Mg−Mn alloys with elongations of higher than 30%. The details are seen in the following chapters.

1.4.1 Mg−RE alloys

The addition of RE elements can largely improve the strength, ductility, heat resistant, and crossing abilities of Mg alloys. Y and Gd elements are the most often used ones in wrought magnesium alloys, with the maximum solid solution of 12.4 and 23.5 wt.%, respectively. After they are dissolved in the Mg matrix, the axis ratio of Mg crystalline is effectively decreased, and the ductility is enhanced. For example, Mg−Y−Nd−Zr alloy has a high elongation of more than 20% after extrusion and heat treatment, and Mg−Gd−Zr alloys can achieve an elongation as high as more than 40% (Details can be seen in Chapter 3: Ultra High Plasticity Mg−Gd−Zr Alloy).

Among Mg−Gd alloys, Mg−Gd−Y−Zr alloys have attracted much attention. Table 1.4 lists the room temperature strengths and elongations of Mg−9Gd−4Y−0.6Zr alloy after different heat treatments and deformation processes. It can be seen that all

Table 1.4 Strengths and elongations at room temperature of Mg−9Gd−4Y−0.6Zr alloy after different heat treatments and deformation processes.

Alloy state	UTS/MPa	YS/MPa	Elo./%
As−Cast	256	209	2.3
Cast-T4	228	177	4.6
Cast-T5	301	262	2.1
Cast-T6	327	279	3.3
As-extruded	312	274	4.8
Ext-T4	238	187	5.7
Ext-T5	370	319	4
Ext-T6	347	293	5.2
Ext+0%pre-def	295	186	26
Ext+4% pre-def	298	265	20.4
Ext+8% pre-def	310	295	15
Ext+12% pre-def	312	305	12.7
Ext+0% pre-def +T5	335	225	16
Ext+4% pre-def +T5	395	330	14.5
Ext+8% pre-def +T5	410	365	12.8
Ext+12% pre-def +T5	405	350	8

Table 1.5 The maximum strengths and elongations at room temperature of Mg−Gd alloys after ageing treatment (Reported in recent years).

Alloys	UTS/MPa	YS/MPa	Elo./%	Alloy State
Mg−7Gd−xY(x=0∼5%)	145−258	81−167	5.2−8.4	T6
Mg−13.5Gd−0.4Zr	298	231	17.1	Ext+T5
Mg−8Gd-xZn−0.4Zr (x=0%−3%)	253−314	171−217	11−17	Ext+T5
Mg−7Gd−4Y−0.6Zr	342	291	6.1	Ext+T5
Mg−9Gd−4Y−0.6Zr	370	319	4	Ext+T5
Mg−12Gd−3Y−0.4Zr	457.6	342.8	3.8	Ext+T5
Mg−15Gd−5Y−0.5Zr	276.9	254	0.5	T6
Mg−8.31Gd−1.12Dy−0.38Zr	355	261	3.8	T6
Mg−9Gd−4Y−0.65Mn	336	310	11.2	Ext+T6
Mg−18.6Gd−1.9Ag−0.24Zr	383.5	291	1.17	T6
Mg−9Gd−3Y−0.6Zn−0.5Zr	430	375	9.5	Ext+T5
Mg−12Gd−4Y−2Nd−0.3Zn−0.6Zr	310	280	2.8	T6
Mg−2Gd−0.6Zr	206	150	36.8	Ext
Mg−4Gd−0.6Zr	207	145	43.4	Ext
Mg−6Gd−0.6Zr	237	168	33.4	Ext
Mg−6Gd−0.6Zr	243	175	31.7	Ext+T5

elongations, including those of the as-cast alloys and the extruded ones, are no more than 6%. The extruded alloy following with pro-deformation and T5 ageing treatment, the strength, and plasticity of Mg−9Gd−4Y−0.6Zr alloy can be enhanced simultaneously, and the maximum elongation achieves a high value of 26%.

Table 1.5 lists the mechanical properties of some Mg−Gd alloys. Mg−xGd−0.6Zr alloys show quite high elongations with the values all beyond 30% attributed to the SSSD effects and the grain refinement effects introduced by the Gd element. (Details can be seen in Chapter 3: Ultra High Plasticity Mg−Gd−Zr Alloy.)

1.4.2 Mg−Li alloys

The addition of Li makes Mg−Li alloys unique to be lighter than Mg metal itself. Mg−Li alloys are of the lowest density in metallic structural materials used in engineering applications. When the Li content is lower than 5.3 wt.%, Li dissolves in α-Mg phase with a hexagonal-close packed (HCP) crystalline structure. Li addition decreases the c/an axis ratio of HCP structure of α-Mg phase. Besides the {0001} <11$\bar{2}$0> basal slip in Mg, the {10$\bar{1}$0} <11$\bar{2}$0> prismatic slip can be activated, which leads to the <c+a> prismatic slip and results in the improvement of plasticity and deformability of Mg alloys. As the Li content is higher than 5.3 wt.%, Li-based β phase, with a body-centered cubic (BCC) crystalline structure, appears in Mg−Li alloys. At the Li content above 10.7 wt.%, Mg−Li alloys contain only β-Li phase and can be deformed even at room temperatures.

Mg—Li alloys present good ductilities, even at an ultra-low temperature of 4K. With the addition of 1—5wt.% Li, the basal texture density of the AZ31 extruded plates can be obviously weakened, and the pole axis of the basal plans declines clearly. The isotropy and the elongation of AZ31 alloy are improved. Mg—Li dual phase alloys often appear to be ductile than Mg—Li alloys of single α-Mg or β-Li phase. At room temperatures, Mg-8~9Li alloys present elongations higher than 50%. Mg—9Li—1Y alloy sheet, with the thickness decreased to 0.6 mm, has an appreciated ductility with the Erichsen value of 9 mm, the limiting drawing ratio of 2.15 bore-expanding ratio of 80%.

Mg—Li dual phase alloys often present to be super-plastic, such as an elongation as high as 1780% at 473K after equal channel angle extrusion (ECAE). Alloying can enhance the ductility of duplex Mg—Li alloys. Y element of no more than 3 wt.% can increase the plasticity of Mg—Li dual phase alloy, without processing such as ECAE. For example, when the temperature is 350°C and the tensile strain rate is $2 \times 10^{-4} \, \text{s}^{-1}$, Mg-8.5Li alloy presents an elongation of 90%. In comparison, Mg-8.5Li-Y alloy manifests an elongation of 390% at a higher strain rate of $4 \times 10^{-3} \, \text{s}^{-1}$.

1.4.3 Mg—Sn alloys

The research of Mg—Sn alloy endures a quite long history because it ascends from the 1960s. The Sn solubility in Mg is low at room temperature, and while its maximum solubility is as high as 14.8 wt.%. Thus the precipitation of Mg_2Sn can easily be seen during the ageing heat treatment of Mg—Sn alloys. The orientation angle of Mg_2Sn precipitates and the Mg matrix often changes with the variation of aging temperature. During the ageing at 130°C—200°C, the orientation relationship of Mg_2Sn and the matrix is $(111)_p//(0001)_m$. As the ageing temperature increases to 200°C~250°C, this relationship changes to $(110)_p//(0001)_m$, which means that Mg_2Sn precipitates are parallel to the Mg matrix and introduces little enhancement to the strength of Mg alloy.

The addition of the Zn element can evidently improve the microstructure and mechanical properties of Mg—Sn alloys. Zn addition can enhance the creep resistant ability of Mg—Sn alloys to be almost zero and bring a steady creep state at 170°C and 60 MPa. The addition of Zn also changes the orientation angle of Mg_2Sn precipitates and the Mg matrix. After the addition of Zn in -9.8 parallel to $(101)_p//(0001)_m$、 $<111>_p//<112(11\bar{2}0)0>_m$, which can markedly increase the age strengthening ability of Mg—Sn alloy. The addition of 1 wt.% Zn improves the peak hardness to as three times as that of Mg—9.8Sn alloy.

With the additions of 1 wt.% Al and 0.5 wt.% Zn, Mg—9.8Sn alloy presents an UTS of 354 MPa, a yield strength of 308 MPa, and an elongation of 12%. Also, Mg—9.8Sn—1Al—0.5—Zn alloy can be extruded at a low temperature of 250°C. Although in this alloy, the Mg_2Sn phase is parallel to the Mg matrix, the synergistic effects of Sn element, with fairly high content, and the microalloying Al and Zn elements have decreased the stacking fault energy (SFE) of the alloy, which resulted in the enhancement of both strength and ductility. As the Sn content is decreased to 8%, the ductility has been improved with the sacrifice of the strength. For example, after the extrusion at 250°C, the UTS, YS, and elongation of Mg-8Sn-Al-Zn alloy are 310, 250 MPa, and 18.5%, respectively. At Mg−3Sn−Al alloy with a low Sn content, the strength has decreased to between 200−300 MPa. Whilst, for Mg−3Sn−Al alloy, the total rolling reduction ratio can reach about 80%, with a single reduction ratio as high as 50%. This situation manifests the possibility of the development of wrought Mg alloy with high ductility.

In recent years, Fusheng Pan's research team from Chongqing University has paid much attention to wrought Mg−Sn alloys. They proposed the "SSSD" theory for Mg alloys. The application of this theory, with the controlling of the Mg_2Sn phase, has resulted in developing a series of Mg−Sn alloys with elongations higher than 20%.

1.4.4 Processing technologies of high plastic magnesium alloys

Wrought magnesium alloy products have outstanding advantages over the cast ones in strength, ductility, reliability, etc. It is quite important to improve the plasticity of wrought magnesium alloys. Besides developing wrought magnesium alloys with high ductility, the processing technologies are also effective to enhance the ductility. High plasticity and formability of magnesium alloys are needed for secondary processing such as punching, drawing, bending, etc. In recent years, research works for these processing technologies is mainly focused on improving the processing technology of magnesium plates and strips. Nowadays, the improvements of the isotropy and the formability of the magnesium plates are hot spots for researching the processing technologies of wrought magnesium alloys. Progress is achieved at the producing processes of magnesium plates and the controlling of their textures.

Generally, the processing technology of magnesium plates is mainly hot breaking-down rolling. With the development and application of magnesium alloys in the latest years, more processing technologies, such as continuous casting and rolling, extrusion breaking-down, etc., are developed. Hot rolling breaking-down and rolling is the most often used processing technology to produce magnesium plates for all thicknesses. To acquire thin magnesium plates, the cast ingots are rolled to plates of 8−10 mm in thickness by hot-rolling, and then to 1−3 mm thickness by hot-rolling with annealing several times. This technology contains many processings, and the finished product ratio is not appreciated, which pushes up the costs of magnesium plates. In recent years, many technologies, such as heavy reduction rolling at elevated temperatures, rolling with pressures inside directions, etc., are developed to decrease the costs and processings, but not fit for the production of AZ31 thin films of 1−3 mm in thickness.

Continuous cast-rolling breaking-down and rolling is a composite technology. Magnesium plates of 3−8 mm in thickness are directly produced from magnesium melts in continuous cast-rolling units and then rolled to 1−3 mm in thickness by in-line hot-rolling or traditional rolling techniques. Continuous cast-rolling and rolling technology for magnesium plates are newly developed in the latest 10 years by researching units from Germany, Australia, Turkey, Korea, Japan, etc. For this technology, many colleges such as Northeast University, Chongqing University, Central South University, Nanchang University, etc., focus on the key technologies, and many companies such as Chinalco Luoyang Copper Processing Co. Ltd., Shanxi Wenxi Yinguang Magnesium Co. Ltd., Fujian Huamei Advanced Materials Co. Ltd., etc., are developing the application technologies. Compared with the hot-rolling breaking-down technology, continuous cast-rolling technology has the advantages of a shortened process. The magnesium plate rolls can be acquired in-line. However, the basal texture of magnesium plates remains unsolved, and the acquired plates show typical solidification microstructures. This situation restricts the enhancement of the total reduction ratio during the following rolling processes and the improvement of the microstructures, quality, and mechanical properties, which makes this technology still under developing.

Extrusion breaking-down is performed in the extruding machine with the deformation mold. Magnesium cast ingots are extruded through symmetric extrusion molds to form plates, with the size of usually 100−500 in width and 1−4 mm in thickness, determined by the specification of the extruding machine. During this process, magnesium ingots are under the pressure stress of three dimensions applied by the extrusion container. The magnesium plates are formed when the ingots are extruded out from the die. The extrusion process is continuous, and magnesium rolls can be acquired by extrusion. Compared with the hot-rolling breaking-down technology, extrusion breaking-down can decrease the number of processes markedly to produce magnesium plates, making the costs of these plates largely decreased, and the basal texture weakened. However, the basal texture can not be removed by the extrusion breaking-down technology, and the following processes, such as punching, etc., are still difficult to be applied.

In these years, many new technologies and methods are applied to control the texture of magnesium plates, such as pre-deformation, asymmetrical rolling, single bending, equal channel rolling, etc. These technologies bring inhomogeneous strains into magnesium plates to improve the basal textures. They are under rapid development and achieve many progresses, and introduce positive effects to improving the basal textures of magnesium plates.

The above technologies are of symmetric deformation and homogeneous strain, which are likely to introduce strong (0002) basal texture to magnesium plates. Although the asymmetric deformation processes, such as pre-deformation, asymmetrical rolling, single bending, equal channel rolling, etc., can weaken the basal texture, they are of high costs and still under developing.

Fusheng Pan's research team from Chongqing University has developed a new asymmetric deformation technology. This technology improves the texture distribution and weakens the (0002) basal texture of magnesium plates, leading to the improvement of the isotropy and the formability. It is a new method of a new principle to weaken the texture with low costs. It has attracted significant attention and has been applied in the industry. This method bases on the design of the model structure and the controlling of the extrusion process, makes the ingot fluid along the inner surfaces of the model with certain speed differences, leading to an asymmetric extrusion of magnesium plates and resulting in graded strains in the thickness direction of these plates. The combination of the short-process extrusion of the plates with the texture controlling of the graded strains can make the c-axes of the grains in magnesium plates to title toward the extrusion direction, and also enable the controlling of both the low-cost extrusion of the plates and the weakening of the textures to be simultaneous.

The design of the structure of the new extrusion model, seen in Fig. 1.1, introduces the extrusion shearing area shown in Fig. 1.1D. This area leads to the speed gradation of the flowing metal in the ingot from the upper surface to the bottom one, in Fig. 1.2A, like the cross shear zone in asymmetrical rolling, and results in graded shearing strains between the upper and the bottom surfaces, for example, in Fig. 1.2B. In the electron back-scattered diffraction (EBSD) results, the high-angle boundaries of the grain in the AZ31 magnesium plates made by this new asymmetric model, from 40 degree to 90 degree, are enhanced in percentage from 29.7% to 54.4%, seen in Fig. 1.3, which means the grains have titled evidently under the graded strains. From both EBSD and X-ray diffraction (XRD) texture results, it can be seen that, compared with the AZ31 plates made by traditional extrusion methods, those plates made by this new method possess an obviously weakened basal texture with the value being decreased from 22.6 to

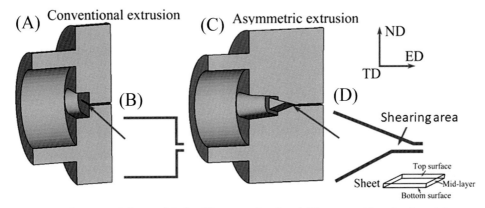

Figure 1.1 Schematic of the models for (A) conventional and (B) asymmetric extrusion plates.

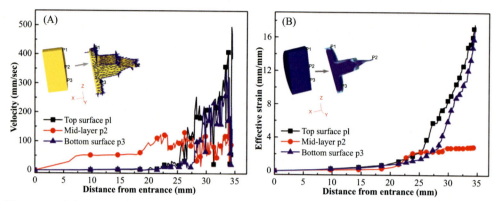

Figure 1.2 Schematic of the simulation of the models for (A) conventional and (B) asymmetric extrusion plates.

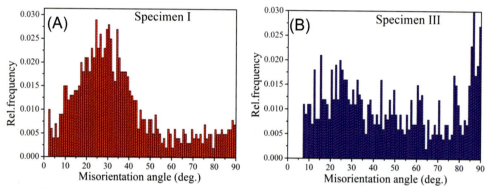

Figure 1.3 Grain orientation distributions for the models for (A) conventional and (B) asymmetric extrusion plates.

15−18, and the *c*-axes titled for 12 degree towards the extrusion direction (ED), seen in Fig. 1.4, which is similar to the variation of high-angle boundaries.

Compared with traditional extrusion methods, the grains of those magnesium plates made by the asymmetric model with graded strains are more homogeneous in their sizes, and the grain orientations title evidently, which improve the isotropy of the plates, seen in Fig. 1.5. Summarily, during the extrusion process, the new asymmetric extrusion technology of the magnesium plates introduces graded strains along the normal directions of the plates to weaken the (0002) basal texture and markedly improve the formability of these extruded magnesium plates.

1.4.5 Influence factors to the plasticity of magnesium alloys

Commercial magnesium alloys are almost all polycrystals with grains, different in orientation to each other, and grain boundaries locating between the grains, and hard

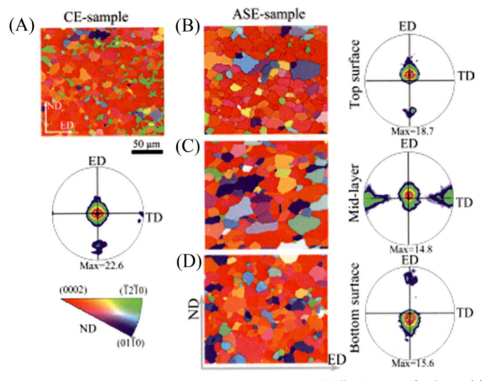

Figure 1.4 (0002) basal pole figures and electron back-scattered diffraction maps for the models for (A) conventional and (B) asymmetric extrusion plates.

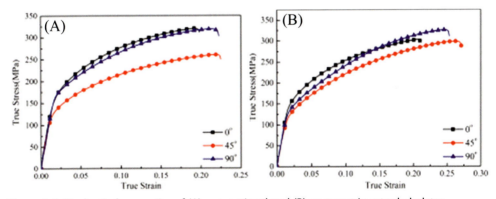

Figure 1.5 Mechanical properties of (A) conventional and (B) asymmetric extruded plates.

second phases distributing in the grains and at the grain boundaries. At the same time, most magnesium alloys are of HCP crystalline structure, which is very different from the cubic structures of iron (Fe) and aluminum (Al) metals. The deformation of magnesium shows a complex mechanism and relates to many aspects. The plasticity of magnesium alloy correlates to the slips and twinning in the matrix, the grain boundaries, and the misorientation angles between the adjacent grains as well. Other factors, such as deformation temperature, deformation speed, deformation state, strain, discontinuous deformation, deformation object size, etc., can also significantly influence the plasticity of magnesium alloys.

There are two interactions between the grains and the grain boundaries during the plastic deformation processes. One is the difference in deformation between the grains. The most evident difference between the adjacent grains lies in the grain orientation. During the deformation of magnesium alloys under force, the grains of different orientations often show different shear stresses at slip directions in the slip planes. When the CRSS is achieved, the slip starts, and then the plastic deformation follows. Some grains are of hard orientations, and their shear stress is too low, that is, lower than the CRSS, to make the slips and the deformation to be started. However, the force introduces torques to this kind of grains and drives them to rotate to coordinate the deformation of the adjacent grains. After the rotation to proper orientations, slips also start in the rotated grains to participate in the deformation.

The other one is the grain boundaries themselves, which are very important to the deformation. The grains on the sides of the grain boundaries are of different grain orientations. The atoms consisting of the grain boundaries usually are irregularly arranged, and the matrix distortion is quite high, which makes the strength of the grain boundaries higher than inner the grains. Therefore besides the deformation mechanisms of slipping and twinning, the plasticity of polycrystallines is also related to grain and their boundaries directly. Accordingly, all aspects related to the deformation mechanisms and the interaction between the grains and their boundaries can affect the plasticity of magnesium alloys. Generally, these aspects are mainly the grain size, the SFE, the axis ratio, and texture. The CRSS, related to SFE and axis ratio, is most important in the complex aspects.

1.4.5.1 Grain size

It is demonstrated by theories and practices that in magnesium alloys with small grains, grain boundaries present to coordinate deformations well to lead to a high plasticity. The causes are mainly of three aspects listed below.

Firstly, a small grain size means a high grain number in a certain space, and the certain deformation strain subjected to the alloy disperses to more grains, and the deformation is more homogeneous in the macro scale. The decrease of grain size decreases the distance of the dislocation slip and the degree of the stress concentration, making the total deformation higher. Besides, smaller grain size can increase the tortuosity of the grain boundary, and the difficulty of crack propagation within the grain will also increase. The macroscopic performance is that the alloy has high plasticity.

Secondly, during the plastic deformation, small grains need low torques to rotate and little energy to slip, making it easy for the grains to rotate and the grain boundaries to slide. If the slip systems are of hard orientations compared to the stress conducted to the grains, the grains are reluctant to make the slips to be started herein. However, they are inclined to rotate to some degree under the drive by the stress and change to be softened in orientations. Then, slips can be activated to enhance the plasticity of the alloy.

Thirdly, during the deformation processes, the prismatic and pyramidal slip systems in the small grains of magnesium alloys are easier to be activated than those in the big grains. The nonbasal slips are relatively easier seen besides grain boundaries. In magnesium grains with the size of no more than 100 μm, nonbasal slips present within the areas 10 μm away from grain boundaries. In those grains with no more than 10 μm in size, nonbasal slips exist throughout the whole grains. When the grain size of magnesium alloys is between 10 and 100 μm, smaller grains mean a higher fraction of those areas with nonbasal slips and a higher deformation ability. For example, the research of Feng Xiaoming et al. demonstrates that after five times of ECAE, the grain size of AZ31 magnesium alloy decreased from 120 to 9 μm. Thus the elongation increased from 28% to 58%. Koike et al. found that the AZ31B magnesium alloy with small grains possessed a room temperature elongation as high as 47%. When the nonbasal slips in the magnesium alloys with small grains at the tensile strain of 2% are observed by transmission electron microscope (TEM), these slips are demonstrated to balance stress at the grain boundaries.

1.4.5.2 Stacking faulty energy

Stacking faulty (SF) is a natural characteristic commonly seen in crystals. Stacking faulty energy (SFE) is the energy of the stacking fault formed by the relative slip dislocation. It is discovered that the SFE variations of magnesium alloys are related to the starting of the nonbasal slips. Thus the SF and SFE of magnesium alloys are closely related to the deformation mechanisms. The studies of the effects of alloying elements on the SFE are essential guides for developing wrought magnesium alloys with high elongations. The calculations of the effects of solid solution atoms on the SF/SFE of the basal planes of the magnesium matrix reveal that Al and Zn atoms can enhance the SFE. The atoms of RE elements, such as Pr, Nd, Gd, Tb, Dy, Y, etc., decrease the steady and unsteady SFEs of the magnesium matrix. The calculations of the unsteady SFEs of $\{0002\}$ basal plane and $\{10\bar{1}1\}$ pyramidic plane in Mg–X (X=Al, Ca, Ce, Gd, Li, Si, Sn, Zn, Zr) binary alloys by the first principle manifest that the solid-solutioned atoms in the magnesium matrix can improve the nuclearance of $<c+a>$ dislocation and the slips by affecting the ratio of the unsteady SFE of $\{0002\}$ basal and $\{10\bar{1}1\}$ pyramidic planes.

The observation of Mg—Y binary alloy further confirms that SF can act as heterogeneous nucleation cores of $<c+a>$ dislocations to improve the activity inherited from the nucleating of the nonbasal dislocations, which benefits to the improvement of random texture. Furthermore, by the using of the relation of SFE and Peierls-Nabarro model, the effects of the solution atoms of alloying elements in Mg—X (X=Al, Ca, Ce, Gd, Li, Si, Sn, Zn, Zr) binary alloys are calculated, as well as the effects of the alloying atoms to the unsteady SFEs of $\{0002\}$ basal and $\{10\bar{1}1\}$ pyramidic planes. The ratios of these two calculated results can be used as the signals of the deformation ability of these binary alloys, which is confirmed in a certain degree. This situation can offer important references for improving the formability of magnesium alloys by the design of the alloys' compositions.

Summarily, the variations of SFEs seem to be related to the starting of the nonbasal slips and also the formabilities of the alloys. However, there are some questions in the research of the formabilities of magnesium alloys by controlling SFEs.

First, the existing research only reports the influence of some alloying elements on the SFE of magnesium-based binary alloys. Although these elements are essential for developing new wrought magnesium alloys, systematic and in-depth research has not been carried out. There is a lack of SFE research on magnesium-based ternary alloys commonly used in the industry. Recent studies have found that a single Y atom reduces the SFE of magnesium solid solution, and a single Zn atom increases the SFE of magnesium solid solution. However, when Y and Zn are added simultaneously, the SFE decreases more than when a single Y is added. It shows that certain alloying elements that cause an increase in SFE in magnesium binary alloys will cause the opposite effect in magnesium ternary alloys. The SFE of magnesium alloy ternary alloys is not a binary purely algebraic superposition. The research results of binary alloy SFE can only reference engineering alloy design and lack a reliable basis. Therefore further systematic research on the SFE of magnesium-based binary alloys and multi-element alloys, and establishing a theoretical approach between magnesium alloy composition design and SFE control are critical fundamental issues that need to be urgently resolved for the development of new high-formability wrought magnesium alloys.

The second is that there is no agreement on the mechanisms of the effects of SFEs on the formability of magnesium alloys. Whether it is that the alloying elements activate the $<c+a>$ nonbasal slips, to enhance the formability of magnesium alloy, by decreasing the steady SFE of basal planes or by the decrease of the unsteady SFE ratios of the basal and pyramidic planes and the enhancement of the nuclearings and dislocations of $<c+a>$ slips, or it affects the deformation by essential effects on the hindrances of the dislocations. It is urgent to carry out systematic studies of calculations and researches to achieve agreement on this question.

The third is that the basic cause of the activation of nonbasal slips is decreasing the CRSSs of these slips to be close to that of basal slips and making nonbasal slips easier to be activated. Previous research on the mechanisms of the SFEs on the formability of magnesium alloys does not concern the effects of the SFEs on the CRSSs of basal and nonbasal slips, which focuses on only the intermediate factors. In fact, the SFEs and CRSSs herein are closely related to each other. Yasi et al. established a relationship between the SFEs of some solid solution magnesium binary alloys and their basal CRSSs by modifying the Fleischer model, which is a quite good reference for researchers to discover the effects of the SFEs to the formability of magnesium alloys from the relations of the SFEs and the CRSSs.

1.4.5.3 Axis ratio

At room temperature with the standard atmospheric pressure, the a and c lattice parameters of the HCP structure in pure Mg are 0.32092 and 0.52105 nm, respectively. The c/a axis ratio 1.6236 of pure Mg is very close to 1.632, which is the theoretical value of the HCP structure. When the temperature is altered, the lattice parameters will change, and the c/a axis ratio will also variate correspondingly. For example, the c/a ratio of Mg increases with the increase of temperature. The effects of the c/a ratio variation on the mechanical properties are based on the change of the slip resistance.

The c/a axis ratio of Mg can be decreased by adding alloying elements, such as Li, Y, In, Ag, etc. As the Li content increases up to 5 wt.%, the axis ratio decreases from 1.625 to 1.610. The c/a axis ratio value of 1.625 will drop to 1.620 by adding 3 wt.% Y element. The decline of axis ratio will benefit to the activation of prismatic slips, like $\{10\bar{1}0\}<11\bar{2}0>$, which leads to the improvement of the plasticity of Mg alloy. Furthermore, Li and Y elements can decrease the activation energy of $<c+a>$ pyramidal slip systems of Mg. Thus the strain coordination in the direction of the c axis will be enhanced as well as the ductility of the alloy during compression. Moreover, different c/a axis ration values will lead to different types of twinning (Fig. 1.6).

At room temperature, the solid solubility of Li is 5.5 wt.% in Mg, which means the influence of Li element on Mg alloy can reach a quite high level. The c/a axis ratios of Mg alloy with different Li contents are listed in Table 1.6. It can be seen that with the increment of Li element to 5 wt.%, the axis ratio decreases from 1.624 in

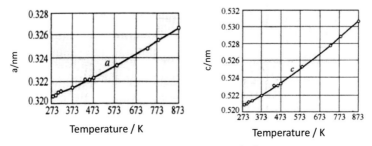

Figure 1.6 The relationship of a and c lattice parameters with the temperature.

AZ31 alloy to 1.608 in LAZ531 alloy. Since the c/a axis ratio decreases, the activation of prismatic $<a>$ slip becomes easier than before. During the deformation, the declination of the basal axes towards the transverse direction (TD) can be mainly attributed to the activation of prismatic slips. It is estimated to be the cause of the prismatic-like texture in the extrusion plates of Mg-4.6Li alloy. Similar to the situations in Ti and Zr alloys, prismatic slips in Mg–Li alloys are very important, even the $<c+a>$ slips also appears(Fig. 1.7—Fig. 1.10).

Table 1.6 The variation of lattice parameters of AZ31 alloy after the addition of Li element.

Alloy	a (Å)	c (Å)	c/a
AZ31	3.2044 ± 0.002	5.2055 ± 0.004	1.6245 ± 0.001
LAZ131	3.1990 ± 0.004	5.1876 ± 0.006	1.6216 ± 0.002
LAZ331	3.1934 ± 0.004	5.1487 ± 0.002	1.6170 ± 0.002
LAZ531	3.1864 ± 0.004	5.1278 ± 0.004	1.6082 ± 0.001

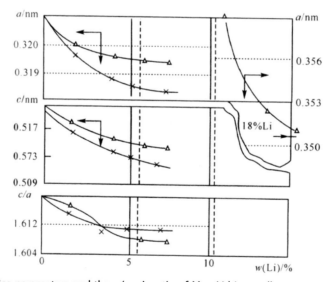

Figure 1.7 Lattice parameters and the a/c axis ratio of Mg–Li binary alloy.

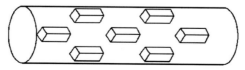

Figure 1.8 Schematic of thread texture.

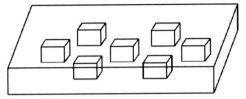

Figure 1.9 Schematic of plate texture.

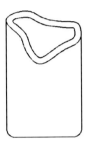

Figure 1.10 Schematic of earring resulted from anisotropy.

1.5 Texture

The grains of polycrstalline metallic materials have structural characteristics of single crystals in some degree, such as anisotropy. The commonly used technologies, such as cast and deforming, affect the metallic materials in certain directions and make the isotropic grains not randomly distributed in the alloys. In these alloys, the direction or those directions that the majority of the grain orientations concentrated are called the preferred orientation(s). In polycrystallines, the orientation structure with preferred orientation(s) is called texture. During deformation processes, the alloys are subjected to forces, and slip planes in the alloys incline to be those with a shear stress of zero, making the grain orientations in these alloys parallel or vertical to the forces applied to the alloys. For example, during drawing, some certain orientation is forced to be parallel to the drawing direction, which leads to a fiber texture; during rolling, easy slip planes of the grains incline to be parallel to the rolling plane, and some certain orientation presents to be parallel to the main deforming direction, resulting in a plate texture.

Magnesium has a HCP crystalline structure. The hexagonal structure only has two independent slip systems at room temperature. This is far away from the demand of von-Mises criterion of five independent slip systems for homogeneous plastic deformation. Therefore magnesium and common magnesium alloys usually present low ductilities. On the other hand, the hexagonal magnesium alloys incline to form preferred textures during plastic deformations. For example, basal slips often occur during the deformations of magnesium alloys. Most of the basal planes of magnesium grains tend to orientate to parallel or vertical directions to the force direction, usually making wrought magnesium alloys show a strong basal texture. During the deformation, anisotropy will cause the metal material to have different deformations in different directions, making the processed materials or devices uneven edges and uneven wall thickness. This will result in the "earing" phenomenon and increase the difficulty of subsequent processing. Thus the scrap rate is increased.

1.5.1 Deformation mechanisms of magnesium alloys

1.5.1.1 Slip

Under applied stress, one part of the crystal and the other part move relative to each other along a certain crystal plane and crystal direction, which is called slip. Essentially, slip is the movement of dislocation. In the unit cell of magnesium, the slip that is easiest to start lies at the most close-packed $<11\bar{2}0>$ direction with the Burger's vector of $a/3<11\bar{2}0>$, which is a $<a>$ dislocation of magnesium crystal. Since the (0001) basal plane, three $\{10\bar{1}0\}$ prismatic planes and six $\{10\bar{1}1\}$ pyramidal planes are all of those planes being able to generate $<a>$ dislocations, the most often seen slip systems in magnesium alloys are basal $<a>$ slip systems, prismatic $<a>$ ones and pyramidal $<a>$ ones. The other kind of perfect dislocation in magnesium is the $<c+a>$ dislocation, which lies in the $<11\bar{2}3>$ direction with the Burger's vector of $\sqrt{c^2 + a^2}<11\bar{2}0>$. The main planes that $<c+a>$ dislocations can occur in are pyramidal $\{10\bar{1}1\}$, $\{11\bar{2}1\}$, $\{10\bar{1}2\}$ and $\{11\bar{2}2\}$ planes. Fig. 1.11 shows the schematic diagrams of the main slip systems in magnesium alloys.

Slip systems in magnesium alloys can be classified by two kinds of methods. One is that by the slip directions to be the $<a>$ slips and the $<c+a>$ ones. The other is that by slip planes to be basal slips and the nonbasal ones. A certain slip plane and a certain slip direction in this plane together can form a slip system. Table 1.7 lists the most common slip systems in magnesium alloys.

The deformation mode easy to be started in magnesium alloys is basal slips. In the view of crystallography, basal slip systems are able to afford two independent slip systems for the plastic deformation of magnesium alloys, and the third basal slip system can be combined by these two independent ones. Similarly, prismatic slip systems can also afford two independent slip systems. Even the basal and prismatic slip systems in magnesium alloys are started simultaneously, there are only four independent slip systems, and can not fulfill the von-Mises criterion, which calls for five independent slip systems for an arbitrary deformation. On the other hand, basal and prismatic slips are all $<a>$ ones, with the slip directions vertical to the c-axis and parallel to the $<11\bar{2}0>$ direction in the basal plane and without coordinating the strains in the directions of the c-axis. Therefore the deformation ability of common polycrystalline magnesium alloys can be improved only if pyramidic $<c+a>$ slip systems are fully started, or the twinning is activated in the deformation.

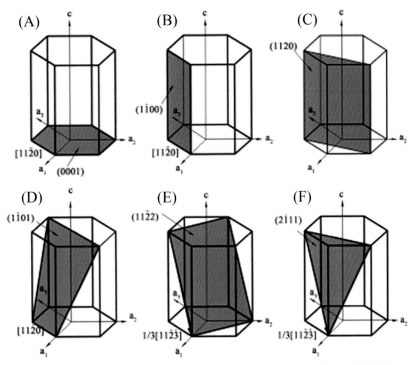

Figure 1.11 Schematic diagram of main slip systems in magnesium alloys: (A) (0001) $<11\bar{2}0>$ basal slip; (B) $\{10\bar{1}0\}<11\bar{2}0>$ prismatic slip; (C) $\{11\bar{2}0\}<11\bar{2}0>$ prismatic slip; (D) $\{10\bar{1}1\}<11\bar{2}0>$ pyramidal slip; (E) $\{11\bar{2}2\}<11\bar{2}3>$ pyramidal slip; (F) $\{11\bar{2}1\}<11\bar{2}3>$ pyramidal slip.

Table 1.7 Main slip systems in common magnesium alloys.

Slip type		Slip plane	Slip direction	Slip character	Number of independent slip systems
Basal slip		(0001)	$<11\bar{2}0>$	a slip	2
Nonbasal slip	Prismatic slip	$\{10\bar{1}0\}$ $\{11\bar{2}0\}$	$<11\bar{2}0>$	a slip	2
	Pyramidic slip	$\{10\bar{1}1\}$ $\{11\bar{2}1\}$ $\{10\bar{2}2\}$	$<11\bar{2}0>$ $<11\bar{2}3>$	a slip $c+a$ slip	4 5

It is shown by many researches that alloying can improve the slips in magnesium alloys. Firstly, alloying elements can affect the deformation behaviors of magnesium alloys by altering the SFE, which is the added energy that needed to form stacking faults in a unit area and can be used to describe the movement of dislocations. Materials with lower SFE have less resistance to dislocation movement, so they have higher forming properties. The first-principles calculations illustrate in previous research that many alloying elements, such as Al, Gd, Sn, Y, Zn, etc., can decrease the SFE of magnesium. Secondly, other researches have discovered that some elements, such as Li, Gd, Y, Zn, etc., can make nonbasal slip systems start at room temperatures, which enhanced the plasticity of magnesium alloys. For example, Mg-(1.0~4.4)wt.%Li alloys show a $<c+a>$ pyramidic slip even at a super low temperature range of 77K−293K. Moreover, added alloying elements could change the crystalline of magnesium and its deformation ability. For example, Li element can introduce β-Li phase, that is, Mg-containing Li-based phase, with a BCC structure, and improve the dislocation and deformation behaviors of magnesium. BCC structure possesses 12 independent slip systems and can easily coordinate the strains during deforming, which gives an outstanding deformation ability to Mg−Li alloys containing β-Li phase. Furthermore, alloying elements such as Li, In, Zn, Y, etc., can also decrease the c/an axis ratios of magnesium alloys, which improves the symmetry of the crystal lattice and activates the nonbasal slip systems.

On the other hand, the hindering effect of the second relative dislocation movement in the alloy can increase the critical shear stress. Alloying elements can also change the structure and distribution of the second phase of the magnesium alloy, thereby affecting the plastic deformation behavior of the magnesium alloy. For example, most second phases in Mg−3wt.%Nd alloy are lamellar and locate in prismatic planes, while those lamellar ones in Mg−3wt.%Nd−1.35wt.%Zn alloy are in the basal planes. The second phases in the basal planes are of a low hindrance for $<a>$ slips. Therefore the age strengthening effects in Mg−3wt.%Nd−1.35wt.%Zn alloy are obviously lower than those in Mg−3wt.%Nd alloy.

During the slipping of metallic crystals, the slip plane slides in the slip direction relatively. Certain shear stress is needed in the direction. Only if the stress achieves a certain value, the slip is able to start. The shear stress needed to drive the certain dislocation in the crystalline slip plane is called the CRSS of the slip system.

When a single crystal is under the effect of external tensile force, F, at the axis direction, the normal direction of the slip plane and the axis direction of the tensile force form an angle of φ, and the slip direction and the axis direction of the tensile force form the angle of λ, which are demonstrated in Fig. 1.12. It can be seen in this figure that the shear stress, τ, at the slip direction in the slip plane is:

$$\tau = \frac{F\cos\lambda}{A/\cos\varphi} = \frac{F}{A}\cos\lambda\cos\varphi \qquad (1.1)$$

where A represents the sectional area of the sample in the plane vertical to F, and $\sigma = F/A$ is the positive tensile stress conducted on the cross-section of the single crystal.

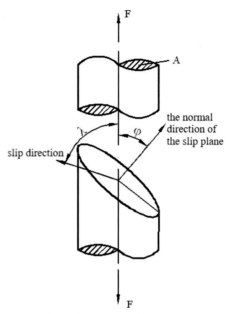

Figure 1.12 Resolved shear stress of tensile force on slip system.

When τ reaches the critical value, τ_c, slips are activated, and at the macro-scale, yield occurs and the positive tensile stress presents to be the limitation of the yield of single crystal. Thus Eq. (1.1) can be deduced to be:

$$\tau_c = \sigma_s \cos\lambda\cos\varphi \qquad (1.2)$$

where τ_c is the CRSS of metallic crystal and also the minimum shear stress to activate the slip system. CRSS, τ_c, is the intrinsic value of crystal, and not relative to the direction and value of the external force, as well as the grain orientations of polycrystalline. The $m = \cos\lambda \cdot \cos\varphi$, also called Schmid factor (SF), is the orientation factor of the slip system relating to the external force, which presents the orientation relationship between the slip plane/direction and the component stress. In practice, the SF value can be determined by the combination of grain orientation, (spatial feature of) slip system, and the external force's state. When both the normal direction of the slip plane and the slip direction are 45 degrees to the external force, these three directions are on one plane, and the SF shows the maximum value of 0.5. In this situation, slips and deformations are the easiest to be activated and the material illustrates the lowest yield strength, and such orientation is called the "soft orientation." When the external is parallel or vertical to the slip plane, when $\varphi = 90$ degrees or $\lambda = 90$ degrees, respectively, SF values are both zero. No matter how is the CRSS, the yield strength is infinitely great, and slip systems can not be activated, and the orientations are hard ones.

1.5.1.2 Twinning

Except for pyramidal $<c+a>$ slips, basal and prismatic slips in magnesium alloys are all $<a>$ ones, with the slip directions of vertical to c-axis and parallel to basal $<11\bar{2}0>$ direction, which are not able to coordinate strains in c-axis direction during the deformation. Twinning, another kind of deformation mechanism, acts as an important role to coordinate the strains in the c-axis direction. Twin deformation refers to the shear deformation of a part of the atoms of the crystal along a certain crystal plane (twin plane) and a certain crystal direction (twin direction) under the action of shear stress. After twin deformation, the deformed part (twin) of the crystal and the undeformed part (parent grain) show a mirror symmetry relationship. There is a certain orientation difference between the twin crystal and the parent grain.

The most common twinning types in magnesium alloys are $\{10\bar{1}2\}$ tension and $\{10\bar{1}1\}$ compression twins, whose basal factors are listed in Table 1.8. It can be seen in this table that both the CRSS and the shear ratio of $\{10\bar{1}2\}$ tension twin are lower than those of the compression one. Therefore during rolling deformation, the twinning in common magnesium alloys is usually the $\{10\bar{1}2\}$ tension one. However, $\{10\bar{1}2\}$ tension twinning is not taking place in magnesium alloy at any circumstance. It only occurs when the c-axis of magnesium unit is under tension or compression.

Fig. 1.13 shows the activating requirement of HCP metals under different shearing stresses. In this figure, positive slopes represent compression twins, and the negative is tension ones. The conditions for compression twinning are: the c-axis of the unit cell is shortened, that is, the c-axis is pressed or stretched perpendicular to the c-axis; the condition for the occurrence of tensile twinning is: the c-axis of the unit cell is elongated, that is, it is pulled along the c-axis of the crystal grain. The presence of the twin is closely relative to the c/a axis ratio of the crystal. Magnesium possesses an axis ratio of 1.6236. Therefore in magnesium, $\{10\bar{1}2\}$ and $\{11\bar{2}1\}$ twins are tension ones. $\{10\bar{1}1\}$ and $\{11\bar{2}2\}$ twins are compression ones.

Table 1.8 Basic factors of main twins in magnesium alloys (r=c/a).

Twin	$\{10\bar{1}2\}$	$\{10\bar{1}1\}$
Type	Tension	Compression
CRSS (MPa)	2−2.8	76−753
Orientation difference	86 degrees	56 degrees
Twin direction/η	$<10\bar{1}1>$	$<10\bar{1}2>$
	$<\bar{1}01\bar{1}>$	$<30\bar{3}2>$
Shearing plane/P	$\{12\bar{1}0\}$	$\{12\bar{1}0\}$
Shearing ratio/s	$\frac{r^2-3}{r\sqrt{3}}=-0.13$	$\frac{4r^2-9}{4r\sqrt{3}}=0.138$
First undistorted plane K_1	$\{10\bar{1}2\}$	$\{10\bar{1}1\}$
Second undistorted plane K2	$\{10\bar{1}2\}$	$\{10\bar{1}3\}$

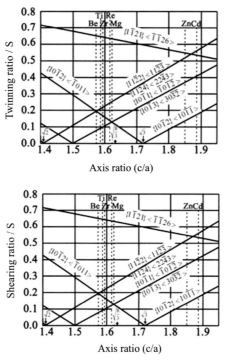

Figure 1.13 Twinning types of hexagonal close-packed structure with different c/a ratios.

The CRSS of twinning in crystals is higher than those of slips. It is hard for twins to be activated in face center cubic and BCC crystals with plenty of independent slip systems. For magnesium of HCP structure, fewer slip systems are readied for being activated, and twins act as important roles during the deformation processes at room temperatures.

During the deforming of magnesium alloys, the shear strain of twinning is usually lower than the strains resulted from slips, and the direct contribution of twinning to the deformation of magnesium alloys is not quite high. While, twinning can alter the orientation of grains, seen in Fig. 1.14, and introduces convenience to the slips and further plastic deformations of magnesium alloys. It is demonstrated that after the introducing of $\{10\bar{1}2\}$ tension twins by the predeforming of AZ31 magnesium plates, the ductility of the annealed plates has been obviously improved, and the IE value is enhanced from 3.2 to 5.3.

Slip and twinning are two kinds of plastic deformation modes with the relationship of both coordinating and competing. Although they both enable the materials to deform, they are in evident differences. Firstly, they have different deformation zones. The deformation zones of slips are usually concentrating on slip planes with an inhomogeneous shearing, while the shearing of twins relates to the whole twinning zone.

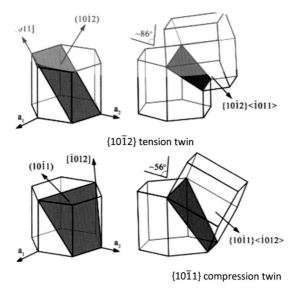

Figure 1.14 Schematic figures of {10$\bar{1}$2} tension and {10$\bar{1}$1} compression twins in magnesium alloys.

Secondly, the distances that the atoms cover are different. During slipping, the distances are integer multiples to the distance of atoms in the slip directions, while during twinning, it is usually only fractions of that distance. Thirdly, the grain orientations are different. During slipping, the orientations of both sides of the slip plane are not changed. While mother grains and their twins are mirror symmetry along the twinning plane, which means the orientations have been changed after twinning. Fourthly, the directions of deformations are different. During slipping, the dislocations can move along the slip directions, or the inverse directions. Whist, during twinning, the movement of dislocations is limited in the twinning directions. Fifthly, the dislocation types are different. During slipping, the dislocations can be perfect ones or not. But, during twinning, they usually are perfect dislocations.

Although there are obvious differences between slipping and twinning, their synergistic effect can effectively coordinate the strains during the deformations. When magnesium crystal is compressed in the vertical direction of the basal plane or tensed in the parallel direction of the basal plane, the grain orientations are not fit for slipping or twinning, and magnesium alloys are difficult to deform. If the crystal is tensed in the vertical direction of the basal plane or compressed in the parallel direction of the basal plane, grain orientations are hard one for $<a>$ slips. Basal and prismatic $<a>$ slips are almost not able to be activated, while twinning can take place. Twinning can change the orientations of the twinning zones in the grains and make the basal planes deviate from hard orientations, enabling the twinned grains to fulfill the starting factors to be activated. When the shearing occurs along the dislocation plane in the grains, the stress conducted on the grains can make the grains rotating along the tension direction and finally changes the orientations of the slip planes. Therefore with the proceeding of twinning and slipping, the final result of the plastic deformation of the crystal is the slip plane parallel to the direction of the tension axis. Neither the slip nor twin can be activated.

After twinning occurs in the magnesium alloy matrix, further twinning may occur if it continues to deform. Twinning occurs again within the existing twins. The original twins are called primary twins or parent twins. The twins that appear inside the primary twins are called secondary twins. For example, the compression twins in the AZ31 alloy appear inside the tensile twins, namely, the secondary twins. The angle with the matrix is 38.5 degrees. Besides, there are some reports on third-order twins.

In addition to the further twinning phenomenon of secondary twinning, the twins in magnesium alloys will also revert back to the matrix under certain conditions, resulting in the phenomenon of reduction or disappearance of twins, that is, the phenomenon of untwining, called detwinning.

1.5.1.3 Movement of grain boundaries

Both dislocation and twinning are deformation mechanisms in the magnesium matrix. In polycrystalline magnesium alloys, especially wrought magnesium alloys, besides the matrix, grain boundaries are of a high fraction and important to the plastic deformation. The most important deformation mechanism between the grains of magnesium alloys is the grain boundary sliding (GBS), which is the relative slide of the magnesium grains. The main function of GBS is coordinating the deformation. One is coordinating the strains resulted from the reluctant deformations of the grains with hard orientations. The other is to release the stress concentration from the strains of basal and prismatic $<\alpha>$ slips at c-axis directions.

Most GBSs in common magnesium alloys occur in deformations of low strains at elevated temperatures, during which the grains roughly maintain their shapes even at large strains and the orientations. The increase in temperature leads to a high contribution of GBS to the plasticity. High strains are often the results of the enhancement of deformation by GBS. The superplastic deformation of magnesium alloys usually occurs at elevated temperatures, with GBSs being the main deformation mechanism.

1.5.2 Characterization of plastic deformation of magnesium alloys

Most deformed magnesium alloy products are profiles and plates. Since the magnesium profiles are near to the final products, they are widely applied. Magnesium plates often need secondary processing to become final products. During processing, such as rolling and extruding, strong anisotropy of magnesium alloy is formed. The earrings and rags occur during the following processes, reducing the yield of final products, which highly obstructs the applications of magnesium plates. Therefore the research of the plasticity of magnesium alloys and the related microstructural characteristics are important works for the developments of wrought magnesium alloys and the secondary processing technologies, such as punching and bending. It is quite beneficial for processing the magnesium plates to near-final products more efficiently and expanding the applications of magnesium alloys.

The plastic characteristics needed to be manifested in the tensile and compression deformation ratios in the single dimension, the anisotropy, and the Erishen value related to two or more dimensions. Among these characteristics, the anisotropy and the Erishen value of magnesium alloys are important references to further processing, such as punching. Since magnesium alloys have strong textures, they often present evident anisotropy. Therefore among the characterizations of those characteristics relating to plasticity, the texture of magnesium alloys is fairly important. There is nearly no parameter proposed to characterize the plastic deformation ability properly. To propose one or two outstanding parameters to characterize the plastic deformation ability is very important for the researchers.

1.5.2.1 Characterization of texture

Texture can be examined in both macroscopic and microcosmic scales. In the macroscopic scale, the preferred orientations of grains in the whole polycrystalline are primarily examined and summed up by using XRD in the present. In the microcosmic scale, the orientation distributions of grains in microareas of polycrystallines, including the distributions of the orientations and their differences for the individual grain and the individual grain boundary, are usually examined by the EBSD method based on the scanning electron microscope (SEM). For samples with the microstructure of a favorable homogeneity, theoretically, different observation planes are different only in their observation angles. Their texture can be exchanged through the relations of their angles. However, in practical plastic deformation products, especially in plates, the microstructures of three areas, that is, on the upper surface, near the lower surface, and in the middle, are usually different, and even of obvious differences. Therefore it should be careful to choose the observation plane during texture measurements and place the sample in the proper position.

1.6 Macrotexture

As the alloy sample is detected by X-ray, the reflections of X-ray for the grains with different orientations are different. By rotating the sample, all reflected X-rays can reach the detector and be recorded by the counter. The basal data for the macro-texture of the sample is acquired by combining the direction of the sample with the results in the counter. The basal data can be transformed to the common data and the pole figures. For example, Fig. 1.15 is an intuitional pole figure that is commonly used.

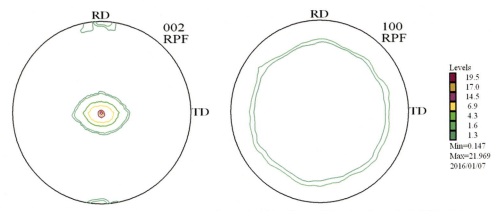

Figure 1.15 Pole figure of macro texture determined by X-ray diffraction for rolled AZ31 plate.

The main information including in the pole figure of Fig. 1.15 is listed below.

1. The *hkl* values of the plane demonstrated in the pole figure. For example, the 002 in the top right corner of the left picture of Fig. 1.15 represents that this pole figure is for the orientation distributing characteristic of {0002} basal plane, and the right picture is for the {10$\bar{1}$0} prismatic plane.

2. The set direction of the sample during the test. Usually, the center of the circle represents the normal direction (ND) of the plane that the sample belongs to, such as the {0002} basal plane in Fig. 1.15, and also the vertical direction of the rolling or extrusion plane; the RD/ED in the peak means that the upper and bottom directions of the circle are the rolling/extrusion direction of the plane that the sample belongs to; the TD at the right of the circle implies that the left and right directions of this circle are vertical to the rolling/extrusion direction of the plane that the sample belongs to. The RD/TD-TD-ND directions are vertical to each other in the sample, and also the top/bottom, the right/left, and the center positions of the circle. During the test, the directions in the sample should be in accordance with the set ones.

3. The closed curves are isodense, during which the density of the grain orientations are identical. The density value of the isodense is distinguished by their colors, which are given in the multicolor square locating on the right of the circle. The maximum and minimum values of the orientation density are also shown with the multicolor square. For different alloy samples, the demarcating criterions are often not the same, which affects the maximum and minimum values and the density values of the isodenses. Therefore it should be unifying the criterion during the processing of the data, the exporting of the pole figure, and the comparing of the textures of different alloys with each other.

In particular, Fig. 1.15 is a typical plate texture for magnesium alloys after the rolling process. Most grain tilts towards the ND and the near directions of the rolling plane, with a peak density value of 21.969 and a little grain to the rolling direction. Schmid factors of both orientations for the basal planes are close to zero. Whilst, the prismatic planes are fairly homogeneously distributed in every direction in the rolling plane of the plate. As the reasons are concerned, the external force is applied to the above and down directions of the alloy plate. The rolling and normal directions of the plate, which are parallel and vertical to the rolling direction, respectively, are the directions with minimum Schmid factor values, very near to zero, and minimum CRSSs. Basal dislocations, easiest to be activated, are reluctant to slip when they move to these two directions, therefore the orientations of basal planes are mainly locating to these two directions and those directions nearby. The normal direction of the plate is that the external force is applied, and thus basal orientations concentrate on this direction, which forms the most evident characteristic for plate texture. There is no external force in the rolling plane, and the prismatic slips distribute homogeneously in this plane after slipping.

1.7 Microtexture

Microtexture, determined by EBSD, is often used to study the plastic deformation mechanisms. The preparation of EBSD samples is similar to that of optical microscopy. After the cutting of the sample, it is grounded to be flat and smooth, and polished to remove the scratches on sand paper of above 1000#. The following step is the most important to delete the stress on the sample surface introduced by the grounding and polishing. After removing the layer with stress by methods such as mechanical polishing, electrolytic polishing, focus ion beam, Ar ion polishing, etc., thus deleting the stress on the sample surface, the sample can be observed and detected by SEM with the EBSD function.

The texture data acquired by the EBSD method are usually the rough shape and size of the grains, the orientations of the grains, the texture of the microareas, and the orientation differences (also called the angle of the grain boundary). The EBSD mapping in Fig. 1.16 is most often seen, since it can directly give the grain orientations and the shapes of the corresponding grains. Major grains in Fig. 1.16 are marked by blue color, which means that these grains are of $\{10\bar{1}0\}$ orientation, that is, the normal directions of their $\{10\bar{1}0\}$ prismatic planes are vertical to the testing plane. There are thin white lines throughout some grains, which are sub-grain or grain boundaries with a grain boundary angle no more than 10 degrees, and this angle varies according to the settings. If the grain orientation in the EBSD measurement area is counted, the texture of the alloy microarea and a pole figure similar to the macro-texture can be obtained. The plastic deformation behavior of the measured area during the deformation process can be further analyzed. Besides, it can also be combined with sample deformation, especially the in situ EBSD test of the sample, which can visualize the evolution of the microtexture during the deformation of the sample.

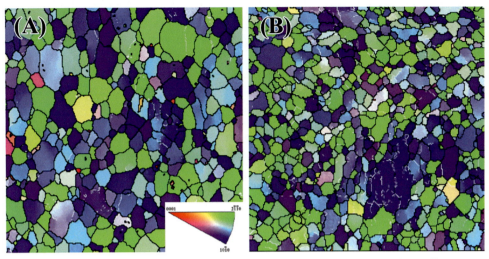

Figure 1.16 Electron back-scattered diffraction maps of M1 alloy rods extruded at different temperatures: (A) 250°C and (B) 300°C.

1.7.1 Static tensile test

The static tensile test is the most basic and commonly used plastic characterization method in engineering and materials. It is used to characterize the basic mechanical properties of materials. Generally, a universal testing machine is used to uniaxially stretch the sample of the magnesium alloy, which can easily obtain the plasticity of the alloy sample in a single dimension. According to GB/T228.1-2010 and ISO6892-1−2009《Metallic materials—Tensile testing—Part 1: Method of test at room temperature》, the plastic deformation ability is represented by the elongation δ and the reduction of area φ during the tensile test along the single-dimensional direction. In this book, the elongation δ is used as the index of plasticity. The gauge length of the sample is measured by an extensometer with standard methods. The following equation determines the elongation:

$$\delta = (L - L_0)/L_0 \times 100\% \tag{1.3}$$

where L_0 is the original gauge length of the sample before the tensile test, L is the gauge length after the fracture of the sample.

Similar to the tensile test, the samples are also able to be compressed by a universal test machine to determine the mechanical properties when they are suffering pressures. The compression plasticity can also be determined by the equation of tensile test, with the result to be a negative value, where the minus sign means the result is the deformation of compression.

The value of the reduction of area, φ, is determined by the following equation:

$$\varphi = (F - F_0)/F_0 \times 100\% \tag{1.4}$$

where F_0 is the original area of the cross-section of the sample before the tensile test, F is the area of the cross-section of the sample after the fracture of the sample.

The reduction of area is reluctant to be affected by factors and shows to be in the range of 0%–100%, with no relation to the gauge length of the sample and the ratio of length for the uniform deformation zone and the concentrated deformation zone as well. Although the reduction of area is advantageous in representing the ductility, the elongation is more usual to be used since the measurement of the parameters of the elongation is more simple and convenient.

1.7.2 Evaluation of anisotropy

In narrow meaning, anisotropy mostly refers to the anisotropy of unidirectional tensile or compressive mechanical properties, namely, the difference in uniaxial tensile or compressive mechanical properties of magnesium alloy sheets in all directions. Anisotropy is mainly used to evaluate the mechanical properties of the sheets prepared by extrusion and rolling. General anisotropy of magnesium alloy also includes the asymmetry of tension and compression.

1.8 Asymmetry of tension and compression

When magnesium alloys deform under the tension or compression force, the mechanical properties of these two situations often show large differences, which is unequality of tension and compression. Since the asymmetry can increase the difficulty of the deformation of magnesium alloys, it quite important to control the asymmetry.

In general, the tension and compression asymmetry of magnesium alloys focus on the inconsistency of the yield strength of magnesium alloys in tension and compression. The tension-compression asymmetry of magnesium alloys is expressed by the ratio of compression yield strength to tensile yield strength. The closer the ratio is to 1, the weaker the asymmetry.

The origin of the asymmetry lies in the different deformation mechanisms of tension and compression processes. During the initial stages of the tension and compression processes, the deformation mechanisms are quite different, leading to different yield strengths in these two processes. Furthermore, the contributions of twinning are also quite not the same during the tension and compression processes.

1.9 Anisotropy of mechanical properties

The objects of the studies of the anisotropy of magnesium alloys' mechanical properties are often magnesium plates. Anisotropy usually refers to the differences of mechanical properties in different directions of magnesium plates, such as the differences of mechanical properties in the directions that are 0 degrees, 45 degrees, and 90 degrees to the rolling direction, that is, the directions parallel, 45 degrees and vertical to the rolling direction, respectively. Usually, it can be evaluated by the ratios of the values of mechanical properties in these directions. The more the ratios are close to 1, the weak the anisotropy is. During the deformation processes, the plates with high anisotropy are not homogeneously deformed in different directions, resulting in earrings.

1.10 Plastic strain ratio and plane anisotropy coefficient

Besides the different directions in the plate plane, the difference in the plasticity of magnesium plates also lies in the directions of width and thickness. During the tension of plate samples, the ratio of the strain in the width, ε_b, to that in the thickness, ε_t, is called the plastic strain ratio, r-value. The r-value is often used to show the ability to resist the thinning or thickening in the plane of the magnesium plate, and represents the difficulty of the deformation of the material in the directions of width and thickness. The r-value can be calculated by the equation listed below:

$$r = \frac{\varepsilon_b}{\varepsilon_t} = \frac{\ln(b/b_0)}{\ln(t/t_0)} \tag{1.5}$$

The material with the r-value higher than 1 shows a higher strain in width than that in thickness. During deformation, this material is easier to be narrowed in width. In those materials with the r-value lower than 1, the strain in the thickness direction is relatively higher, which means they incline to be decreased in thickness during deforming. The plates with higher r-values are easier to be processed to cups with higher depths.

During the process of plates to be deformed to cup-like cylindrical products, the earing is easy to occur in anisotropic plates, which is that the convex lugs/ears exist at the mouth of the product. A higher height of the ears means a higher anisotropy of the plate, a higher difference of the deformation in the different parts of the product, a higher heterogeneity of the thickness, lower quality of the finished product, more material to be abandoned at the mouth of the product, and a larger size of the original plate as well.

To show the tendency of the formation of the ears directly, the plane anisotropy coefficient, Δr, is used to illustrate the average difference of the ratios of thickness in different directions of the plate plane. The Δr value can be determined by the equation listed below.

$$\Delta r = \frac{r_0 + r_{90} - 2r_{45}}{2} \tag{1.6}$$

The *r*-values are the average value of the measured values of the samples machined from 0 degree, 45 degrees and 90 degrees directions of the sample, and can be calculated by the following equation.

$$\bar{r} = \frac{r_0 + 2r_{45} + r_{90}}{4} \tag{1.7}$$

1.10.1 Erichsen drawing test

Cup drawing test, also called Erichsen drawing test or Erichsen test, is used to illustrate the plastic deformation ability of plate samples. Usually, when some certain spherical punches or steel balls are used to press the plate samples being circularly clamped, the pressed depth (mm), as the crack is seen in the sample, is the drawing depth, also called cupping test value or Erichsen value (IE value). Cup drawing test is used to evaluate the punching ability of alloy plates, such as bulged type molding, drawing process of complex surface, etc., and often used at the metallic plates with the thickness below 2 mm. The diameter of the punch or the steel ball is usually 8−20 mm to fit for samples of different sizes.

The main factors that affect the results of Erichsen test are the thickness tolerance of the plate sample, the lubrication of the test, and the judgment of the crack by the tester. Among these factors, the effects of both the thickness tolerance and the lubrication can be avoided by enhancing the controlling as precise as possible, while the judgment of the crack by the tester presents is occasional. Therefore some automatic judging methods, such as using 5% decrease of the load as the criterion, are developed.

Although great efforts are made to research the formability of magnesium alloys, there is no scientific method and index for the quantitative characterization of magnesium alloy formability. More efforts should be summoned to focus on this area.

Further reading

Agnew S, Horton J, Yoo M. Transmission electron microscopy investigation of <c+a> dislocations in Mg and α-solid solution Mg-Li alloys. Metallurgical and Materials Transactions A 2002;33:851−8.

Agnew SR, Yoo MH, Tome CN. Application of texture simulation to understanding mechanical behavior of Mg and solid solution alloys containing Li or Y. Acta Materials 2001;49(20):4277−89.

Alok S, Nakamura M, Watanabe M, et al. Quasicrystal strengthened Mg-Zn-Y alloys by extrusion. Scripta Materialia 2003;49:417−22.

Ando S, Tanaka M, Tonda H. Pyramidal slip in magnesium alloy single crystals. Materials Science Forum 2003;419:87−92.

Bae DH, Lee MH, Kim KT, et al. Application of quasicrystalline particles as a strengthening phase in Mg−Zn−Y alloys. Journal of Alloys and Compounds 2002;342:45−450.

Bae JH, Rao AKP, Kim KH, et al. Cladding of Mg alloy with Al by twin-roll casting. Scripta Materialia 2011;64:836−9.

Barnett M, Sullivan A, Stanford N, et al. Texture selection mechanisms in uniaxially extruded magnesium alloys. Scripta Materialia 2010;63(7):721−4.

Barnett MR, Keshavarz Z, Beer AG, et al. Non-schmid behavior during secondary twinning in a poly-crystalline magnesium alloy. Acta Materials 2008;56:5−15.

Boehlert C, Chen Z, Chakkedath A, et al. In situ analysis of the tensile deformation mechanisms in extruded Mg-1Mn-1Nd (wt%). Philosophical Magazine 2013;93(6):598−617.

Bohlen J, Yi S, Letzig D, et al. Effect of rare earth elements on the microstructure and texture development in magnesium−manganese alloys during extrusion. Materials Science and Engineering: A 2010;527(26):7092−8.

Borkar H, Gauvin R, Pekguleryuz M. Effect of extrusion temperature on texture evolution and recrystallization in extruded Mg−1%Mn and Mg−1%Mn−1.6%Sr alloys. Journal of Alloys and Compounds 2013;555(5):219−24.

Borkar H, Hoseini M, Pekguleryuz M. Effect of strontium on flow behavior and texture evolution during the hot deformation of Mg−1wt% Mn alloy. Materials Science and Engineering: A 2012;537 (1):49−57.

Borkar H, Hoseini M, Pekguleryuz M. Effect of strontium on the texture and mechanical properties of extruded Mg−1% Mn alloys. Materials Science and Engineering: A 2012;549(15):168−75.

Borkar H, Pekguleryuz M. Effect of extrusion conditions on microstructure and texture of Mg-1%Mn and Mg-1%Mn-1.6%Sr alloys. Magnesium Technology 2012;2012:461−4.

Borkar H, Pekguleryuz M. Effect of extrusion parameters on texture and microstructure evolution of extruded Mg-1%Mn and Mg-1%Mn-Sr alloys. Metallurgical and Materials Transactions A 2015;46 (1):488−95.

Borkar H, Pekguleryuz M. Microstructure and texture evolution in Mg−1%Mn−Sr alloys during extrusion. Journal of Materials Science 2013;48(4):1436−47.

Brömmelhoff K, Huppmann M, Reimers W. The effect of heat treatments on the microstructure, texture and mechanical properties of the extruded magnesium alloy ME21: dedicated to Prof. Dr.-Ing. Heinrich Wollenberger on the occasion of his 80th birthday. International Journal of Materials Research 2011;102(9):1133−41.

Byrne J. Plastic deformation of Mg−Mn alloy single crystals. Acta Metallurgica 1963;11(9):1023−7.

Cao YL. Effect of Ca on as-cast microstructure and mechanical properties of AZ40M and ZK61M magnesium alloys. Nonferrous Metal Process 2014;(6):22−6.

Carter JT, Krajewski PE, Verma R. The hot blow forming of AZ31 Mg sheet: formability assessment and application development. Journal of the Minerals Metals & Materials Society 2008;60(11):77−81.

Celikin M, Kaya A, Pekguleryuz M. Effect of manganese on the creep behavior of magnesium and the role of α-Mn precipitation during creep. Materials Science and Engineering: A 2012;534(1):129−41.

Celikin M, Kaya A, Pekguleryuz M. Microstructural investigation and the creep behavior of Mg−Sr−Mn alloys. Materials Science and Engineering: A 2012;550(30):39−50.

Celikin M, Pekguleryuz M. The effect of Ce addition on the creep resistant Mg-Sr-Mn alloys. Materials Science Forum 2014;784:358−62.

Celikin M, Sediako D, Pekguleryuz M. Texture change in pure Mg and Mg-1.5wt% Mn casting alloy during compressive creep-deformation. Magnesium Technology 2010;2010:249−51.

Chakkedath A, Bohlen J, Yi S, et al. Behavior of extruded Mg-1Mn (wt pct) at temperatures between 298 K and 523 K (25°C and 250°C). Metallurgical and Materials Transactions A 2014;45 (8):3254−74.

Chakkedath A, Bohlen J, Yi S, et al. The effect of Nd on the tension and compression deformation behavior of extruded Mg-1Mn (wt pct) at temperatures between 298 K and 523 K (25°C and 250°C). Metallurgical and Materials Transactions A 2014;45(8):3254−74.

Chang TC, Wang JY, Chu CL, Lee S. Mechanical properties and microstructures of various Mg-Li alloys. Materials Letters 2006;60:3272−6.

Chen B, Feng LP, Zhong H, et al. Microstructures and properties of a Mg-Li-Al-Zn wrought alloy. Journal of Beijing University of Aeronautics and Astronautics 2004;30(10):976−9.

Chen L, Jiang N, Meng LG, et al. Effect of RE element Nd on microstructure and mechanical properties of AZ80 magnesium alloy. Specification Cast & Nonferrous Alloy 2017;37(5):576−80.

Chen ZH, Yan HG, Chen JH, et al. Magnesium alloys. Beijing: Chemical Industry Press; 2004.

Chen ZH. Wrought magnesium alloys. Beijing: Chemical Industry Press; 2005.

Christian JW, Mahajan S. Deformation twinning. Progress in Materials Science 1995;39:1−157.

Dai Q, Lan W, Chen X. Effect of initial texture on rollability of Mg-3Al-1Zn alloy sheet. Journal of Engineering Materials and Technology 2014;136(1):011005.

Dharmendra C, Rao KP, Prasad YVRK, et al. Hot workability analysis with processing map and texture characteristics of as-cast TX32 magnesium alloy. Journal of Materials Science 2013;48:5236−46.

Ding PD, Jiang B, Wang J, et al. Status and development of magnesium alloy thin strip casting. Materials Science Forum 2007;546−549:361−4.

Ding PD, Pan FS, Jiang B, et al. Twin-roll strip casting of magnesium alloys in China. Transactions of Nonferrous Metals Society of China 2008;18(1):7−11.

Dong DQ. Effect of rare-earth La on microstructure of Mg-4.5% Zn as-cast magnesium alloy. Light Alloy Fabrication Technology 2008;36(10):11−13.

Duan CY. Research about microstructures and mechanical properties of AM60-Nd/Ce/Sc magnesium alloys. Chongqing: Chongqing University of Technology; 2015.

Dudamell N, Hidalgo-Manrique P, Chakkedath A, et al. Influence of strain rate on the twin and slip activity of a magnesium alloy containing neodymium. Materials Science and Engineering: A 2013;583(20):220−31.

Fang C, Zhang J, Liao AL, et al. Hot compression deformation characteristics of Mg−Mn alloys. Transactions of Nonferrous Metals Society of China 2010;20(10):1841−5.

Fang DQ, Ma N, Cai K, et al. Age hardening behaviors, mechanical and corrosion properties of deformed Mg−Mn−Sn sheets by pre-rolled treatment. Materials & Design 2014;54:72−8.

Fang XY, Yi DQ, Nie JF, et al. Effect of Zr, Mn and Sc additions on the grain size of Mg-Gd alloy. Journal of Alloys and Compounds 2009;470(1):311−16.

Feng XM, Ai TT, Zhang H. The microstructure and mechanical properties of AZ31 Mg alloy processed by equal channel angular extrusion. Special-cast Non-ferrous Alloy 2008;28(7):499−501.

Furui M, Kitamura H, Anada H, Langdon TG. Influence of preliminary extrusion conditions on the superplastic properties of a magnesium alloy processed by ECAP. Acta Materials 2007;55:1083−91.

Furui M, Xu C, Aida T, Inoue M, Anada H, Langdon TC. Improving the superplastic properties of a two-phase Mg-8%Li alloy through processing by ECAP. Materials Science and Engineering: A 2005;410−411:439−42.

Gall S, Coelho R, Mller S, et al. Mechanical properties and forming behavior of extruded AZ31 and ME21 magnesium alloy sheets. Materials Science and Engineering: A 2013;579(1):180−7.

Gall S, Huppmann M, Mayer H, et al. Hot working behavior of AZ31 and ME21 magnesium alloys. Journal of Materials Science 2013;48(1):473−80.

Gandel D., Birbilis N., Easton M., et al. Influence of manganese, zirconium and iron on the corrosion of magnesium. In: Proceedings of corrosion & prevention; 2010. p. 875−5.

Gandel D, Easton M, Gibson M, et al. Calphad simulation of the Mg−(Mn, Zr)−Fe system and experimental comparison with as-cast alloy microstructures as relevant to impurity driven corrosion of Mg−Alloys. Materials Chemistry and Physics 2014;143(3):1082−91.

Gandel D, Easton M, Gibson M, et al. Influence of Mn and Zr on the corrosion of Al-free Mg alloys: part 2-impact of Mn and Zr on Mg alloy electrochemistry and corrosion. Corrosion 2013;69(8):744−51.

Gandel D, Easton M, Gibson M, et al. Influence of Mn and Zr on the corrosion of Al-free Mg alloys: part 1-electrochemical behavior of Mn and Zr. Corrosion 2012;69(7):666−71.

Ganeshan S, Shang SL, Wang Y, et al. Effect of alloying elements on the elastic properties of Mg from first-principles calculations. Acta Materials 2009;57:3876−84.

Gąsior W, Moser Z, Zakulski W, et al. Thermodynamic studies and the phase diagram of the Li-Mg system. Metallurgical and Materials Transactions A 1996;27(9):2419−28.

Gu SW, Hao H, Zhang AM, et al. Effects of Gd on microstructure and properties of as-cast AZ31 magnesium alloy. Specification Cast & Nonferrous Alloy 2011;31(5):472−5.

Guan SK, Wang YX. Research development of automotive high temperature magnesium alloy. Automobile Technology & Material 2003;(4):3−8.

Hang HY, Wang HY, Wang C, et al. First-principles calculations of generalized stacking fault energy in Mg alloys with Sn, Pb and Sn+Pb dopings. Materials Science and Engineering: A 2013;584:82−7.

Hauser FE, Landon PR, Dorn JE. Deformation and fracture of alpha solid solutions of lithium in magnesium. Transactions of the American Society Meteorological 1958;50:856−81.

Henes S, Gerold VZ. Radiographical analysis of segregation process in Mg-Pb and Mg-Sn welded alloys. Zeitschrift für Metallkunde 1962;53:743−51.

Hidalgo-Manrique P, Yi S, Bohlen J, et al. Effect of Nd additions on extrusion texture development and on slip activity in a Mg−Mn alloy. Metallurgical and Materials Transactions A 2013;44 (10):4819−29.

Hou ZS, Lu GX. Fundamental of metal. Shanghai: Shanghai Scientific & Technical Publishers; 1989.

Hu GS, Zhang DF, Zhao DZ, et al. Microstructures and mechanical properties of extruded and aged Mg−Zn-Mn-Sn-Y alloys. Transactions of Nonferrous Metals Society of China 2014;24(10):3070−5.

Huang ZH, Liu-Wang HB, Qi WJ, et al. Effects of Sm on microstructure and mechanical property of AZ91D alloy. Chinese Journal of Nonferrous Metals 2015;10:2649−55.

Huppmann M, Gall S, Mller S, et al. Changes of the texture and the mechanical properties of the extruded Mg alloy ME21 as a function of the process parameters. Materials Science and Engineering: A 2010;528(1):342−54.

Illkova K, Dobron P, Chmelik F, et al. Acoustic emission study of the deformation behavior of Mg−Mn alloys containing rare earth elements. Acta Physica Polonica Series A: General Physics 2012;122 (3):634−8.

Ji GQ. Effect of Sm and Ti compound inoculation on microstructure and mechanical properties of AM60 alloy. Taiyuan: Taiyuan University of Technology; 2011.

Jian WW, Kang ZX, Li YY. Effect of hot plastic deformation on microstructure and mechanical property of Mg-Mn-Ce magnesium alloy. Transactions of Nonferrous Metals Society of China 2007;17 (6):1158−63.

Jiang B, Gao L, Huang GJ, et al. Effect of extrusion processing parameters on microstructure and mechanical properties of as-extruded AZ31 sheets. Transactions of Nonferrous Metals Society of China 2008;18:160−4.

Kang JJ. Study on strenging and toughening of AZ61 magnesium alloy with Pr and Ti additions. Taiyuan: Taiyuan University of Technology; 2012.

Kaya AA, Duygulu O, Ucuncuoglu S, et al. Production of 150 cm wide AZ31 magnesium sheet by twin roll casting. Transactions of Nonferrous Metals Society of China 2008;18(1):s185−8.

Kim HL, Bang WK, Chang YW. Deformation behavior of as-rolled and strip-cast AZ31 magnesium alloy sheets. Materials Science and Engineering: A 2011;528:5356−65.

Kleiner S, Uggowitzer PJ. Mechanical anisotropy of extruded Mg-6%Al-1%Zn alloy. Materials Science and Engineering: A 2004;379:258−63.

Koike J. Enhanced deformation mechanisms by anisotropic plasticity in polycrystalline mg alloys at room temperature. Metallurgical and Materials Transactions A 2005;36:1689−96.

Koike J, Kobayashi T, Mukai T, et al. The activity of non-basal slip systems and dynamic recovery at room temperature in fine-grained AZ31B magneium alloys. Acta Materials 2003;51(7):2055−9.

Kojima Yo, Inoue Makoto, Osamu Tanno. Superplasticity of Mg-Li alloy. The Japan Institute of Metals 1990;54(3):354−5.

Lentz M, Klaus M, Coelho RS, et al. Analysis of the deformation behavior of magnesium-rare earth alloys Mg-2 pct Mn-1 pct rare earth and Mg-5 pct Y-4 pct rare earth by in situ energy-dispersive X-ray synchrotron diffraction and elasto-plastic self-consistent modeling. Metallurgical and Materials Transactions A 2014;45(12):5721−35.

Li CQ. Effects of Sm, Ca and Sb on mechanical properties of AE51 magnesium alloy. Luoyang: Henan University of Science & Technology; 2011.

Li DS, Li D, Tong YY, et al. Effect of Sr on microstructure and mechanical properties of AM50. Materials Reviews 2010;24(10):47−9.

Li JF, Zhen ZQ, Tao GY. Review on super light Mg-Li alloy. Light Alloy Fabrication Technology 2004;32(10):35−8.

Li JH, Jie WQ, Yang GY. Effect of gadolinium on microstructure and the mechanical properties of Mg-Zn-Zr cast magnesium alloys. Rare Metal Materials and Engineering 2008;37(9):1587−91.

Li Q, Wang QD, Wang YG, et al. Effect of Nd and Y addition on microstructure and mechanical properties of as-cast Mg−Zn-Zr alloy. Journal of Alloys and Compounds 2007;427:115−23.

Li RH, Pan FS, Jiang B, et al. Effect of Li addition on the mechanical behavior and texture of the as-extruded AZ31 magnesium alloy. Materials Science and Engineering: A 2013;562:33−8.

Li RH, Jiang B, Li FJ. Preparation and processing of high performance Mg-Li Alloy sheet. Changchun: Jilin university Press; 2014.

Li YG, Duan JH, Liu HL, et al. Effect of yttrium on mechanical properties and microstructure of Mg-Zn - Zr alloy. Mod Machinery 2003;5:86−8.

Liang D, Cowley CB. The twin-roll strip casting of magnesium. Journal of the Minerals Metals & Materials Society 2004;5:26−8.

Liu Q. Research progress on plastic deformation mechanism of Mg alloys. Acta Metallurgica Sin 2010;46:1458−72.

Liu WJ. Effect of Sm and Nd on the microstructure and mechanical properties of AZ81 magnesium alloy. Luoyang: Henan University of Science & Technology; 2014.

Liu XH, Wu RZ, Niu ZY, Zhang JH, Zhang ML. Superplasticity at elevated temperature of a Mg-8% Li-2%Zn alloy. Journal of Alloys and Compounds 2012;541:372−5.

Liu Z, Zhang K, Zeng XQ. Theory and application of magnesium matrix light mass alloy. Beijing: China Machine Press; 2002.

Liu ZL, Ding WJ, Yuan G, et al. Survey of the Mg−Al based elevated temperature magnesium casting alloys. Materials for Mechanical Engineering 2001;(11):1−4.

Liu ZM, Xing SM, Bao PW, et al. Deep drawing of twin-roll cast AZ31B Mg alloy sheets at warm temperatures by pre-forming. Chinese Journal of Nonferrous Metals 2010;20(4) 688−94.

Lou XY, Li M, Boger RK, Agnew SR, Wagoner RH. Hardening evolution of AZ31B Mg sheet. International Journal of Plasticity 2007;23:44−86.

Luo D, Wang HY, Chen L, et al. Strong strain hardening ability in an as-cast Mg-3Sn-1Zn alloy. Materials Letters 2013;94:51−4.

Luo ZP, Zhang SQ, Lu LQ, et al. Influence of heat treatment on microstructure and mechanical properties of as-extruded Mg-Nd-Zr magnesium alloy. Journal of the Chinese Rare Earth Society 1994;12 (2):183−5.

Ma CJ, Zhang D, Qin JM, et al. Mechanical properties and damping capacity of Mg-Li-Al alloys. Chinese Journal of Nonferrous Metals 2000;10(1):10−14.

Mackenzie LWF, Pekguleryuz M. The influences of alloying additions and processing parameters on the rolling microstructures and textures of magnesium alloys. Materials Science and Engineering: A 2008;480(1−2):189−97.

Martin É, Capolungo L, Jiang L, et al. Variant selection during secondary twinning in Mg-3%Al. Acta Materials 2010;58:3970−83.

Masoumi M. Microstructure and texture studies on magnesium sheet alloys. Canada: McGill University Montreal; 2011.

Masoumi M, Pekguleryuz M. Effect of Sr on the texture of rolled Mg−Mn-based alloys. Materials Letters 2012;71(15):104−7.

Mendis CL, Bae JH, Kim NJ, et al. Microstructures and tensile properties of a twin roll cast and heat-treated Mg-2.4Zn-0.1Ag-0.1Ca-0.1Zr alloy. Scripta Materialia 2011;64:335−8.

Mendis CL, Bettles CJ, Gibson MA, et al. An enhanced age hardening response in Mg-Sn based alloys containing Zn. Materials Science and Engineering: A 2006;435−436:163−71.

Mendis CL, Oh-Ishi K, Hono K. Enhanced age hardening in a Mg-2.4at.% Zn alloy by trace additions of Ag and Ca. Scripta Materialia 2007;57(6):485−8.

Metenier P, González-Doncel G, Ruano OA, et al. Superplastic behavior of a fine-grained two-phase Mg-9wt.%Li alloy. Materials Science and Engineering: A 1990;125:195−202.

Mu SJ, Jonas JJ, Gottstein G. Variant selection of primary, secondary and tertiary twins in a deformed Mg alloy. Acta Materials 2012;60:2043−53.

Nie JF. Effects of precipitate shape and orientation on dispersion strengthening in magnesium alloys. Scripta Materialia 2003;48(8):1009−15.

Pan FS, Han EH. High-performance wrought magnesium alloys and its processing technology. Beijing: Science Press; 2007.

Park S, You B. Low-temperature superplasticity of extruded Mg-Sn-Al-Zn alloy. Scripta Materialia 2011;65(3):202−5.

Pekguleryuz M, Celikin M. Creep resistance in magnesium alloys. International Materials Reviews 2010;55(4):197−217.

Planken J. Precipitation hardening in magnesium-tin alloys. Journal of Materials Science 1969;4(10):927−9.

Proust G, Tomé CN, Jain A, Agnew SR. Modeling the effect of twinning and detwinning during strain-path changes of magnesium alloy AZ31. International Journal of Plasticity 2009;25:861−80.

Roberts CS. Magnesium and its Alloys. Wiley; 1960.

Ryuuji Nininmiya, Koichi Miyake. A study of superlisht and superplastic Mg-Li alloy. Journal of the Japan Institute of Metals 2001;51(10):509−13.

Sandlöbes S, Friák M, Zaefferer S, et al. The relation between ductility and stacking fault energies in Mg and Mg-Y alloys. Acta Materials 2012;60:3011−21.

Sandlöbes S, Zaefferer S, Schestakow I, et al. On the role of non-basal deformation mechanisms for the ductility of Mg and Mg-Y alloys. Acta Materials 2011;59:429−39.

Sang J, Kang ZX, Li YY. Corrosion resistance of Mg-Mn-Ce magnesium alloy modified by polymer plating. Transactions of Nonferrous Metals Society of China 2008;18:s374−9.

Sasaki T, Oh-Ishi K, Ohkubo T, et al. Effect of double aging and microalloying on the age hardening behavior of a Mg-Sn-Zn alloy. Materials Science and Engineering: A 2011;530:1−8.

Sasaki TT, Ju JD, Hono K, et al. Heat-treatable Mg-Sn-Zn wrought alloy. Scripta Materialia 2009;61 (1):80−3.

Sasaki TT, Yamamoto K, Honma T, et al. A high-strength Mg-Sn-Zn-Al alloy extruded at low temperature. Scripta Materialia 2008;59(10):1111−14.

Shen F, Dong Q. Effect of rare earth on microstructure and mechanical properties of AZ31 magnesium alloy. Foundry Technology 2016;(8):1572−4.

Shi XB. Effect of Ca addition on microstructures and mechanical properties of AZ-series magnesium alloy. Taiyuan: Taiyuan University of Science and Technology; 2016.

Skjerpe P, Simensen C. Precipitation in Mg−Mn alloys. Meat Science 1983;17(8):403−7.

Song B. Study on the plastic deformation behavior and anisotropy of precipitation and twinning strengthening magnesium alloys. Chongqing: Chongqing University; 2013.

Song FQ, Zhang QL, Xu YC, et al. Grain refinement of wrought magnesium alloy and microstructures and properties in annealing states. Met Form Technology 2004;(3):46−8.

Song XJ. Effects of Y and Gd on microstructure and properties of AZ61 magnesium alloy. Luoyang: Henan University of Science & Technology; 2014.

Stanford N. The effect of calcium on the texture, microstructure and mechanical properties of extruded Mg−Mn−Ca alloys. Materials Science and Engineering: A 2010;528(1):314−22.

Stanford N, Atwell D, Beer A, et al. Effect of microalloying with rare-earth elements on the texture of extruded magnesium-based alloys. Scripta Materialia 2008;59(7):772−5.

Stanford N, Barnett M. Effect of composition on the texture and deformation behavior of wrought Mg alloys. Scripta Materialia 2008;58(3):179−82.

Stanford N, Sha G, La Fontaine A, et al. Atom probe tomography of solute distributions in Mg-based alloys. Metallurgical and Materials Transactions A 2009;40(10):2480−7.

Stratford D, Beckley L. Precipitation Processes in Mg−Th, Mg−Th−Mn, Mg−Mn, and Mg−Zr Alloys. Meat Science 1972;6(1):83−9.

Styczynski A, Hartig C, Bohlen J, et al. Cold rolling textures in AZ31 wrought magnesium alloy. Scripta Materialia 2004;50(7):943−7.

Suresh K, Rao KP, Prasad YVRK, et al. Study of hot forging behavior of as-cast Mg-3Al-1Zn-2Ca alloy towards optimization of its hot workability. Materials & Design 2014;57:697−704.

Suzuki K, Chino Y, Huang XS, et al. Elastic and damping properties of AZ31 magnesium alloy sheet processed by high-temperature rolling. Materials Transactions 2011;52(11):2040−4.

Takuda H, Kikuchi S, Yoshida N, et al. Tensile properties and press formability of a Mg-9Li-Y alloy sheet. Materials Transactions 2003;44(11):2266−70.

Wang C, Zhang HY, Wang HY, et al. Effects of doping atoms on the generalized stacking-fault energies of Mg alloys from first-principles calculations. Scripta Materialia 2013;69:445−8.

Wang HL, Li R, Yang JJ. Effect of RE and heat treatment on microstructure and mechanical properties of AZ91D. Mod Bus Trades India 2016;37(32):190−2.

Wang HY, Nan XL, Zhang N, et al. Strong strain hardening ability in an as-cast Mg-3Al-3Sn alloy. Materials Chemistry and Physics 2012;132(2−3):248−52.

Wang HY, Xue ES, Xiao W, et al. Influence of grain size on deformation mechanisms in rolled Mg−3Al−3Sn alloy at room temperature. Materials Science and Engineering: A 2011;528 (29−30):8790−4.

Wang HY, Zhang N, Wang C, et al. First-principles study of the generalized stacking fault energy in Mg-3Al-3Sn alloy. Scripta Materialia 2011;65(8):723−6.

Wang J, Jiang B, Ding PD, et al. Study on solidification microstructure of AZ31 alloy strips by vertical twin roll casting. Materials Science Forum 2007;546−549:383−6.

Wang J, Lu R, Qin D, et al. A study of the ultrahigh damping capacities in Mg−Mn alloys. Materials Science and Engineering: A 2013;560(10):667−71.

Wang J, Liao R, Wang L, et al. Investigations of the properties of Mg-5Al-0.3Mn-xCe (x=0−3, wt. %) alloys. Journal of Alloys and Compounds 2009;477(1):341−5.

Wang LG, Zhang BF, Guan SK, et al. Effects of RE, B on microstructure and properties of AM60 alloy. Rare Metal Materials and Engineering 2007;36(1):59−62.

Wang YN, Huang JC. The role of twinning and untwinning in yielding behavior in hot-extruded Mg−Al−Zn alloy. Acta Materials 2007;55:897−905.

Wang YX, Fu JW, Yang YS. Effect of Nd addition on microstructures and mechanical properties of AZ80 magnesium alloys. Transactions of Nonferrous Metals Society of China 2012;21(6):1322−8.

Wilson R, Bettles CJ, Muddle BC, et al. Precipitation hardening in Mg-3 wt% Nd (-Zn) casting alloys. Materials Science Forum 2003;419:267−72.

Wu AR, Gu Y, Xia CQ. Study of microstructure and properties of Mg-RE(Ce Nd Y)-Zn-Zr alloys. Hot Working. Technology 2004;12:21−3.

Wu D, Chen RS, Han EH. Excellent room-temperature ductility and formability of rolled Mg-Gd-Zn alloy sheets. Journal of Alloys and Compounds 2011;509:2856−63.

Wu YF, Du WB, Yan ZJ, et al. Effects of Nd on tensile properties of Mg-6Al as-cast alloy. Rare Metal Materials and Engineering 2010;39(10):60−4.

Xiao Y, Zhang XM, Chen BX, et al. Mechanical properties of Mg-9Gd-4Y-0.6Zr alloy. Transactions of Nonferrous Metals Society of China 2006;16(s3):1669−72.

Xiao Y, Zhang XM, Chen JM, et al. Microstructures and mechanical properties of extruded Mg-9Gd-4Y-0.6Zr-T5 at elevated temperatures. Chinese Journal of Nonferrous Metals 2006;16(4):709−14.

Xin YC, Zhou XJ, Liu Q. Suppressing the tension−compression yield asymmetry of Mg alloy by hybrid extension twins structure. Materials Science and Engineering: A 2013;567:9−13.

Xu DF. Microstructure and properties of Mg-4Al-RE-xCa heat-resistant magnesium alloys. Spec Cast & Nonferrous Alloy 2015;35(4):419−22.

Xu TY. Effect of alloying elements and solidification rate on microstructure and mechanical properties of AM60 alloy. Chongqing: Chongqing University; 2016.

Yan H, Chen RS, Han EH. A comparative study of texture and ductility of Mg-1.2Zn-0.8Gd alloy fabricated by rolling and equal channel angular extrusion. Mater Charact 2011;62:321−6.

Yan YQ, Den J, Zhang YJ, et al. Effects of Ce element on microstructure and tensile properties at room temperature of ZK40 alloy. Hot Working Technology 2004;(4):4−6.

Yang G. Effects of Y, Bi elements on microstructure and properties of Sn-containing AZ80 magnesium alloys. Chongqing: Chongqing University; 2015.

Yang J, Wang JL, Wang LD, et al. Microstructure and mechanical properties of Mg-4.5Zn-xNd (x=0, 1 and 2, wt%) alloys. Materials Science and Engineering 2008;A479:339−44.

Yang L, Hou H, Zhao YH, et al. Strengthening effect of heat treatment on AZ80 alloy. Hot Working Technology 2014;43(24):179−81.

Yang QS. Study on microstructures and mechanical properties of As-extruded magnesium alloys plate. Chongqing: Chongqing University; 2013.

Yang QS, Jiang B, Dai JH, et al. Mechanical properties and anisotropy of AZ31 alloy sheet processed by flat extrusion container. Journal of Materials Research 2013;28:1148−54.

Yang QS, Jiang B, Huang XY, et al. Influence of microstructural evolution on mechanical behavior of AZ31 alloy sheet processed by flat extrusion container. Materials Science and Technology 2013;29:1012−16.

Yang QS, Jiang B, Tian Y, et al. A tilted weak texture processed by an asymmetric extrusion for magnesium alloy sheets. Materials Letters 2013;100:29−31.

Yoo MH. Slip, twinning, and fracture in hexagonal close-packed metals. Metallurgical and Materials Transactions A 1981;12:409−18.

Yoshida Y, Cisar L, Kamado S, Kojima Y. Low temperature superplasticity of ECAE processed Mg-10% Li-1%Zn alloy. Materials Transactions 2002;43(10):2419−23.

Yuasa M, Hayashi M, Mabuchi M, et al. Improved plastic anisotropy of Mg−Zn-Ca alloys exhibiting high-stretch formability: a first-principles study. Acta Materials 2014;65:207−14.

Yuasa M, Miyazawa N, Hayashi M, et al. Effects of group II elements on the cold stretch formability of Mg−Zn alloys. Acta Materials 2015;83:294−303.

Zeng Y, Jiang B, Zhang MX, et al. Effect of $Mg_{24}Y_5$ intermetallic particles on grain refinement of Mg-9Li alloy. Intermetallics 2014;45:18−23.

Zhang EL, He WW, Du HL. Microstructure, mechanical properties and corrosion properties Mg−Zn-Y alloys with low Zn contet. Materials Science and Engineering 2008;A488:102−11.

Zhang H, Huang GS, Li JH, et al. Influence of warm pre-stretching on microstructure and properties of AZ31 magnesium alloy. Journal of Alloys and Compounds 2013;563:150−4.

Zhang H, Huang GS, Wang LF, et al. Improved formability of Mg−Al−1Zn alloy by pre-stretching and annealing. Scripta Materialia 2012;67:495−8.

Zhang HJ, Meng J, Tang DJ. Investigation,exploitation and application of magnesium-rare earth alloy as a structure material. Journal of the Chinese Rare Earth Society 2004;22(1):40−7.

Zhang J, Fang C, Yuan F, et al. A comparative analysis of constitutive behaviors of Mg−Mn alloys with different heat-treatment parameters. Materials & Design 2011;32(4):1783−9.

Zhang J, Zhang D, Zheng T, et al. Microstructures, tensile properties and corrosion behavior of die-cast Mg-4Al-based alloys containing La and/or Ce. Materials Science and Engineering: A 2008;489 (1−2):113−19.

Zhang JY. Study on strengthening and toughening as-cast AM60 magnesium alloy. Taiyuan: Taiyuan University of Technology; 2008.

Zhang JZ, Ma ZX, Li DF. Effect of Pre-deformation on the mechanical properties of Mg-Gd-Y-Zr alloy. Chinese Journal of Rare Metals 2008;32(4):522−5.

Zhang XP, Yuan GY, Liu Y, et al. Effects of alloying elements on the microstructure and mechanical properties of Mg−Zn-Gd alloys. Special Casting & Nonferrous Alloys 2008;28(11):882−5.

Zhao F. Effect of Nd 和 Ca on microstructure and mechanical properties of AZ63 magnesium alloy. Taiyuan: Taiyuan University of Science and Technology; 2012.

Zhou HT, Zeng XQ, Liu WF, et al. Effect of Ce on microstructures and mechanical properties of AZ61 wrought magnesium. The Chinese Journal of Nonferrous Metals 2004;1:99−104.

CHAPTER 2

"Solid solution strengthening and ductilizing" theory for magnesium alloys

As a HCP crystal structure, magnesium alloys have only two independent sliding systems, both of which are on the same slip plane from a crystallographic point of view. In addition, the critical resolved shear stress (CRSS) in basal slip is much lower than that in prismatic slip and pyramidal slip, which causes poor plastic formability and poor uniform formability due to the formation of the basal texture. To further improve the plasticity and formability of magnesium alloys, new theory and approach should be developed from the basic factor that limits the plasticity and uniform formability to improve strength and plasticity simultaneously.

It is an important way to modulate the mechanical properties of alloys by adding appropriate alloying elements for solution and subsequent precipitation treatment. The plasticity in cubic metals, especially in steel and aluminum alloy materials, is generally reduced by solution strengthening and precipitation strengthening. The plasticity reduction due to precipitation strengthening has been validated from a large number of experiments in magnesium alloys. However, some new phenomena are found during solution treatment, that is, some elements can improve the plasticity and strength simultaneously. Based on this phenomenon, a new principle of alloy design is proposed through experimental verification and theoretical application.

2.1 Alloy design theory of solid solution strengthening and ductilizing

Since 2002, Fusheng Pan's research group in Chongqing University has performed significant research on the effects of alloy elements on the strength and ductility of magnesium alloys, such as Ag, Nd, Zn, Mn, Sn, Er, Al, Y, Ca, Ce, Li, Gd, Sc, and Sr. Among these studies, a common phenomenon of precipitation strengthening is that the strength of magnesium alloys increases while the ductility decreases. The magnesium alloy research team in Chongqing University focuses on how the changes of the resistance to basal and nonbasal slip systems affect ductility after alloying elements are dissolved in magnesium alloys. It is found that some specific elements can reduce the difference of slip resistance between basal and nonbasal slip systems, which is

High Plasticity Magnesium Alloys
DOI: https://doi.org/10.1016/B978-0-12-820110-7.00002-1

beneficial to the activation of nonbasal slip and thus improves the ductility of magnesium alloys. Around 2005, Fusheng Pan and his collaborators proposed an alloy design theory of "solid solution strengthening and ductilizing" in combination with research work at home and abroad. The main idea is shown in Fig. 2.1.

Fig. 2.1 shows that the solid solution of alloy elements in magnesium can increase or decrease the slip resistance to basal and nonbasal slip systems. In this case, the relationship between $\Delta\tau'$ and $\Delta\tau$ can be divided into three types, $\Delta\tau \approx \Delta\tau$, $\Delta\tau' > \Delta\tau' < \Delta\tau$, where $\Delta\tau'$ is the difference of slip resistance between basal and nonbasal slip systems in solid solution alloys, $\Delta\tau$ is the difference of slip resistance between basal and nonbasal slip system in pure magnesium.

1. $\Delta\tau' \approx \Delta\tau$ (see Fig. 2.1A and D): Alloy element solution has little effect on the difference of slip resistance between basal and nonbasal slip systems, which cannot facilitate the activation of nonbasal dislocation slip and then improve the ductility of magnesium alloys.
2. $\Delta\tau' > \Delta\tau$ (see Fig. 2.1C, F, and G): Alloy element solution increases the difference of slip resistance between basal and nonbasal slip systems, which is difficult to start the nonbasal dislocation slip and is not good to the improvement of ductility.

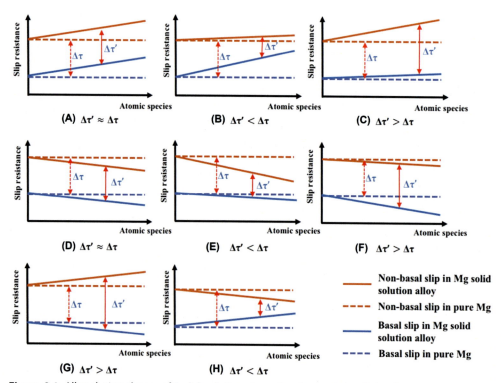

Figure 2.1 Alloy design theory of "solid solution strengthening and ductilizing."

3. $\Delta\tau' < \Delta\tau$ (see Fig. 2.1B, E, and H): Alloy element solution reduces the difference of slip resistance between the basal and nonbasal slip systems, which is beneficial to the initiation of the nonbasal dislocation slip and the improvement of the homogeneous plastic deformation ability of magnesium alloys. However, the decrease of the resistance to the basal dislocation slip is not beneficial to the improvement of strength of magnesium alloys. As shown in Fig. 2.1E, though the ductility can be improved under that condition, the strength can be lost to some extent, which limits the industrial application of magnesium alloys.

Fig. 2.1B and H shows the development direction that should be pursued in alloy design, and that in Fig. 2.1H is the most beneficial way to improve both strength and ductility of magnesium alloys simultaneously. Therefore the alloy elements, which can increase the resistance to the basal dislocation slip and decrease the difference of slip resistance between the basal and nonbasal dislocations, should be chosen to design alloy composition. Such element can not only enhance the strength of magnesium alloys, but also improve the ductility of magnesium alloys by promoting the nonbasal dislocation slip. In this regard, the strength and ductility of magnesium alloys can be improved simultaneously, achieving the purpose of the alloy design idea "solid solution strengthening and ductilizing."

2.2 Theoretical calculation of solid solution strengthening and ductilizing

The previous studies show that the change of c/a axial ratio for magnesium with a close packed hexagonal structure could improve the ductility of magnesium alloys by changing the atomic spacing and activating the nonbasal dislocation slip. More and more studies have found that the increase or decrease of the axial ratio has no corresponding relationship with the difficulty of the nonbasal dislocation activity. Therefore the change of the axial ratio is not the key factor to stimulate the activation of the nonbasal dislocation slip.

Stacking fault is a common intrinsic characteristic in crystal materials, and stacking fault energy is the energy corresponding to the stacking fault formed by the specific relative sliding displacement. In recent years, a large number of studies have shown that the change of stacking fault energy is related to both the activation of the basal and nonbasal dislocation slips and the ability of the plastic deformation. Fusheng Pan's research group in Chongqing University has studied the effects of various solid solution atoms on the stacking fault energy of magnesium alloys. The results show that the alloying elements, such as Al, Bi, Ca, Dy, Er, Ga, Gd, Ho, In, Lu, Nd, Pb, Sm, Sn, Y, and Yb, can significantly reduce the stacking fault energy of I_1, and the alloying elements, such as Ca, Dy, Er, Gd, Ho, Lu, Nd, Sm, Y, and Yb, can greatly reduce the unstable stacking fault energy of the prismatic slip system and are beneficial to the reduction of the CRSS of the prismatic dislocation. The alloying elements, such as Ag, Al, Ca, Dy, Er, Ga, Gd, Ho, Li, Lu, Nd, Sm, Y, Yb, and Zn, are beneficial to activate the pyramidal dislocation slip and can improve the intrinsic ductility of magnesium alloys.

Based on the first principles calculation, Yasi and his group established the quantitative relationship between the stacking fault energy and CRSS of the basal slip for some binary magnesium solid solution alloys, and proved the direct correlation between the stacking fault energy and CRSS. In recent years, with the development of computer technology and methods, the stacking fault energy of the close packed crystal structure metals can be reliably calculated via the theoretical model based on the local density functional theory (DFT). Such calculation is even more accurate than the experimental results in many cases. Important results and progress have been made by using the first principles calculation method, which provides practical conditions for theoretical calculation of high plasticity magnesium alloys.

2.2.1 Effect of alloying elements on stacking fault energy

The close packed hexagonal structure is stacked by the close packed plane (0001) in a sequence of •••ABABAB•••. For convenience, the symbol "$\triangle$" is used to denote the stacking sequence of AB, BC and CA, and the symbol "∇" is used to denote the stacking sequence of BA, AC and CB. Therefore the stacking sequence of the close packed hexagonal structure can be expressed as •••$\triangle\nabla\triangle\nabla$•••. If the atoms on a certain layer are all dislocated, a plane defect is formed, and the stacking sequence is changed to •••$\nabla\triangle\nabla\nabla\triangle\nabla$••• or •••$\nabla\triangle\nabla\nabla\nabla\triangle\nabla$•••, which is called stacking fault.

The formation of stacking fault breaks the integrity and periodicity of the crystal, which increases the crystal energy. The stacking fault energy (γ) is defined as the increased energy per unit area caused by stacking fault. If the energies of the perfect crystal and the crystal with stacking fault are E_0 and E_γ, the stacking fault energy is expressed as:

$$\gamma = \frac{E_\gamma - E_0}{A} \tag{2.1}$$

where A is the area of the close packed plane (0001).

As long as the energy of the perfect crystal (E_0) and the energy of the crystal with stacking fault (E_γ) are calculated via the first-principle method, the stacking fault energy can be calculated by Eq. (2.1).

When the relative shear displacement of two parts occurs, the stacking fault is formed. By making the X and Y coordinates being the shear displacement and the system energy with unit area change respectively, the obtained curve is the generalized stacking fault energy (GSFE) curve.

2.2.2 Stacking fault energy (I_2) for the basal plane of Mg

The calculation model consists of $2 \times 2 \times 6$ supercell lattices containing 24 Mg atoms with a stacking sequence of BABABABABABA. The close packed plane A represents the basal plane (0001), and B represents the basal plane (0002), as shown in Fig. 2.2A. To calculate the change of the stacking fault energy of the pure magnesium, a stacking fault is introduced in the basal plane by making all the atoms in and above a certain basal plane translate a certain distance of $n \times 0.1a$ in a linear combination of $1/3 \left[\bar{1} \, \bar{1} 20 \right]$ and $\left[\bar{1} 100 \right]$ (a is the lattice constant of magnesium cell), as shown in Fig. 2.2B.

Fig. 2.3 shows the typical interstitial positions in a close packed hexagonal crystal, which can be represented by A, B and C respectively. It can be seen from the stacking sequence of the close packed hexagonal crystal (Fig. 2.3) that the atomic stacking sequence is ABABABABABA. The formation of I_2 type stacking fault can be regarded as that certain close packed plane (A) of the close packed hexagonal crystal is taken off, and then all the atoms in and above this plane are sheared along the $1/3 \left[\bar{1} 010 \right]$ vector. Therefore the atoms in the plane B above the plane which is taken off moves to the C-interstitial position, and the atoms in the plane A above the plane which is taken off moves to the B-interstitial position. The stacking sequence after the shear displacement becomes BABABABC'ACACAC, as shown in Fig. 2.3. According to the definition of stacking fault, a typical stable intrinsic stacking fault I_2 of the close packed hexagonal crystal is constructed in the basal plane (0001).

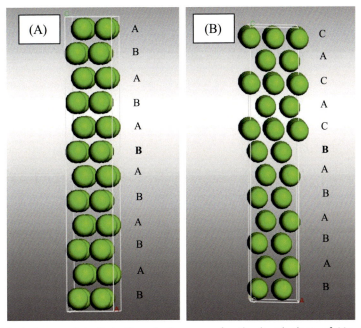

Figure 2.2 Calculation model of stacking fault energy for the basal plane of Mg: (A) $2 \times 2 \times 6$ supercell, (B) $2 \times 2 \times 6$ supercell with I_2 stacking fault.

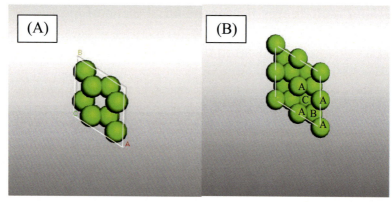

Figure 2.3 Stacking sequence in HCP structure (Projection) (A) perfect HCP structure (B) HCP structure with stacking fault l_2.

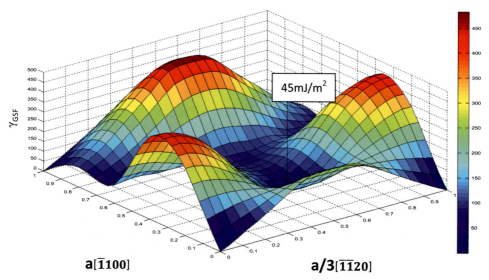

Figure 2.4 Generalized stacking fault energy surface for the basal plane of Mg from first-principle calculation.

Fig. 2.4 shows the GSFE surface for the basal plane of Mg from the first-principle calculation, in which the X-coordinate is along $a/3 \left[\bar{1}\,\bar{1}20\right]$, and the Y-coordinate is along $a \left[\bar{1}100\right]$. The stacking fault energy of the pure Mg changes greatly along the crystal direction.

Figs. 2.5 and 2.6 show the stacking fault energy curve (γ_{GSF}) along $\left[\bar{1}010\right]$ and $1/3 \left[\bar{1}\,\bar{1}20\right]$ in the basal plane of Mg. In Fig. 2.5, the stable stacking fault energy (γ_{is}) along [-1010] is 45 mJ m^{-2}, and its unstable stacking fault energy (γ_{us}) is 101 mJ m^{-2}. In Fig. 2.6, there is no stable stacking fault along $1/3 \left[\bar{1}\,\bar{1}20\right]$ in the basal plane of Mg. γ_{us} is

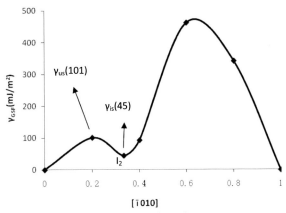

Figure 2.5 Stacking fault energy curve (γ_{GSF}) along $\left[\bar{1}010\right]$ in the basal plane of Mg.

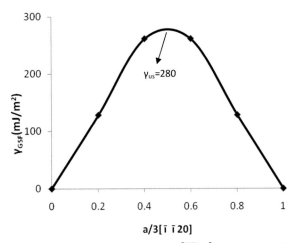

Figure 2.6 Stacking fault energy curve (γ_{GSF}) along 1/3 $\left[\bar{1}\bar{1}20\right]$ in the basal plane of Mg.

as high as 280 mJ m^{-2}, indicating that producing full dislocations along 1/3 $\left[\bar{1}\,\bar{1}20\right]$ in the magnesium basal plane needs to overcome a high energy. It also shows that the total dislocations generated along 1/3 $\left[\bar{1}\,\bar{1}20\right]$ are unstable from the thermodynamic point of view. Similarly, the dislocations generated along $\left[\bar{1}010\right]$ in the magnesium basal plane have high stability. Therefore the dislocations along the direction of a higher energy will spontaneously propagate to the dislocations with lower energy. It can be inferred that during the dislocation motion, the dislocation reaction of the total dislocation $a/3\left[\bar{1}\,\bar{1}20\right]$ may satisfy:

$$\frac{1}{3}\left[\bar{1}\bar{1}20\right] \rightarrow \frac{1}{3}\left[\bar{1}010\right] + \frac{1}{3}\left[0\bar{1}10\right] \tag{2.2}$$

It is consistent with the classical dislocation theory of the close packed hexagonal structure.

In addition, the direction of I_2 stacking fault is $1/3\left[1\,\overline{1}20\right]$, and the stable stacking fault energy is 25 meV m^{-2} (45 mJ m^{-2}), which is in good agreement with the experimental result (23 meV m^{-2}). All these calculations on the stacking fault energy for the basal plane of Mg proves the feasibility and reliability of the first-principle calculation, which lays a good foundation for evaluating the effect of alloying elements on the stacking fault energy.

2.2.3 Effect of alloying elements on generalized stacking fault energy

As shown in Fig. 2.7, $2 \times 2 \times 6$ supercell with 48 atoms are established with VASP to replace one Y or Zn atom (i.e., the solid solubility is 2.08 at.%) respectively. The influence of alloying elements (Y and Zn) on the stacking fault energy (I_2) of Mg is calculated by introducing I_2 type stacking fault (all the atoms in and above (1000) plane are sheared along the $1/3\left[\overline{1}010\right]$ vector). Table 2.1 shows the calculated stacking fault energy in the basal plane of Mg with addition of 2.08% Y or Zn.

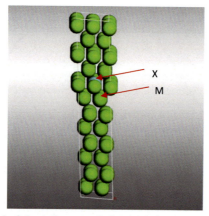

Figure 2.7 Calculation model of the influence of alloying elements (Y and Zn) on the stacking fault energy in the basal plane of Mg.

Table 2.1 Stacking fault energy and chemical misfit in the basal plane of Mg with addition of 2.08% Y or Zn.

Element	γ_{12}/ (mJ m^2)	$\Delta\gamma_{12}/\gamma_{12}$/%	ε_{SFE}
Mg	45.0	0	—
Y	35.0	-22.2	-0.88
Zn	45.3	$+0.67$	$+0.04$

The calculated stacking fault energy for pure Mg is 45 mJ m^{-2}, and addition of 2.08% Y sharply decreases the stacking fault energy of the alloy by 22.2% from 45 to 35 mJ m^{-2}. And addition of 2.08% Zn just slightly increases the stacking fault energy by 0.67% from 45 to 45.3 mJ m^{-2}. "+" means that the stacking fault energy is "increased" by alloying elements, and "$-$" means that the stacking fault energy is "decreased" by alloying elements.

In order to validate the calculated value of the stacking fault energy with addition elements, the chemical misfit (ε_{SFE}) is calculated from the supercell energies:

$$\varepsilon_{SFE} = \frac{E_{displaced}(solute) - E_{undisplaced}(solute) - 2\sqrt{3}a^2\gamma_{12}}{\gamma_{12}\sqrt{3}a^2/2} \tag{2.3}$$

where γ_{12} is the stacking fault energy of pure Mg, and the "displaced" and "undisplaced" geometries corresponded to the supercell with and without an I_2 intrinsic stacking faults, and $\sqrt{3}a^2/2$ is the basal plane area. The term $E_{displaced}(solute) - E_{undisplaced}(solute)$ represents the interfacial energy introduced by the stacking faults. The term $E_{displaced}(solute) - E_{undisplaced}(solute) - 2\sqrt{3}a^2\gamma_{12}$ represents the influence of alloying elements on the stacking fault energy. $\varepsilon_{SFE>0}$ indicates that the alloying elements can increase the stacking fault energy, and vice versa. The absolute value of ε_{SFE} represents the magnitude of the stacking fault energy of Mg caused by alloying elements.

Table 2.1 shows the calculated results of the chemical misfit (ε_{SFE}) of Y and Zn. The chemical misfits (ε_{SFE}) of Y and Zn are -0.88 and $+0.04$ respectively, indicating that the stacking fault energy of Mg decreases with the addition of Y, while that of Mg increases with the addition of Y. The variation range of the stacking fault energy of Mg caused by Y is significantly larger than that of Mg caused by Zn. Yasi also reported that ε_{SFE} of Y and Zn is -1.70 and $+0.32$ respectively, indicating that the stacking fault energy of Mg alloy can be reduced by adding Y and increased by adding Zn. The variation range of the stacking fault energy caused by Y is significantly greater than that of Zn. Generally speaking, the lower the stacking fault energy, the higher the probability of stacking fault. It is observed that the precipitation of LPSO phase in Mg$-$Zn$-$Zr$-$Y alloy is accompanied by the stacking fault, indicating that the remarkably decreased stacking fault energy due to the addition of Y can easily form the stacking fault.

2.2.4 Effect of alloying elements on critical resolved shear stress

The CRSS of Mg for nonbasal slip at room temperature (RT) is much higher than that for basal slip, resulting in the difficulty to activate nonbasal slip. Thus Mg and its alloys show poor formability at RT due to the lack of independent slip systems. Previous researches indicated that the nonbasal slip could be activated with the addition of certain alloying elements. For example, nonbasal slip could be activated with the addition of Li, Y, Zn, Ca, Gd, which makes the magnesium alloy show good formability at RT.

Akhtar et al. found the CRSS for Mg basal slip rose while that for pyramidal decreased when Zn was added to Mg alloys. Herrera-Solaz et al. combined experimental method with crystallography finite element simulation and demonstrated that CRSSs for all slips in Mg increased while the ratio of CRSSs for nonbasal slip and twinning decreased with increasing addition of Nd. However, few systematic investigations focus on the effect of the alloying elements on the CRSS for both basal slip and nonbasal slip.

First, the site preferences of different solute atoms (Al, Gd, Li, Mn, Sn, Y, Zn) in Mg crystal are calculated by the first-principle calculation, and the feasibility of this calculation method is verified by experiments. Then, the elements with the same doping preference as Li atoms are found out, and the changes of theoretical critical shear strength (τ_{max}) for basal slip and nonbasal slip are calculated after these solute atoms (including Li atoms) are doped into the Mg crystal. The specific calculation process is designed as follows: First, the structure relaxation of Mg supercell doped with solute atoms is carried out to obtain the equilibrium crystal structure, and the most stable site preferences is found out. At the same time, the calculation results are verified by experiments. Second, the GSFE for the basal and nonbasal slip of Mg crystal after doping is calculated when the atoms which are in the same site preferences as Li are found out. Finally, τ_{max} of the corresponding slip system is calculated through the generalized fault energy curve. The ratio of τ_{max} for the nonbasal slip to the basal slip is compared before and after doping the solute atoms.

2.2.5 Calculation methods for critical resolved shear stress

Generally, the CRSS can be experimentally measured by compression test of single crystal. There are some approaches to prepare single crystal, such as Ohno continuous casting, Bridgeman Process and Czochralski Method. However, because of the limitation of single crystal preparation conditions, it is difficult to prepare metal and alloy single crystals with specific composition or special phase state (such as single-phase solid solution).

Because of the limited solid solubility of alloying elements in Mg, the magnesium alloy usually has a second phase. The existence of second phase and the random orientations of grains before deformation make it hard to obtain Mg alloys singe crystal for experimental CRSS test. Therefore simulation is an alternative method to qualitatively or semiquantitatively investigate the effects of alloying elements on CRSS of Mg alloys. The simulation not only provides a convenient and effective method to evaluate the influence of various alloying elements on the formability of magnesium alloy, but also has an important significance for in-depth understanding of plastic deformation behavior of magnesium alloy, and provides theoretical guidance for the experimental selection and design of alloy elements.

According to the Peierls—Nabarro model, the resistance F of dislocation movement is periodic, and the dislocation can move only when the external shear stress exceeds F_{max}. The critical shear stress is defined as the external shear stress which makes the dislocation move over the dislocation energy barrier (lattice resistance), and it is defined as σ_{P-N}. The critical shear stress is also called Peierls—Nabarro force, which is equivalent to the critical shear stress required to move a dislocation to overcome the lattice resistance in an ideal crystal. σ_{P-N} can be calculated by the following equation:

$$\sigma_{P-N} = \frac{2\mu}{q} \exp\left(-\frac{2\pi a}{qb}\right) \tag{2.4}$$

where a is the dislocation width, b is the Burgers vector, μ is the shear modulus, and q is 1 for the screw dislocation and $1 - \nu$ for the edge dislocation (ν is the Poisson's ratio).

Dislocation slip is most easily carried out along the direction with the smallest σ_{P-N}. According to Eq. (2.4), the shear modulus μ and q are constant for the same material, and the close packed plane is accompanied with smaller dislocation width and smaller Bergers vector. Therefore the dislocation slip deformation in the crystal is easily carried out along the close packed direction in the close packed plane.

B.Joós made it capable to quantitatively forecast by amending Eq. (2.4). The Peierls force can be evaluated directly by the GSFE surface, which is valid for both wide dislocations and narrow dislocations. Based on the continuum theory, the stress field generated at a certain point x' at the dislocation line is proportional to that at the point $1/(x - x')$, and the integral-differential formula for P—N model is:

$$\frac{K}{2\pi} \int_{-\infty}^{+\infty} \frac{1}{x - x'} \frac{df(x')}{dx'} dx' = F_b(f(x)) \tag{2.5}$$

where K is the elastic constant which is dependent on the type of dislocation and the direction of Burgers vector, and $F(f)$ is the GSFE. A simple sinusoidal shear stress is introduced into the original P—N model, and the concept of theoretical shear strength is introduced. The theoretical critical shear strength of a part of a crystal slipping along a certain crystal direction is an important parameter to measure the critical shear stress in the crystal, which is defined as τ_{max}. For simplifying analytical solution, the sinusoidal restoring force is assumed as follow:

$$F_b(f(x)) = \tau_{max} \sin \frac{2\pi f(x)}{b} \tag{2.6}$$

According to the approximate formula from Eq. (2.6), Eq. (2.5) has a soliton solution:

$$f(x) = \frac{b}{\pi}\tan^{-1}\frac{x}{\xi} + \frac{2}{b} \tag{2.7}$$

where $\xi = Kb/4\pi\tau_{max}$ is the half-width of the dislocation.

In the absence of dislocations, the interplanar distance of atoms in the x direction is defined as a'. A dislocation is introduced at the u position, and the upper half part of the crystal which is perpendicular to the dislocation line is ma'. This plane will replace the crystal surface of the lower half along the Burgers vector, and the formula is $f(ma' - u)$. The mismatch energy is the sum of the misfit energies of each part of the atomic surface, which can be written as:

$$W(u) = \sum_{-\infty}^{+\infty}\gamma\big(f\big(ma' - \mu\big)\big)a' = \gamma(f(-\mu))a' = \frac{Kb^2a'}{4\pi^2}\frac{\xi}{\xi^2 + \mu^2} \tag{2.8}$$

and hence,

$$\sigma(\mu) = \frac{1}{b}\frac{dW}{d\mu} = \frac{a'}{b}\frac{d\gamma(f(-\mu))}{df}\frac{df}{d\mu} \tag{2.9}$$

Therefore the maximum value of Peierls force is:

$$\sigma_{P-N} = \frac{3\sqrt{3}}{8}\tau_{max}\frac{a'}{\pi\xi} \tag{2.10}$$

A more general form is obtained under this boundary condition, which can be applied to nonsinusoidal shear forces. Eq. (2.9) can be written as:

$$\sigma(u) = \frac{1}{b}\frac{dW}{du} = \frac{a'}{b}\frac{d\gamma(f(-u))}{df}\frac{df}{du} \tag{2.11}$$

Similar to Eq. (2.7), by introducing a $\tan^{-1}$ in Eq. (2.11), Eq. (2.11) can be written as:

$$\sigma(\mu) = -\frac{a'}{\pi}\frac{d\gamma(f(-\mu))}{df}\frac{\xi}{\mu^2 + \xi^2} \tag{2.12}$$

With combination of Eqs. (2.7) and (2.12), σ_{P-N} can be calculated by the following equation:

$$\sigma_{P-N} = \frac{a'}{\pi\xi}\max\left\{\frac{d\gamma(f)}{df}\sin^2\frac{\pi f}{b}\right\} \tag{2.13}$$

The theoretical shear strength (τ_{max}) of the ideal crystal sliding along a certain direction to the dislocation on a certain crystal plane can be obtained by Eqs. (2.10) and (2.13):

$$\tau_{max} = \frac{8\sqrt{3}}{9} \max\left\{ \frac{d\gamma(f)}{df} \sin^2 \frac{\pi f}{b} \right\} \tag{2.14}$$

where $d\gamma(f)/df$ is the derivative of the GSFE curve when the crystal slips along a certain slip system.

Theoretical critical shear strength (τ_{max}) is the minimum required shear strength for lattice dislocations to start slipping along the slipping direction on the slip plane during the sliding deformation of the perfect crystal, which is an intrinsic parameter based on electronic structure and chemical bonds of crystals. In real crystals, there are various defects inside in different dimensions (such as point, line and plane), and these defects have accumulated energy in crystal before slip systems start. As a result, the value of τ_{max} tend to be higher than that of CRSS, but the variation tendencies of them are in accordance. Therefore the variation rule of CRSS in Mg crystal can be qualitatively investigated by the calculation of τ_{max}. Theoretically, the decreasing value of the ratio of the nonbasal slip τ_{max} to the basal slip τ_{max} can illustrate that the CRSS ratio also decreases, further indicating that the nonbasal slip systems in this alloy is easier to be activated than those in pure Mg.

2.2.6 Site preferences of alloying elements in Mg crystal

1. The first-principle calculation

Based on the crystal structure symmetry, a small part of the large system (Fig. 2.8B) is selected from the center of the magnesium supercell (Fig. 2.8A) to calculate the site preference of solute atoms in magnesium crystals, and the doping site of the solute atom in a large system is inferred by using the calculation results of the small system. A $3 \times 3 \times 2$ Mg supercell is constructed by the first principles calculation, which contains 36 atoms, as shown in Fig. 2.8C. In order to use the doping position of the solute atoms to determine the preferred crystal plane, each supercell contains three solute atoms (three atoms can form an atomic plane). Based on the slip system, the solute atoms occupy the basal plane $\{0001\}$, the prismatic plane $\{10\bar{1}0\}$, $\{11\bar{2}0\}$, and the pyramidal plane $\{11\bar{2}1\}$, $\{11\bar{2}2\}$, $\{10\bar{1}1\}$, $\{10\bar{1}2\}$ of the magnesium crystal, as shown in Fig. 2.9. In the model of $Mg_{33}X_3$ solid solution, the atomic concentration of solute is 8.33 at.%.

The first-principle calculations are performed by using the Vienna *ab initio* simulation package (VASP) which is based on the DFT. The electron exchange and the correlation are calculated by the generalized gradient approximation (GGA) with the advantage of high calculation accuracy and fast convergence rate. The interaction between electrons and ions is described by the Projector Augmented Wave pseudo potential method. The cutoff energy is set to 380 eV. The Brillouin zone is sampled

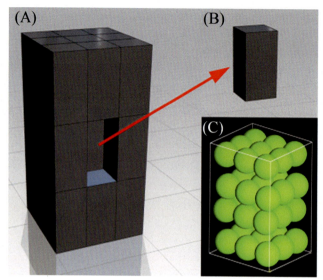

Figure 2.8 Mg supercell based on the crystal structure symmetry: (A) Mg supercell of a large system; (B) a small part of the center of the Mg supercell; (C) Mg supercell containing 36 atoms.

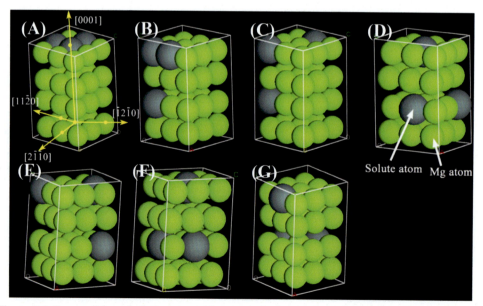

Figure 2.9 Supercell for calculating site preferences of solute atoms in Mg crystal (A) {0001} plane, (B) {10$\bar{1}$0} plane, (C) {11$\bar{2}$0} plane, (D) {11$\bar{2}$1} plane, (E) {11$\bar{2}$2} plane, (F) {10$\bar{1}$1} plane and (G) {10$\bar{1}$2} plane.

using a $7 \times 7 \times 5$ Monkhorst-Pack k-point mesh. The convergence accuracy of energy calculation is 1.0×10^{-5} eV atom^{-1}.

VESTA (visualization for electronic and structure analysis) software package is used for modeling and processing related crystallographic data of Mg crystal. The software can build crystal model accurately, and it is easy to operate. It can not only generate coordinate files for VASP, but also process output files after structure optimization for VASP.

Fig. 2.10 shows the calculated cohesive energies of Mg crystal with different alloying elements at different doping sites. Sites 1−7 refer to the planes of {0001}, {10$\bar{1}$0}, {11$\bar{2}$0}, {11$\bar{2}$1}, {11$\bar{2}$2}, {10$\bar{1}$1}, and {10$\bar{1}$2}, respectively. The lower the total energy of the system is, the more stable the system is. The crystallographic structure with the lowest cohesive energy corresponds to the preferred site of the solute atoms. Site preferences of solute atoms in Mg crystal are listed in Table 2.2. The Gd, Sn, Zn, Li atoms prefer to dope in {11$\bar{2}$0} plane of Mg crystal, while the Al, Mn, Zn atoms prefer to dope in {10$\bar{1}$0}, {0001}, and {11$\bar{2}$2} planes respectively, as shown in Table 2.2.

The electronic structures of Gd, Li, Sn and Y are 4f75d16s2, 2S1, 5s25p2 and 4d15s2, respectively. It can be seen that only Gd contains f-layer electrons. The f-layer electrons is the strongly correlated to electrons, but the GGA and LDA exchange correlation potentials in the traditional DFT calculation ignore its strong correlation characteristics. However, the existence of the electrons in f orbital might contribute to an absence of convergence during the GGA process of first-principle calculations. Therefore only Li, Sn, and Y are selected for the calculation except Gd.

2.3 Experimental verification for the site preferences of alloying elements by X-ray diffraction

The atoms in a crystal structure have a regular arrangement, so it is convenient to analyze the diffraction of crystal by taking the atom as an individual scattering source. When analyzing the diffraction of crystal materials, the X-ray with wavelength λ and intensity I_0 is incident on the polycrystalline sample. The unit cell volume is V_0, and the volume of irradiated crystal is V. If the {hkl} diffraction of a crystal plane occurs in the direction of 2θ from the incident angle, the integral strength with a unit length of L is detected at R from the sample plane, which is the absolute strength. However, only the relative value of the strength should be considered in the experiment. For the diffractive rays corresponding to the same phase in the same diffraction pattern, only the relative integrated intensity I of each diffraction pattern needs to be considered:

$$I = P|F_{HKL}|^2 \frac{1 + \cos^2 2\theta}{\sin^2 \theta \cos \theta} A(\theta) e^{-2M} \qquad (2.15)$$

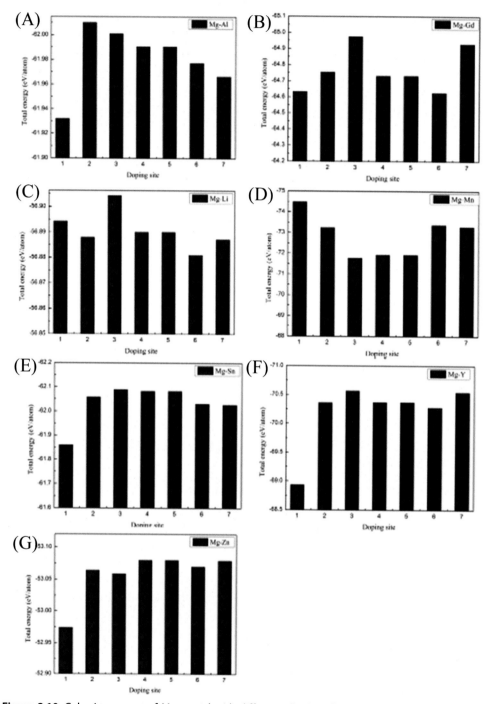

Figure 2.10 Cohesive energy of Mg crystal with different alloying elements in different substituted sites: (A) Mg−Al; (B) Mg−Gd; (C) Mg−Li; (D) Mg−Mn; (E) Mg−Sn; (F) Mg−Y; (G) Mg−Zn. Sites 1−7 refer to the planes of {0001}, {10$\bar{1}$0}, {11$\bar{2}$0}, {11$\bar{2}$1}, {11$\bar{2}$2}, {10$\bar{1}$1}, {10$\bar{1}$2}, respectively.

Table 2.2 Site preferences of solute atoms in Mg crystal.

Solute atom	Al	Gd	Li	Mn	Sn	Y	Zn
Site preferences	{10$\bar{1}$0}	{11$\bar{2}$0}	{11$\bar{2}$0}	{0001}	{11$\bar{2}$0}	{11$\bar{2}$0}	{11$\bar{2}$2}

where P is the multiplicity factor, 2θ is the diffraction angle, $A(\theta)$ is the absorption factor, e^{-2M} is the temperature factor, and $|F_{HKL}|^2$ is the structure factor.

The multiplicity factor P is related to the crystallographic symmetry and the indices of the lattice plane. In a crystal, the arrangement of all the planes of the same crystal plane group is the same, and the spacing between the crystal planes is equal. In polycrystalline diffraction, the diffraction angles of all planes of the same crystal plane group are the same, that is, 2θ. Therefore the crystal plane of the same group (called equivalent crystal planes) will overlap on the same diffraction ring. When the number of equivalent planes on a certain crystal plane group increases, the diffraction probability of the crystal plane increases, and the corresponding diffraction peak increases. The multiplicity factor refers to the number of equivalent plane in a plane group. For the Mg−Sn alloys, the HCP structure and the crystallographic symmetry of Mg do not change after the doping of Sn. Hence, the P values in certain crystal planes (cylindrical plane {11$\bar{2}$0}) of the Mg−Sn alloys and pure Mg are the same. In the X-ray diffraction (XRD) patterns of Mg, the incident angle corresponding to the crystal plane is 57.28 degrees. The solid solution of Sn atom does not change the HCP structure of Mg, so the diffraction angle is also a definite value. For the cylindrical sample, when θ changes, the change of the absorption factor $A(\theta)$ is opposite to that of the temperature factor e^{-2M}, and the effects of the two factors can be counteracted. Furthermore, the diffraction intensity of the unit cell can be characterized by the structure factor $|F_{HKL}|^2$ which reflects the effect of atomic species, amount and site on the diffraction intensity in the (HKL) plane.

To summarize, the structure factor $|F_{HKL}|^2$ may be the only coefficient influencing the peak intensity in an individual crystal plane during the XRD analysis of magnesium crystal, and it can be obtained by Eq. (2.16):

$$|F_{HKL}|^2 = \left[\sum_{j=0}^{n} f_j \cos 2\pi(Hx_j + Ky_j + Lz_j) \right]^2 + \sum_{j=0}^{n} f_j \sin 2\pi(Hx_j + Ky_j + Lz_j)]^2 \quad (2.16)$$

where j refers to the atom in the unit cell with the real-space coordinates (x_j, y_j, z_j), f_j is the atomic scattering factor, and (HKL) are the crystal plane indices.

Fig. 2.11A shows the crystal structure in a $3 \times 3 \times 2$ supercell of the Mg-based solid solution with 36 atoms, where 3 solute atoms are dissolved in the (11$\bar{2}$0) plane (Fig. 2.11B). For easy calculation, the orthogonal coordinates with the original point of

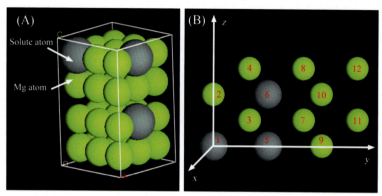

Figure 2.11 Distribution of solute atoms substituted in the (11$\bar{2}$0) plane of a Mg crystal (A) Solute atoms in a Mg supercell and (B) Atom configuration of a Mg crystal with solute atoms doping in the (11$\bar{2}$0) plane.

a solute atom are constructed to calculate the structure factor, as shown in Fig. 2.11B. In the (11$\bar{2}$0) plane shown in Fig. 2.11B, 3 solute atoms are labeled as No. 1, 5 and 6 and the others are Mg atoms. Table 2.3 lists the corresponding coordinates of the 12 atoms in Fig. 2.11 (a and c refer to the lattice constant of Mg-based solid solution).

In the coordinate system, the values x_j of all the atoms are zero and (HKL) of the plane of (11$\bar{2}$0) is (110), as shown in Table 2.3. Accordingly, $Hx_j + Ky_j + Lz_j = Ky_j$, and $|F_{HKL}|^2$ can be deduced as:

$$
\begin{aligned}
|F_{HKL}|^2 &= \left[\sum_{j=0}^{n} f_j \cos 2\pi (Ky_j)\right]^2 + \left[\sum_{j=0}^{n} f_j \sin 2\pi (Ky_j)\right]^2 \\
&= \left\{\sum_{j=1}^{2} f_j + \sum_{j=3}^{4} f_j \cos 2\pi \left(2a/\sqrt{3}\right)\right] + \sum_{j=5}^{6} f_j \cos 2\pi \left(\sqrt{3}a\right)\right] \\
&\quad + \sum_{j=7}^{8} f_j \cos 2\pi \left(5a/\sqrt{3}\right)\right] + \sum_{j=9}^{10} f_j \cos 2\pi \left(2\sqrt{3}a\right)\right] + \sum_{j=11}^{12} f_j \cos 2\pi \left(8a/\sqrt{3}\right)\right]\right\}^2 \\
&\quad + \left\{\sum_{j=3}^{4} f_j \sin 2\pi \left(2a/\sqrt{3}\right)\right] + \sum_{j=5}^{6} f_j \sin 2\pi \left(\sqrt{3}a\right)\right] + \sum_{j=7}^{8} f_j \sin 2\pi \left(5a/\sqrt{3}\right)\right] \\
&\quad + \sum_{j=9}^{10} f_j \sin 2\pi \left(2\sqrt{3}a\right)\right] + \sum_{j=11}^{12} f_j \sin 2\pi \left(8a/\sqrt{3}\right)\right]\right\}^2
\end{aligned}
\tag{2.17}
$$

X-ray diffractive rays are obtained by bombarding copper target with electron beam. The wavelength of X-ray produced by copper target is $\lambda = 0.15406$, and the incident angle is 57.28 degrees, $\lambda^{-1}\sin\theta = 5.46$. For easy calculation, the solute atom is assumed to be Sn in Fig. 2.12. According to the manual of XRD, the atomic

Table 2.3 Corresponding coordinates of the 12 atoms in the $(11\bar{2}0)$ plane in Fig. 2.11B.

No.	1	2	3	4	5	6	7	8	9	10	11	12
x_j	0	0	0	0	0	0	0	0	0	0	0	0
y_j	0	0	$2a/\sqrt{3}$	$2a/\sqrt{3}$	$\sqrt{3}a$	$\sqrt{3}a$	$5a/\sqrt{3}$	$5a/\sqrt{3}$	$2\sqrt{3}a$	$2\sqrt{3}a$	$8a/\sqrt{3}$	$8a/\sqrt{3}$
z_j	0	c	$2/c$	$3c/2$	0	c	$2/c$	$3c/2$	0	c	$2/c$	$3c/2$

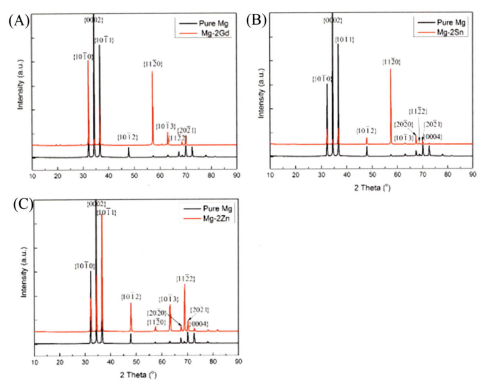

Figure 2.12 X-ray diffraction patterns of solid solution alloys compared with the standard pure Mg: (A) Mg−Gd, (B) Mg−Sn, (C) Mg−Zn.

scattering factor of Mg is 4.8, and that of Sn is 25.6, that is, $f_1 = f_5 = f_6 = f_{Sn} = 25.6$, $f_{2-4} = f_{7-12} = f_{Mg} = 4.8$. Make:

$$
\begin{aligned}
2\pi(2a/\sqrt{3}) &= M1; \\
2\pi(\sqrt{3}a) &= M2; \\
2\pi(5a/\sqrt{3}) &= M3; \\
2\pi(2\sqrt{3}a) &= M4; \\
2\pi(8a/\sqrt{3}) &= M5.
\end{aligned}
\tag{2.18}
$$

$|F_{HKL}|^2$ of Mg–Sn system can be deduced as:

$$|F_{HKL}|^2_{Mg-Sn} = [30.4 + 51.2\cos M2 + 9.6(\cos M1 + \cos M3 + \cos M4 + \cos M5)]^2$$
$$+[51.2\sin M2 + 9.6(\sin M1 + \sin M3 + \sin M4 + \sin M5)]^2$$

(2.19)

In the pure Mg system,

$$|F_{HKL}|^2_{Mg} = [9.6(1 + \cos M1 + \cos M2 + \cos M3 + \cos M4 + \cos M5)]^2 +$$
$$[9.6(\sin M1 + \sin M2 + \sin M3 + \sin M4 + \sin M5)]^2$$

(2.20)

Ganeshan et al. investigated the change of lattice parameters after doping a variety of solute atoms via calculation and experiment. They found that a value ranged from 3.177 to 3.252 in the binary Mg-based solute solutions. Thus the $M(1-5)$ value, which relates to the a value, only changes a little after doping solute atoms, especially compared to the effect of f_j value (related to atomic species). Therefore $|F_{HKL}|^2_{Mg-Sn} > |F_{HKL}|^2_{Mg}$. Except B, C, and N, most of the metal atoms display larger f_j values than Mg. From above analysis, it can be deduced that a crystal plane with abnormal XRD peak intensity, compared to that of the pure Mg, corresponds to the plane which is solid solution by large numbers of solute atoms.

Fig. 2.12 shows a comparison of the XRD patterns of the solid solution alloys to that of the standard pure Mg. The top three peaks in the standard pure Mg appear in the {0002}, {10$\bar{1}$1}, and {10$\bar{1}$0} crystal planes. However, Fig. 2.12A shows that the top three peaks change after solid solution. Table 2.4 lists the corresponding crystal planes of the top three peaks of the pure Mg and solid solution alloys in the XRD patterns. The crystal planes with abnormal peak intensities of Mg–Gd, Mg–Sn, Mg–Zn alloys appear in the {11$\bar{2}$0}, {11$\bar{2}$0}, and {11$\bar{2}$2} planes respectively, indicating that Gd, Sn and Zn atoms prefer to substitute in the {11$\bar{2}$0}, {11$\bar{2}$0} and {11$\bar{2}$2} planes respectively, which agrees well with the calculation results.

2.3.1 Calculation of theoretical critical shear strength

According to the model proposed by B. Joós, the theoretical critical shear strength τ_{max} of an ideal crystal to start dislocation slipping along a certain crystal direction on a certain crystal plane can be expressed as follows:

$$\tau_{max} = \frac{8\sqrt{3}}{9} \max \left\{ \frac{d\gamma(f)}{df} \sin^2 \frac{\pi f}{b} \right\}$$

(2.21)

where f is the instantaneous displacement of dislocation movement, b is the Bergers vector, $d\gamma(f)/df$ is the derivative of the GSFE curve when the lattice dislocations slip along a certain slip system.

Table 2.4 Corresponding crystal planes of top tree peaks in X-ray diffraction pattern of pure Mg and solid solution alloys.

Alloy	Strong peak 1	Strong peak 2	Strong peak 3
Pure Mg	$\{0002\}$	$\{10\bar{1}1\}$	$\{10\bar{1}0\}$
Mg-2Gd	$\{10\bar{1}0\}$	$\{11\bar{2}0\}$	$\{10\bar{1}1\}$
Mg-2Sn	$\{11\bar{2}0\}$	$\{0002\}$	$\{10\bar{1}1\}$
Mg-2Zn	$\{10\bar{1}1\}$	$\{10\bar{1}0\}$	$\{11\bar{2}2\}$

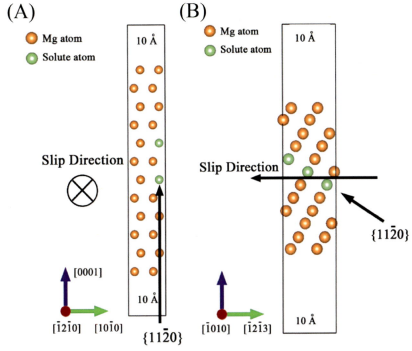

Figure 2.13 Supercells used in the calculations of generalized stacking fault energies. (A) for $\{0001\}$ $<11\bar{2}0>$ slip system, (B) for $\{11\bar{2}2\}$ $<11\bar{2}3>$ slip systems. Vacuum width of 10 Å is added to avoid the interactions due to periodic images.

The GSFE of the basal slip and nonbasal slip of magnesium alloy is calculated by replacing the corresponding Mg atom with the alloying element atom in the solid solution position. τ_{max} of the corresponding system is calculated by combining Eq. (2.8). Finally, the change of the ratio of the nonbasal slip to the basal slip of the Mg crystal before and after doping is compared. The slip systems involved in the calculation are the base slip system $\{0001\}<11\bar{2}0>$ and the pyramidal slip system $\{11\bar{2}2\}<11\bar{2}3>$.

Fig. 2.13 shows the supercell model used in the calculation of GSFE, where the supercell contains 48 atoms, including 3 solute atoms with a solute concentration of

6.25 at.%. Based on the symmetry of crystal, a periodic boundary condition is adopted during calculation, and effects caused by adjacent supercells are excluded by setting up a vacuum gap of 10 Å in both the upper and lower part of the supercell. The coordinate system in the supercell is established and directions of axis are defined as follow: x axis is the slip direction; y axis is perpendicular to the slip direction in the slip plane, and z axis is perpendicular to the slip plane. In order to simulate the slipping process, two parts of a supercell are tangentially moved along the x direction during the calculation of GSFE. The tangential variable at each step is $|\boldsymbol{b}|/50$ ($\boldsymbol{b}$ is the Burgers vector of each slip system). The stacking fault energy is calculated by Eq. 2.1. When calculating the total system energy (E_{tot}), the x direction is fixed, and the relaxation of shape and size is allowed for guaranteeing accuracy during generating stacking faults. Theoretical shear strength curves of the basal slip system and the nonbasal slip system are shown in Fig. 2.14.

Furthermore, the ratios of τ_{max} between nonbasal and basal slips are obtained in Table 2.5. It can be found that the ratio of τ_{max} between nonbasal and basal slips decreases with Li, Sn, or Y addition. These three alloying elements can also reduce the ratio of CRSS for nonbasal to that for basal slip.

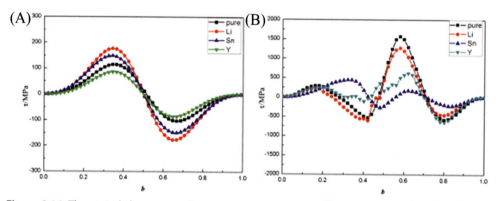

Figure 2.14 Theoretical shear strength curves (A) for {0001} <11$\bar{2}$0> of the basal slip system and (B) for {11$\bar{2}$2} <11$\bar{2}$3> of the pyramidal slip system.

Table 2.5 The τ_{max} ratio of the nonbasal slip to the basal slip for various Mg-based binary alloys.

Alloy	τ_{max} ratio of nonbasal to basal	τ_{max} ratio change rate (%)	Change rate per atom (/at.%)	Change rate per weight (/wt.%)
Pure Mg	13.7	0.0	0.00	0.00
Mg–Li	7.2	47.7	7.63	25.49
Mg–Sn	3.0	78.2	12.51	3.18
Mg–Y	7.5	45.7	7.32	2.33

In order to better connect the experimental data, the ratio change rate of $\tau_{\max}$ per atom and per weight is listed in Table 2.5 after the solute atom is doped into Mg crystal. In the supercell model for calculating the GSFE curve, the atomic concentration is 6.25 at.%, and the weight percent of Li, Sn and Y atoms are 1.87 wt.%, 24.56 wt.%, and 19.61 wt.%, respectively.

The calculation results in Table 2.5 demonstrates that the ability to decrease the $\tau_{\max}$ ratio of the nonbasal and basal slips per atom can be ranked as Sn > Li > Y; while that per weight can be ranked as Li > Sn > Y.

2.3.2 Calculation of modified stacking fault energy and critical resolved shear stress

Most researches on stacking fault energy are based on the first-principle calculation at 0K. In order to better explain the relationship between stacking fault energy and the activation of basal and nonbasal dislocation slip and the deformation ability of magnesium alloys, it is necessary to extend the study of stacking fault energy at 0K to finite temperature.

Fusheng Pan's team describes the interaction between atoms by using the second-nearest neighbor modified embedded-atom method in the molecular dynamics simulation, and they have studied the effects of various alloying elements on the stacking fault energy of magnesium at different temperatures. For Mg−Al, Mg−Zn and Mg−Y alloys with solution content of 0−3 at.%, the GSFE of the basal slip system $<11\overline{2}0>$, the basal slip system $<10\overline{1}0>$, the prismatic slip system $\{10\overline{1}0\}<11\overline{2}0>$ and the pyramidal slip system $\{11\overline{2}2\}<11\overline{2}3>$ are studied at 0K−500K. The effects of temperature and atomic content on stacking fault energy and the relationship between stacking fault energy and micro plastic deformation mode are discussed. In order to better explain the influence of the stacking fault energy on the activation of each slip system, the plastic forming parameter χ proposed by Moitra et al. [1] is modified. The parameters χ_1 is related to the activation of the basal and prismatic slip systems and the parameter χ_2 is related to the activation of basal and pyramidal slip systems. The two parameters are defined to explain the relationship between the stacking fault energy and the deforming ability of magnesium alloys at the same temperature, as shown in Eqs. (2.22) and (2.23).

$$\chi_1 = \frac{\dfrac{\left(\frac{\gamma_{sf}^{B}}{\gamma_{usf}^{B}}\right)_{X}}{\left(\frac{\gamma_{sf}^{B}}{\gamma_{usf}^{B}}\right)_{Mg}}}{\dfrac{\left(\gamma_{sf}^{M}\right)_{X}}{\left(\gamma_{sf}^{M}\right)_{Mg}}} \tag{2.22}$$

$$\chi_2 = \frac{\left(\gamma_{sf}^{B}/\gamma_{usf}^{B}\right)_{X}}{\left(\gamma_{sf}^{B}/\gamma_{usf}^{B}\right)_{Mg}} \Big/ \frac{\left(\gamma_{sf}^{Pyr}/\gamma_{usf}^{Pyr}\right)_{X}}{\left(\gamma_{sf}^{Pyr}/\gamma_{usf}^{Pyr}\right)_{Mg}} \tag{2.23}$$

where γ_{sf}^{B} is the stable stacking fault energy of the basal slip system $\{0001\}$ $<10\bar{1}0>$, γ_{usf}^{B} is the unstable stacking fault energy of the basal slip system $\{0001\}$ $<10\bar{1}0>$, γ_{sf}^{M} is the substitution value of the stable stacking fault energy and unstable stacking fault energy of the prismatic slip system, γ_{sf}^{Pyr} and γ_{usf}^{Pyr} is the stable stacking fault energy γ_{sf} and unstable stacking fault energy γ_{usf}^{I} on the second-order pyramidal plane respectively, and the subscript Mg represents pure Mg and X represents Mg—X alloy. According to Eqs. (2.22) and (2.23), the values of χ_1 and χ_2 of pure magnesium equal 1. If the values of χ_1 and χ_2 are greater than 1, the ductility of the alloy is better. Fig. 2.15 shows the variation of χ_1 and χ_2 values of Mg—Al, Mg—Zn and Mg—Y alloys with the increase of solution atoms at different temperatures. It can be seen that the values of χ_1 and χ_2 of the Mg—Al alloy are mostly less than 1 at each composition and temperature, and both decrease slightly with the increase of Al content. However, the values of χ_1 and χ_2 of Mg—Zn and Mg—Y alloys are mostly greater than 1. The values of χ_1 increase with the increase of solid solution content, which means the tendency of prismatic dislocation slip increases. However, the values of χ_2 increase slightly with the increase of Zn content. The values of χ_1 and χ_2 of the Mg—Y alloy increase with the increase of Y content, which are higher than that of the Mg—Zn alloy. According to the calculation results, the increase of Al content does not significantly improve the activation of the nonbasal dislocation slip at the same temperature, and the increase of Zn content is beneficial to the activation of the nonbasal dislocation slip, while the increase of Y content obviously increases the possibility of the nonbasal dislocation slip in magnesium alloys.

At present, most researches focus on the effect of stacking fault energy on the activation of the basal and nonbasal dislocation slips to explain the ductility of magnesium alloys by studying the stacking fault energy. In fact, the key to activate the nonbasal dislocation slip is to narrow the difference of the slip resistance between the basal and nonbasal dislocation slip, and the stacking fault energy is directly related to the slip resistance of each slip system. Yasi et al. established the relationship between the stacking fault energy and the slip resistance (CRSS) of the basal dislocation in some binary solid solution magnesium alloys at 0K by modifying Fleischer model, as shown in Eq. (2.24).

$$\Delta\tau_{crss(0001)} \approx (38.9\text{MPa})\left\{ \left(\varepsilon_b/0.176\right)^2 + \left(\varepsilon_{SFE}/0.176\right)^2 - \varepsilon_b\varepsilon_{SFE}/2.98 \right\}^{3/2} c_s^{1/2} \quad (2.24)$$

where $\Delta\tau_{crss(0001)}$ is the change of CRSS of the basal dislocation slip in the solid solution alloy compared with pure magnesium (MPa), ε_b is the size misfit caused by solid solution atoms, ε_{SFE} is the chemical misfit caused by solid solution atoms (i.e., the change of stacking fault energy), and c_s is the concentration of the solid solution atom (at.%). The results show that the stacking fault energy and the size misfit caused by solid solution atoms affect

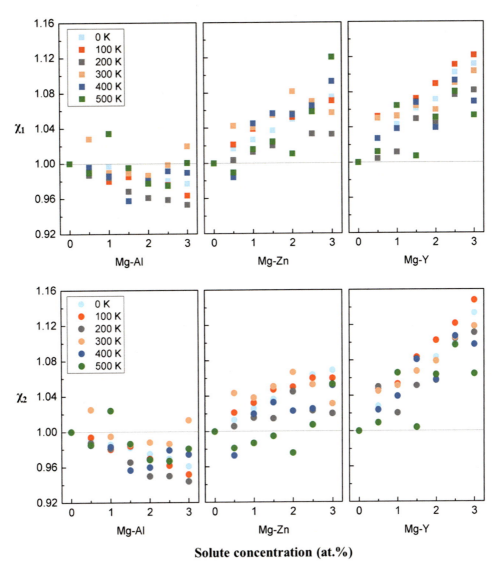

Figure 2.15 χ_1 and χ_2 values of Mg–Al, Mg–Zn and Mg–Y alloys.

the resistance to basal dislocation slip, which lays a theoretical foundation for the quantitative analysis of the relationship between the stacking fault energy and the slip resistance. Accordingly, the effect of some solid solution atoms on the changes of CRSS of the basal dislocation is calculated by Eq. (2.24), including the changes ($\Delta\tau_{\text{crss}}^{M}$ and $\Delta\tau_{\text{crss}}^{1}$) of CRSS of the basal dislocation at the maximum solid solubility of alloying elements and at the solid solution content being 1 at.%. The results are listed in Table 2.6.

Table 2.6 Changes of the resistance to the basal dislocation slip (CRSS) in Mg containing different solid solution elements.

Elements	Maximum solid solubility, c_M/at.%	$\Delta\tau_{crss}^M$/MPa	$\Delta\tau_{crss}^1$/MPa
Al	11.5	7.08	2.09
Zn	2.69	5.23	3.19
Mn	0.996	10.13	(10.04)
Sc	15	6.11	1.58
Si	1.16	4.75	4.41
La	0.14	6.95	/
Sn	3.35	2.34	1.28
Y	3.4	16.71	9.06
Dy	4.83	18.89	8.6
Ag	3.83	8.54	4.37
Ti	0.12	0.49	/
Yb	1.2	11.15	10.18
Ca	0.44	6.85	/
Zr	1.04	1.00	0.98
Er	6.9	19.05	7.25
Fe	0.00043	0.24	/
Gd	4.53	23.22	10.91
Li	17	4.80	1.17

As shown in Table 2.6, Al, Zn, Y, Gd, Mn, Yb, Ag, Dy, Er, and other elements can increase the slip resistance of basal dislocation in Mg, and the values of $\Delta\tau_{crss}^M$ and $\Delta\tau_{crss}^1$ of these elements are positive. According to the values of $\Delta\tau_{crss}^1$, Gd, Yb and Mn have the obvious effect on increasing the slip resistance of basal dislocation in Mg, followed by Y, Dy and Er. Among them, Mn is the lowest cost element, which is a favorable element for alloying to develop high ductility and low-cost magnesium alloy. However, the solid solubility of Mn is limited, and the maximum solid solubility of Mn is only close to 1 at% in Mg. Therefore how to use the maximum solid solubility of Mn to improve the ductility and how to use the fine precipitates of Mn to refine the grain and further improve the ductility are the problems that must be considered simultaneously when Mn is used in magnesium alloys.

The elements Al, Zn and Y can increase the resistance to basal dislocation slip for Mg−Al, Mg−Zn and Mg−Y alloys. Compared with the values of $\Delta\tau_{crss}^1$, Y has the greatest effect on increasing the resistance to basal dislocation slip in magnesium, followed by Zn and Al. The resistance to the basal dislocation slip increases in Mg−Al, Mg−Zn and Mg−Y solid solution alloys, which makes the dislocation movement difficult and causes the solid solution strengthening effect of the basal plane. Thus the yield strength of the alloys increases. Combined with the results of molecular dynamics simulation, the solid solution of Zn and Y is conducive to the activation of the nonbasal dislocation slip, which improves the ductility of magnesium alloys. Therefore the yield strength and ductility of magnesium alloys can be improved simultaneously by solid solution of Zn and Y in Mg, that is, the strengthening and ductilizing is achieved by solid solution of Zn and Y. The detailed results about the effect of Zn and Y on the resistance to nonbasal dislocation slip will be reported later.

2.4 Experimental verification of solid solution strengthening and ductilizing

The alloy design theory of "solid solution strengthening and ductilizing" can be indirectly verified by using the stress—strain curve and visco-plastic self-consistent model. Fig. 2.16A shows the tensile stress—strain curves of Mg-2wt.% Al and Mg-2wt.%Y binary alloys. The yield strength and the fracture elongation of pure magnesium can be significantly improved by adding Al and Y. The tensile strength of the Mg-2Al alloy is higher, and the fracture elongation of the Mg-2Y alloy is higher.

For the pure Mg, Mg-2Al and Mg-2Y alloys, the plastic deformation during tensile process at the RT is simulated by the visco-plastic self-consistent model (as shown in Fig. 2.16B). The influence of solutes Al and Y on the difference of slip resistance between basal and nonbasal slip systems in Mg is analyzed. The results are shown in Table 2.7. $\Delta\tau^{\text{Basal}}$, $\Delta\tau^{\text{Prismatic}}$ and $\Delta\tau^{\text{Pyramidal}}$ are the changes of resistance to basal dislocation slip, prismatic dislocation slip and pyramidal dislocation slip in solid solution alloys respectively. The positive values indicate increase, and the negative values indicate decrease. $\Delta\tau^{Pr-B}$ and $\Delta\tau^{Py-B}$ are the difference of the slip resistance between the prismatic and basal slip systems and that between the pyramidal and basal slip systems respectively.

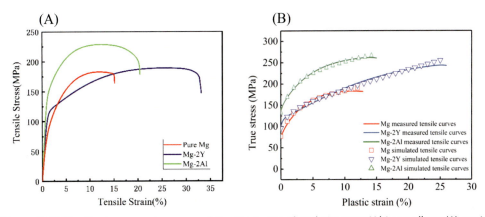

Figure 2.16 Tensile stress—strain curves of the Mg-2wt.% Al and Mg-2wt.%Y binary alloys: (A) engineering stress—strain curve, (B) true stress—strain curve.

Table 2.7 Mechanical properties and slip resistance data of the pure Mg, Mg-2Al, and Mg-2Y alloys.

Sample	Tensile yield strength /MPa	Elongation to failure /%	$\Delta\tau^{Basal}$/MPa	$\Delta\tau^{Prismatic}$/MPa	$\Delta\tau^{Pyramidal}$/MPa	$\Delta\tau^{Pr-B}$ /MPa	$\Delta\tau^{Py-B}$ /MPa
Mg	75.5	15.3	—	—	—	145	180
Mg-2Al	151.6	20.2	13	30	15	162	182
Mg-2Y	110.7	32.6	3	-90	-70	52	167

When Al is dissolved into Mg, the value of $\Delta\tau^{Basal}$ increases, resulting in solid solution strengthening, and the value of $\Delta\tau^{Py-B}$ changes little, which has little effect on the activation of the pyramidal dislocation slip. The value of $\Delta\tau^{Pr-B}$ increases, indicating the difficulty to activate the prismatic dislocation slip. When Y is dissolved into Mg, the value of $\Delta\tau^{Basal}$ increases slightly, but the value of $\Delta\tau^{Py-B}$ and $\Delta\tau^{Pr-B}$ decreases significantly, which is beneficial to the activation of prismatic and pyramidal dislocation slip, resulting in a good deformation ability of the Mg—Y alloy.

Furthermore, the increment of the resistance to the basal dislocation slip of the Mg-2Al alloy is larger than that of the Mg-2Y alloy, so the yield strength of the Mg-2Al alloy is higher than that of the Mg-2Y alloy. The values of $\Delta\tau^{Pr-B}$ and $\Delta\tau^{Py-B}$ of the Mg-2Y alloy is much smaller than that of Mg-2Al alloy, which means that it is more conducive to the activation of the nonbasal slip system. Therefore the ductility of the Mg-2Y alloy is higher than that of the Mg-2Al alloy. The effect of the Al and Y addition on the plastic deformation ability of the pure Mg is the same as that of the molecular dynamics simulation.

Table 2.8 shows the mechanical properties of some binary magnesium alloys from experiments and literatures. The addition of Al, Y, Mn, Gd, Zn, Er, and other alloy elements can improve the tensile ultimate strength, yield strength and elongation of pure magnesium. The tensile strength and elongation of the binary alloys increase with the increasing content of Gd, Zn and other alloy elements under the same process. Despite the influence of grain refinement, precipitation strengthening and purification, the deformation and the heat treatment process of the binary alloy reported in the experiment and literature are all around 400°C, which means that the alloying elements are mostly dissolved in the matrix. Therefore the effect of solution strengthening and ductilizing in magnesium is verified by the improvement of the strength and ductility of pure magnesium simultaneously after the solid solution of alloying elements.

2.5 Application of solid solution strengthening and ductilizing

Based on the alloy design theory of solid solution strengthening and ductilizing and the long-range ordered phase control, the magnesium alloy research team in

Table 2.8 Tensile properties of some binary magnesium alloys.

Alloys/wt.%	Tensile yield strength/ MPa	Ultimate tensile strength/ MPa	Elongation to failure/ %	Processing route
Mg	76	183	15.3	Heat treated at 420°C for 24h–extruded at 350°C
Mg-2Al	152	228	20.2	Heat treated at 420°C for 24h–extruded at 350°C
Mg-3Al	158	241	20.8	Heat treated at 420°C for 24h–extruded at 350°C
Mg-4Al	162	252	20.8	Heat treated at 420°C for 24h–extruded at 350°C
Mg-2Y	111	190	32.6	Heat treated at 520°C for 24h–extruded at 450°C
Mg-2Y	92	189	21	Heat treated at 480°C for 12h–extruded at 420°C
Mg-1.0Mn	178	217	18.3	Heat treated at 500°C for 24h–extruded at 350°C
Mg-0.89Mn	204.3	234.1	38.8	As-extruded
Mg-1.0Gd	80	186	25.5	Heat treated at 520°C for 24h–extruded at 450°C
Mg-3.0Gd	78	187	31.8	Heat treated at 520°C for 24h–extruded at 450°C
Mg-0.75Gd	145	210	12	Hot rolled at 400°C-Annealed for 1 h at 380°C
Mg-2.75Gd	160	205	21	Hot rolled at 400°C-Annealed for 1 h at 380°C
Mg-4.65Gd	165	210	26	Hot rolled at 400°C-Annealed for 1 h at 380°C
Mg-1Zn	126	215	17.3	Heat treated at 400°C for 24h–extruded at 350°C
Mg-2Zn	129	217	20.0	Heat treated at 400°C for 24h–extruded at 350°C
Mg-4Zn	139	242	25.1	Heat treated at 400°C for 24h–extruded at 350°C
Mg-2Er	83	251	19.6	Heat treated at 520°C for 48h–extruded at 400°C-Annealed for 1 h at 400°C
Mg-4Er	80	184	28.4	Heat treated at 520°C for 48h–extruded at 400°C-Annealed for 20 min at 400°C
Mg-8Er	153	260	44	As-extruded

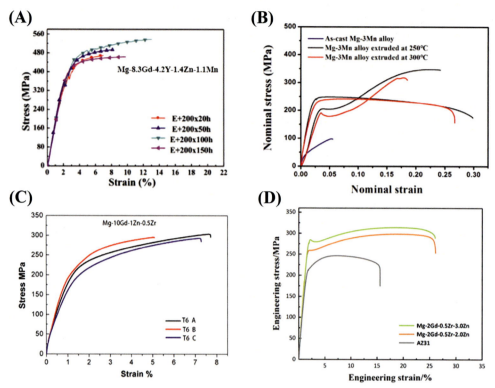

Figure 2.17 The tensile mechanical properties of high-performance magnesium alloys: (A) ultrahigh-strength wrought magnesium alloy, (B) high-ductility magnesium alloy containing manganese, (C) high-strength and -ductility cast magnesium alloy, (D) ultrahigh-ductility magnesium alloy.

Chongqing University has developed more than 20 kinds of new high–performance magnesium alloys, including ultrahigh-strength wrought magnesium alloys, ultrahigh-strength cast magnesium alloys, ultrahigh-ductility magnesium alloys, low–cost and high-ductility magnesium alloys, ultralight magnesium alloys, high electromagnetic shielding properties magnesium alloys and high thermal conductivity magnesium alloy, etc. Fig. 2.17A–D is the tensile mechanical properties curves of ultrahigh-strength wrought magnesium alloy, high-ductility magnesium alloy-containing manganese, high-strength and -ductility cast magnesium alloy and ultrahigh-ductility magnesium alloy developed by the magnesium alloy research team in Chongqing University respectively. In addition, some new high–performance magnesium alloys and their properties developed by Chongqing University are listed in Table 2.9, in which 16 kinds have been approved as national standard alloys and 9 kinds have been approved as international standard alloys.

How to improve the room and low temperature ductility of magnesium alloy is an urgent problem to be solved in the popularization and application of magnesium alloy.

Table 2.9 Some new high-performance magnesium alloys developed by Chongqing University.

Alloys	Series	Ultimate tensile strength/MPa	Elongation to failure/%
Ultrahigh plastic magnesium alloys	Mg–X–Gd	200–250 MPa	50%–63%
High plastic wrought magnesium alloys	Mg–Zn–Zr–Nd (Er)	230–300 MPa	20%–40%
Low-cost wrought magnesium alloys	Mg–Mn–Al	280–330 MPa	20%–23%
Wrought magnesium alloys without RE addition	Mg–Zn–Mn–Sn	380–400 MPa	8%–10%
Ultrahigh strength wrought magnesium alloys	Mg–Gd–Y–Zn	500–550 MPa	10%–13%
High strength cast magnesium alloys	Mg–Gd–Y–Zn	330–380 MPa	9%–12%
Wrought magnesium alloys	Mg–Zn–Zr–Y–Ce	400–420 MPa	9%–12%
Ultralight magnesium alloys	Mg–Li–Al–X	200–230 MPa	20%–25%

The theory of solid solution strengthening and ductilizing of magnesium alloy can provide a new idea of alloy design for the development of the high ductility magnesium alloy, which can improve the strength and ductility of the magnesium alloy simultaneously. In the alloy design idea of solid solution strengthening and ductilizing, the application of Mn is very valuable. On the one hand, the cost of Mn is very low, and on the other hand, the effect of Mn on solid solution ductilizing at low temperature is very significant and Mn also has obvious precipitation effect, which is of great significance for the development of low-cost ultra-fine grain wrought magnesium alloy at low temperature. Moreover, because the damping property of magnesium alloy is closely related to the mobility of dislocations, the theory of solid solution strengthening and ductilizing can also provide a new solution to the contradiction between the improvement of strength and the deterioration of damping property, which means solid solution strengthening and damping property increasing can be realized through the design of solid solution magnesium alloy. In addition, the ductility of titanium alloy, beryllium alloy, zinc alloy and other hexagonal structure metals are relatively poor, so it is worth to explore how to improve the ductility and formability by solid solution strengthening and ductilizing.

2.6 Summary

In this chapter, the phenomenon of the increase in the plasticity of magnesium alloy accompanied with solution strengthening is stated from both theoretical and experimental points of view. The new design principle and theory of solution strengthening and ductilizing for magnesium alloy are proposed.

1. A new design theory of solution strengthening and ductilizing is proposed. The strength and ductility of magnesium alloy can be improved simultaneously under the guidance of this new design theory. The stacking fault energy and CRSS of the Mg—X solid solution (X = Al, Gd, Li, Mn, Sn, Y, and Zn) are calculated by first-principle calculations. The calculation results are verified by experiments and application examples. The results show that the theory of solution strengthening and ductilizing of magnesium alloy is reliable and feasible in both theory and practice.

2. In the alloy design idea of solid solution strengthening and ductilizing, the effect of Mn and Gd is very obvious. The application of Mn is very valuable. On the one hand, the cost of Mn is very low. On the other hand, the effect of Mn on solid solution ductilizing at low temperature is very significant, and Mn also has obvious precipitation effect, which is of great significance for the development of low-cost ultra-fine grain wrought magnesium alloy at low temperature.

3. The theory of solution strengthening and ductilizing of magnesium alloy can provide a new idea of alloy design for the development of high plasticity magnesium alloy. Because the damping property of magnesium alloy is closely related to the mobility of dislocations, the theory of solid solution strengthening and ductilizing can also provide a new solution to the contradiction between the improvement of strength and the deterioration of damping property, which means that the solid solution strengthening and damping property increasing can be realized through the design of solid solution magnesium alloy. In addition, the ductility of titanium alloy, beryllium alloy, zinc alloy and other hexagonal structure metals are relatively poor, so it is worth to explore how to improve the ductility and formability by solid solution strengthening and ductilizing.

4. The accurate application of the principle of solution strengthening and ductilizing in the design and development of new magnesium alloys needs further improvement and development in many aspects. First, it is very difficult to accurately calculate the influence of alloy elements on the resistance in the nonbasal plane at different temperatures, especially under the interaction of multiple elements, which also includes the difficulty of accurate calculation of multi-element stacking fault energy. Second, the validity and accuracy of experimental verification need to be further improved. Third, there is a lack of a large number of accurate multi-element phase diagrams of magnesium alloy, and the change of solid solution content of alloy elements is not completely clear, and several thermodynamic and kinetic studies need to be strengthened. Fourthly, as for the research of solid solution, precipitation and dislocation for magnesium alloy, the disconnection phenomenon of thermodynamics, kinetics, and accurate microstructure prediction is still critical; therefore collaborative research is very important.

Reference

[1] Wu ZX, Curtin WA. The origins of high hardening and low ductility in magnesium. Nature 2015;526:62−7.

Further reading

Agnew S, Horton J, Yoo M. Transmission electron microscopy investigation of <c+a> dislocations in Mg and α-solid solution Mg-Li alloys. Metallurgical and Materials Transactions A 2002;33:851−8.

Akhtar A, Teghtsoonian E. Solid solution strengthening of magnesium single crystals—I alloying behaviour in basal slip. Acta Metallurgica 1969;17:1339−49.

Akhtar A, Teghtsoonian E. Solid solution strengthening of magnesium single crystals—II the effect of solute on the ease of prismatic slip. Acta Metallurgica 1969;17:1351−6.

Akhtar A, Teghtsoonian E. Supplement to Trans. JIM 1968;9:692−7.

Ando S, Tonda H. Non-basal slip in magnesium-lithium alloy single crystals. Materials Transactions JIM 2000;41:1188−91.

Arrabal R, Mingo B, Pardo A, et al. Role of alloyed Nd in the microstructure and atmospheric corrosion of as-cast magnesium alloy AZ91. Corrosion Science 2015;97:38−48.

Chen YJ, Chen Q, Wang ZD, et al. Preparation and microstructure analysis of Al-1%Si single crystal. Journal of University of Science and Technology Beijing 2005;27:50−4.

Chetty N, Weinert M. Stacking faults in magnesium. Physical Review B 1997;56(17):10844.

Chino Y, Kadob M, Mabuchi M. Enhancement of tensile ductility and stretch formability of magnesium by addition of 0.2 wt% (0.035 at%)Ce. Materials Science and Engineering A 2008;494:343−9.

Chino Y, Kadob M, Mabuchi M. Texture and stretch formability of a rolled Mg-Zn alloy containing dilute content of Y. Materials Science and Engineering A 2009;513–514:394−400.

Ding WJ, Li DQ, Wang QD, et al. Microstructure and mechanical properties of hot-rolled Mg-Zn-Nd-Zr alloys. Materials Science and Engineering A 2008;483–484:228−30.

Fang C, Zhang J, Pan FS. First-principles study on solute-basal dislocation interaction in Mg alloys. Journal of Alloys and Compounds 2019;785:911−17.

Hantzsche K, Bohlen J, Wendt J, et al. Effect of rare earth additions on microstructure and texture development of magnesium alloy sheets. Scripta Materials 2010;63:725−30.

He SM, Zeng XQ, Peng LM, et al. Microstructure and strengthening mechanism of high strength Mg-10Gd-2Y-0.5Zr alloy. Journal of Alloys and Compounds 2007;427(1−2):316−23.

Hirth JP, Lothe J. Theory of dislocations. John Wiley and Sons, Inc; 1982. p. 857.

Huang H, Miao HW, Yuan GY, et al. Fabrication of ultra-high strength magnesium alloys over 540 MPa with low alloying concentration by double continuously extrusion. Journal of Magnesium and Alloys 2018;6(2):107−13.

Huang H, Yuan GY, Chu ZH, et al. Microstructure and mechanical properties of double continuously extruded Mg-Zn-Gd-based magnesium alloys. Materials Science and Engineering A 2013;560:241−8.

Huntington H. Modification of the Peierls-Nabarro model for edge dislocation core. Proceedings of the Royal Society of London Series B 1955;68:1043.

Jiang B, Liu WJ, Qiu D, et al. Grain refinement of Ca addition in a twin-roll-cast Mg-3Al-1Zn alloy. Materials Chemistry and Physics 2012;133(2−3):611−16.

Jiang B, Yin HM, Yang QS, et al. Effect of stannum addition on microstructure of as-cast and as-extruded Mg-5Li alloys. Transactions of Nonferrous Metals Society of China 2011;21(11):2378−83.

Jiang B, Zeng Y, Yin HM, et al. Effect of Sr on microstructure and aging behavior of Mg-14Li alloys. Progress in Natural Science: Materials International 2012;22(2):160−8.

Jiang B, Zeng Y, Zhang MX, et al. Effects of Sn on microstructure of as-cast and as-extruded Mg−9Li alloys. Transactions of Nonferrous Metals Society of China 2013;23(4):904−8.

Jiang B, Zhou GY, Dai JH, et al. Effect of second phases on microstructure and mechanical properties of as-cast Mg-Ca-Sn magnesium alloy. Rare Metal Materials and Engineering 2014;43(10):2445−9.

Joos B, Duesbery M. The Peierls stress of dislocations: an analytic formula. Physical Review Letters 1997;78(2):266–9.

Kang F, Li Z, Wang JT, et al. The activation of <c+a> non-basal slip in Magnesium alloys. Journal of Materials Science 2012;47:7854–9.

Koizumi T, Egami M, Yamashita K, et al. Platelet precipitate in an age-hardening Mg-Zn-Gd alloy. Journal of Alloys and Compounds 2018;752:407–11.

Kresse G, Furthmüller J. Efficiency of ab-initio total energy calculations for metals and semiconductors using a plane-wave basis set. Computational Materials Science 1996;6:15–50.

Kresse G, Hafner J. Ab initio molecular dynamics for open-shell transition metals. Physical Review B 1993;48:13115.

Li JC, He ZL, Fu PH, et al. Heat treatment and mechanical properties of a high-strength cast Mg–Gd–Zn alloy. Materials Science and Engineering A 2016;651:745–52.

Li RH, Pan FS, Jiang B, et al. Effect of Li addition on the mechanical behavior and texture of the as-extruded AZ31 magnesium alloy. Materials Science and Engineering A 2013;562:33–8.

Li RH, Pan FS, Jiang B, et al. Effects of yttrium and strontium additions on as-cast microstructure of Mg-14Li-1Al alloys. Transactions of Nonferrous Metals Society of China 2011;21(4):778–83.

Liu GB, Zhang J, Dou YC. First-principles study of solute–solute binding in magnesium alloys. Computational Materials Science 2015;103:97–104.

Liu SJ, Wang K, Wang JF, et al. Ageing behavior and mechanisms of strengthening and toughening of ultrahigh-strength Mg-Gd-Y-Zn-Mn alloy. Materials Science and Engineering A 2019;758:96–8.

Liu TT, Pan FS, Zhang XY. Effect of Sc addition on the work-hardening behavior of ZK60 magnesium alloy. Materials & Design 2013;43:572–7.

Liu TT, Pan FS. Development and application of "solid solution strengthening and ductilizing" for magnesium alloys. The Chinese Journal of Nonferrous Metals 2019;29(9):2050–63.

Miao JS, Sun WH, Klarner AD, et al. Interphase boundary segregation of silver and enhanced precipitation of Mg17Al12 phase in a Mg-Al-Sn-Ag alloy. Scripta Materialia 2018;154:192–6.

Moitra A, Kim SG, Horstemeyer MF. Solute effect on the <a+c> dislocation nucleation mechanism in magnesium. Acta Materialia 2014;75:106–12.

Momma K, Izumi F. VESTA 3 for three-dimensional visualization of crystal, volumetric and morphology data. Journal of Applied Crystallography 2011;44:1272–6.

Monnet G, Pouchon MA. Determination of the critical resolved shear stress and the friction stress in austenitic stainless steels by compression of pillars extracted from single grains. Materials Letters 2013;98:128–30.

Nie JF. Precipitation and hardening in magnesium alloys. Metallurgical and Materials Transactions A 2012;43(11):3891–939.

Note R, Kimura S, Inoue A. Preparation of decagonal Al-Ni-Co single quasicrystal by Czochralski method. Materials Transactions JIM 1997;38:943–9.

Pan FS, Mao JJ, Zhang G, et al. Development of high-strength, low-cost wrought Mg–2.0mass% Zn alloy with high Mn content. Progress in Natural Science: Materials International 2016;26(6):630–5.

She J, Pan FS, Guo W, et al. Effect of high Mn content on development of ultra-fine grain extruded magnesium alloy. Materials & Design 2016;90:7–12.

Shi DK. The fundamentals of materials science. Beijing: China Machine Press; 2003.

Sun G, Wang SH, Zhang XF, et al. Effect of solution treatment and prestretching deformation on microstructure and properties of 2197 Al-Li alloy. Heat Treatment of Metals 2011;36(10):75–8.

Tabache MG, Bourret ED, Elliot AG. Measurements of the critical resolved shear stress for indium-doped and undoped GaAs single crystals. Applied Physics Letters 1986;49:289–91.

Tan J, Sun YH, Xie HB, et al. Atomic-resolution investigation of Y-rich solid solution with an invariable orientation in Mg-Y binary alloy. Journal of Alloys and Compounds 2018;766:716–20.

Tan YX, Hu ZZ. Materials characterization. Beijing: China Machine Press; 2013.

Tong X, You GQ, Wang YC, et al. Effect of ultrasonic treatment on segregation and mechanical properties of as-cast Mg-Gd binary alloys. Materials Science and Engineering A 2018;731:44–53.

Vítek V. Intrinsic stacking faults in body-centred cubic crystals. Philosophical Magazine 1968;18:773–86.

Wang C, Zhang HY, Wang HY, et al. Effects of doping atoms on the generalized stacking-fault energies of Mg alloys from first-principles calculations. Scripta Materiars 2013;69:445−8.

Wang F, Hu T, Zhang YT, et al. Effects of Al and Zn contents on the microstructure and mechanical properties of Mg-Al-Zn-Ca magnesium alloys. Materials Science and Engineering A 2017;704:57−65.

Wang JF, Song PF, Huang S, et al. High-strength and good-ductility Mg−RE−Zn−Mn magnesium alloy with long-period stacking ordered phase. Materials Letters 2013;93:415−18.

Wang JF, Wang K, Hou F, et al. Enhanced strength and ductility of Mg-RE-Zn alloy simultaneously by trace Ag addition. Materials Science and Engineering A 2018;728:10−19.

Wang LF, Huang GS, Quan Q, et al. The effect of twinning and detwinning on the mechanical property of AZ31 extruded magnesium alloy during strain-path changes. Materials & Design 2014;63:177−84.

Wang QH, Shen YQ, Jiang B, et al. A good balance between ductility and stretch formability of dilute Mg-Sn-Y sheet at room temperature. Materials Science and Engineering A 2018;736:404−16.

Wang Y, Chen LQ, Liu ZK, et al. First-principles calculations of twin-boundary and stacking-fault energies in magnesium. Scripta Materialia 2010;62:646−9.

Wang YH. Calculation and study on stacking fault energy for magnesium alloy based on molecular dynamics. Chongqing: Chongqing University; 2019. p. 51−61.

Wang ZJ, Jia WP, Cui JZ. Study on the deformation behavior of Mg-3.6% Er magnesium alloy. Journal of Rare Earths 2007;25(6):744−8.

Wen L, Chen P, Tong ZF, et al. A systematic investigation of stacking faults in magnesium via first-principles calculation. European Physical Journal B 2009;72:397−403.

Wu D, Chen RS, Han EH. Excellent room-temperature ductility and formability of rolled Mg-Gd-Zn alloy sheets. Journal of Alloys and Compounds 2011;509:2856−63.

Wu DH, Hu GX, Inui H, et al. Orientation dependence of CRSS for ⟨011] super lattice slip in γ-TiAl single crystals. Acta Metallurgica Sinica 1999;35(4):337 -106.

Wu L, Pan FS, Yang MB, et al. As-cast microstructure and Sr-containing phases of AZ31 magnesium alloys with high Sr contents. Transactions of Nonferrous Metals Society of China 2011;21 (4):784−9.

Wu ZX, Ahmad R, Yin BL, Sandloebes S, Curtin WA. Mechanistic origin and prediction of enhanced ductility in magnesium alloys. Science 2018;359:447−52.

Xua J, Jiang B, Song JF, et al. Unusual texture formation in Mg-3Al-1Zn alloy sheets processed by slope extrusion. Materials Science and Engineering A 2018;732:1−5.

Yan H, Chen RS, Han EH. A comparative study of texture and ductility of Mg-1.2Zn-0.8Gd alloy fabricated by rolling and equal channel angular extrusion. Materials Characterization 2011;62:321−6.

Yang QS, Jiang B, Jiang W, et al. Evolution of microstructure and mechanical properties of Mg-Mn-Ce alloys under hot extrusion. Materials Science and Engineering A 2015;628:143−8.

Yasi JA, Hector JLG, Trinkle DR. First-principles data for solid-solution strengthening of magnesium: from geometry and chemistry to properties. Acta Materialia 2010;58:5704−13.

Yin HM, Jiang B, Huang XY, et al. Effect of Ce addition on microstructure of Mg-9Li alloy. Transactions of Nonferrous Metals Society of China 2013;23(7):1936−41.

Yu ZJ, Huang YD, Qiu X, et al. Fabrication of a high strength Mg−11Gd−4.5Y−1Nd−1.5Zn−0.5Zr (wt%) alloy by thermomechanical treatments. Materials Science and Engineering A 2015;622:121−30.

Yu ZP, Yan YH, Yao J, et al. Effect of tensile direction on mechanical properties and microstructural evolutions of rolled Mg-Al-Zn-Sn magnesium alloy sheets at room and elevated temperatures. Journal of Alloys and Compounds 2018;744:211−19.

Yuasa M, Hayashi M, Mabuchi M, et al. Improved plastic anisotropy of Mg-Zn-Ca alloys exhibiting high-stretch formability: a first-principles study. Acta Materials 2014;65:207−14.

Yuasa M, Miyazawa N, Hayashi M, et al. Effects of group II elements on the cold stretch formability of Mg-Zn alloys. Acta Materials 2015;83:294−303.

Zeng Y, Jiang B, Li RH, et al. Effect of Li content on microstructure, texture and mechanical properties of cold rolled Mg-3Al-1Zn alloy. Materials Science and Engineering A 2015;631:189−95.

Zhang HY, Wang HY, Wang C, et al. First-principles calculations of generalized stacking fault energy in Mg alloys with Sn, Pb and Sn+Pb dopings. Materials Science and Engineering A 2013;584:82−7.

Zhang J, Dou Y, Dong H. Intrinsic ductility of Mg-based binary alloys: a first-principles study. Scripta Materials 2014;89:13−16.

Zhang J, Dou YC, Liu GB, et al. First-principles study of stacking fault energies in Mg-based binary alloys. Computational Materials Science 2013;79:564−9.

Zhang J, Li WG, Zhang BX, et al. Influence of Er addition and extrusion temperature on the microstructure and mechanical properties of a Mg-Zn-Zr magnesium alloy. Materials Science and Engineering A 2011;528(13−14):4740−6.

Zhang J, Liu M, Dou YC, et al. Role of alloying elements in the mechanical behaviors of an Mg-Zn-Zr-Er alloy. Metallurgical and Materials Transactions A 2014;45(12):5499−507.

Zhang J, Zhang XF, Li WG, et al. Partition of Er among the constituent phases and the yield phenomenon in a semi-continuously cast Mg-Zn-Zr alloy. Scripta Materialia 2010;63(4):367−70.

Zhang JM, Yu CM. Effect of aging on microstructure and mechanical properties of low alloy high strength steel containing Nb. Transactions of Materials and Heat Treatment 2019;40(6):123−9.

Zhang J, Qi MA, Fusheng PAN. Effects of trace Er addition on the microstructure and mechanical properties of Mg−Zn−Zr alloy. Materials & Design 2010;31(9):4043−9.

Zhao YZ, Pan FS, Peng J, et al. Effect of neodymium on the as-extruded ZK20 magnesium alloy. Journal of Rare Earths 2010;28:631−5.

Zhao ZL, Sun ZW, Liang W, et al. Influence of Al and Si additions on the microstructure and mechanical properties of Mg-4Li alloys. Materials Science and Engineering A 2017;702:206−17.

Zhu M, Wang LY, Song GF, et al. Study on Cu-Al-Ni-Be single crystal's processing and it's properties. Journal of Functional Materials 2007;38:1474−7.

CHAPTER 3

Ultrahigh plasticity Mg−Gd−Zr alloy

According to the previous chapter, Gd has a large solid solubility in magnesium alloy, and this is very effective in reducing the CRSS gap between basal and nonbasal slip. Therefore Gd is one of the most effective elements to improve the plasticity of magnesium alloy. In addition, Mg−Gd serial alloys are one of the rare earth magnesium alloys with excellent castability, mechanical properties, and corrosion resistance. In general, Y, Zr, Zn, Mn, Dy, Nd, Sm, and other RE elements are added into Mg−Gd binary alloys to form different types of alloys. Combined with different heat treatment and deformation process, the mechanical properties and corrosion resistance of Mg−Gd−X alloy can be improved even with a low amount of Gd content, and thus the cost is reduced. At present, the mechanical properties of Mg−Gd−Y−Zr alloy have surpassed those of WE54 and WE43 alloys which are generally used in aerospace and other fields both at room temperature and high temperature. The mechanical properties of some alloys are superior or comparable to that of some ultrahigh strength 7000 series aluminum alloys. This chapter mainly introduces the microstructure and mechanical properties of as-cast and wrought Mg−Gd−Zr alloys with low Gd content. The influence of deformation and heat treatment on the microstructure and mechanical properties of the alloy is discussed as well.

3.1 Microstructure and properties of as-cast Mg−Gd−Zr alloy

3.1.1 Effect of Gd content on microstructure and properties

Fig. 3.1 shows the optical microstructure of as-cast Mg−xGd−0.6Zr (VKx1, x = 2, 4, 6, wt.%) alloys. The addition of Gd refines the microstructure, but the grain refining effect is not proportional to the increase of Gd addition. The most uniform and finest microstructure is obtained when the addition of Gd is 4 wt.%. TheVK61 alloy shows the worst homogeneity and the largest average grain size of 55 μm. The VK 21 alloy exhibits a moderate uniform microstructure and grain refinement. Such a phenomenon is due to a combined effect of altered solid solubility of Zr affected by Gd addition and the formation of the second phase. When Gd is added, part of Gd dissolve in the magnesium matrix, reducing the solid solubility of Zr in magnesium and deteriorating the refining effect of Zr on the microstructure. The other part of Gd, which exceeds its solid solubility, will form the second phase with magnesium to improve the strength of the alloy. According to the XRD results of the alloy, as shown in

High Plasticity Magnesium Alloys
DOI: https://doi.org/10.1016/B978-0-12-820110-7.00003-3
83

Fig. 3.2, VK21 and VK41 alloys with relatively low Gd content have no obvious second phase peak. This indicates that Gd mainly exists as solid solution in magnesium alloy. With the increase of Gd content, the peaks of $Mg_{5.05}Gd$、 Mg_3Gd and Mg_2Gd are observed in VK61 alloy. According to the phase diagram of Mg–Gd binary alloys, Mg_3Gd and Mg_2Gd phases appear only when the content of Gd exceeds 53 wt.%. Hence, the formation of Gd containing second phases is due to the serious segregation of Gd during the solidification of the alloy.

The microhardness of the alloys (as shown in Fig. 3.3) confirms the conjecture of second phase strengthening to some extent. Typically, the more concentrated the microhardness values are, the more uniform the microstructure is. With the increase of Gd content, the microhardness of the alloy increases gradually, and the dHV/dx $(x = 4 \sim 6\%)$ is greater than the $dHV/dx(x = 2 \sim 4\%)$. This indicates that the strength

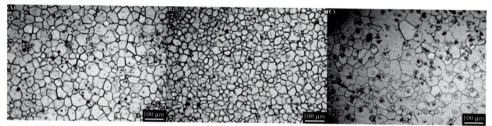

Figure 3.1 OM of as-cast Mg–Gd–Zr alloy, (A) VK21, (B) VK41, (C) VK61.

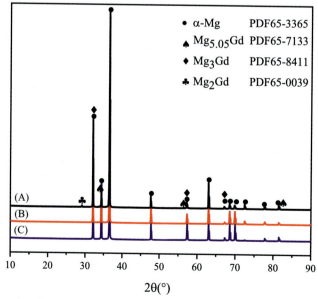

Figure 3.2 XRD results of as cast Mg–Gd–Zr alloy: (A) VK 61, (B) VK 41, (C) VK 21.

of the VK61 alloy is higher than that of the former two alloys. Microhardness measurement is to press the specified indenter into the tested material with a certain load. The hardness of the tested material is evaluated according to local plastic deformation on the surface of the material. A small microhardness means the material is soft, the alloy is easy to undertake plastic deformation, and the alloy has better plastic formability. Therefore it is speculated that the plasticity of VK21 with low Gd content is the best, which is due to solid solution plasticization of Gd. However, with a further increase of Gd content, the formation of a large number of second phases significantly deteriorates the solid solution plasticization of Gd. Hence, VK61 alloy shows the worst plasticity.

3.1.2 Effect of Gd addition on lattice parameters of as cast alloy matrix

The calculation results of the c/a axial ratio of alloy matrix after Gd addition by XRD software are shown in Table 3.1. The solid solution of Gd element increases both the lattice parameters c and a as compared with those of pure Mg. The c/a ratio decreases with the increase of Gd content. Because of the magnesium has a close packed hexagonal structure, the axial ratio (c/a) not only affects the crystal plane spacing, but also affects the crystal plane sliding. The c/a of pure Mg and most of Mg alloys is larger

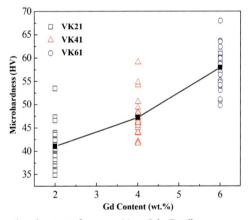

Figure 3.3 Microhardness distribution of as cast Mg–Gd–Zr alloy.

Table 3.1 Lattice parameters of pure Mg and as cast Mg–Gd–Zr alloy matrix.

Alloys	a/Å	c/Å	c/a
Mg	3.20890	5.21010	1.62364
VK21	3.21213	5.21369	1.62313
VK41	3.21405	5.21441	1.62238
VK61	3.21726	5.21256	1.62019

than or close to 1.633, and the close-packed and slip plane are (0001). When the c/a is less than 1.633, because the basal plane spacing is reduced, (0001) plane is no longer the most close-packed plane, the slip may start at the prismatic plane $(10\bar{1}0)$ or pyramidal plane $(11\bar{2}2)$ and others. Thus, prismatic or pyramidal $<c+a>$ slip can be activated. Therefore the decrease of c/a axial ratio can change the CRSS of basal slip and nonbasal slip, and thus promote the activation of nonbasal slip. Especially, the activation of pyramidal $<c+a>$ dislocation slip can coordinate the deformation in c axis, which improve the plasticity of the alloy significantly.

Theoretically, with the increase of Gd content, the c/a axial ratio decreases and results in better plasticity. Hence, the plasticity of VK61 alloy should be the best. However, the actual mechanical properties (Fig. 3.3) show that the VK21 alloy has the best plasticity. Combined with the phase constituent of Mg—Gd alloys, it is considered that the formation of a stable phase with a high melting point does a disservice to plasticity. During the plastic deformation of the alloy, the stable phase will pin the dislocation and hinder the movement of dislocation, which results in difficult plastic deformation and greatly increased strength. Therefore it is expected that to improve the plasticity of the alloy, reducing the content of the second phase and increasing the solid solution in the matrix through high temperature solution treatment is eligible.

3.2 Microstructure and properties of as-quenched alloy

On the premise of ensuring the solid solution of alloy elements, the effect of microstructure coarsening on its properties can be reduced by solution treatment at low temperatures. Therefore it is necessary to explore the solid solution process of the alloy.

3.2.1 Effect of solution temperature on microstructure and properties of Vk61 alloy

It can be seen from the DSC curve of as-cast VK61 alloy in Fig. 3.4 that the alloy is melted at 648°C. The dissolution of the second phase occurs in the corresponding temperature range of two broad endothermic peaks, namely 280°C—375°C and 445°C—523°C. Therefore the solution treatment temperature of the alloy is selected as 400°C and 500°C, and the XRD results after solid solution are shown in Fig. 3.5. There is a large amount of Mg_2Gd,Mg_3Gd in as-cast VK61 alloy. After solution treatment at 400°C, a new Mg_5Gd peak appears near 20 degrees, while the Mg_2Gd peak near 28 degrees becomes weak. This reveals that a new reaction similar to the peritectoid reaction may take place.

$$\alpha - Mg + Mg_2Gd \rightarrow Mg_3Gd$$

$$\alpha - Mg + Mg_3Gd \rightarrow Mg_5Gd$$

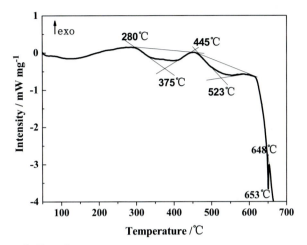

Figure 3.4 DSC curve of VK61 alloy.

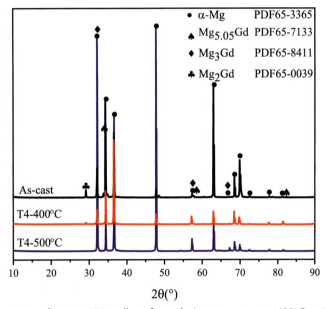

Figure 3.5 XRD pattern of as-cast VK61 alloy after solution treatment at 400°C and 500°C.

Combined with the DSC curve, it can be seen that the reaction is an endothermic reaction which starts at 280°C, and the peak temperature of the reaction is 375°C. After the reaction, Mg_2Gd and Mg_3Gd tend to disappear and a large number of Mg_5Gd phases form. With the increase of temperature, the Mg_5Gd phase starts to dissolve at 445°C and ends at 523°C. The second endothermic reaction occurs at 445°C and reaches its peak at 523°C.

To obtain the best solution temperature of the alloy, the VK61 alloy with the highest Gd content was treated with different solution processes, as shown in Fig. 3.6. The solution treatment process is 300°C (hold for 6 h) + 400°C/420°C/440°C/460°C/480°C/500°C (hold for 10 h) + water quenching. In the picture, the large black areas and gray areas are stain and watermark left by metallographic treatment. With the increase of solution temperature, the content of the second phase in the alloy decreases gradually, and the particle size decreases. The second phase continuously and densely distributed along the grain boundary at 400°C (Fig. 3.6A) gradually changes to discontinuous distribution at 420°C (Fig. 3.6B). As a further increase of solution treatment temperature, there are very few fine second phase particles along the grain boundary and inside the grain. The grains grow slightly at 440°C (Fig. 3.7C). The second phase particles are fully dissolved into the matrix and the grains continue to grow slowly at 460°C (Fig. 3.6D). As the solution temperature of the alloy reaches 480°C and 500°C (Fig. 3.6E, F), the microstructure coarsens and the grains grow through merging each other. Therefore with the same holding time, solution temperature of 440°C and 460°C is relatively better. While the temperature continues to rise, it will cause grain coarsening and slight overburning. As a consequence, for the full dissolution of the second phase, the best solution treatment process of VK61 alloy is 300°C (hold for 6 h) + 460°C (hold for10 h) + water quenching

The microhardness of VK61 alloy after solution treatment at different temperatures were measured and shown in Fig. 3.7. The increase of solution temperature has little effect on the microhardness of the alloy, but there is a certain trend. With the increase of solution temperature, the microhardness of the alloy increases and reaches the peak

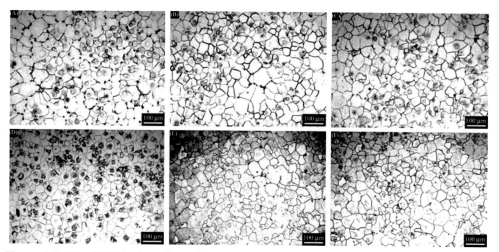

Figure 3.6 OM of VK61 alloy after solution treatment at different temperatures: (A) 400°C, (B) 420°C, (C) 440°C, (D) 460°C, (E) 480°C, (F) 500°C.

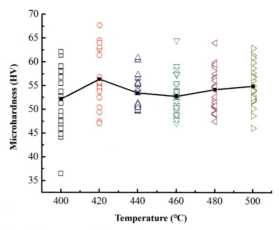

Figure 3.7 Hardness of solution treated VK61 alloy at different temperatures.

value at 420°C. As the temperature continues to increase, the microhardness of the alloy is relatively low at 440°C and 460°C. When the solution temperature increases further, the microhardness increases slightly. When the solution temperature is low, the solution effect is not ideal, The second phase strengthening is the main strengthening (400°C) mechanism in the alloy. With the increase of temperature, the solution strengthening and the second phase strengthening coexist, such as at the solution treatment temperature of 420°C. When the temperature continues to increase, the dissolution of the second phase decreases the hardness of the alloy (440°C and 460°C). The slight fluctuation of microhardness at the higher solution temperature is due to the coarsening of the microstructure. The homogeneity of microstructure after solution treatment is relatively good at 440°C and 460°C. Therefore the best solution treatment process of VK61 alloy can be further determined at 300°C (hold for 6 h) + 460°C (hold for 10 h) + water quenching.

3.2.2 Effect of Gd content on microstructure and properties of as-quenched Mg—xGd—0.6Zr alloy

Fig. 3.8 shows the optical microstructures of Mg—xGd—0.6Zr (x = 2, 4, 6, wt.%) alloy after solution treatment. The solution process is 400°C (hold for 6 h) + 500°C (hold for 10 h) + water quenching. With the increase of Gd content, the microstructure of solid solution alloy is slightly refined. Compared with as-cast alloy, the microstructure of VK21 and VK41 with lower Gd content is coarsened. Because of the existence of the second phase in VK61 alloy, the grain growth is hindered, which leads to the refined microstructure of VK61 alloy, and the homogeneity of the microstructure is improved significantly. In the figure, in addition to the grain boundary, there are also some black dots. Other large black areas and gray areas are stain and watermark left by metallographic treatment.

To determine the solid solution effect of the alloy at high temperature, EDS analysis was carried out for VK61 with the highest Gd content, as shown in Fig. 3.9. There are only a few fine Gd containing second phase and the Zr-riched phase in the alloy. This indicates that the second phase has dissolved at high temperature, and the high temperature solid solubility accelerates the solid solution behavior of Gd in the matrix. The solid solubility of Gd in magnesium can reach 22 wt.% at 500°C, which is far more than the addition amount of Gd in the alloy. However, the distribution of Gd in the matrix is not completely uniform, some enrichment occurs at the grain boundary with lower energy. Solid solution treatment at high temperature only diffuses the Gd near the grain boundary.

Fig. 3.10 shows the XRD pattern of Mg$-x$Gd$-$0.6Zr ($x = 2$, 4, 6, wt.%) alloy in solid-solution state. All three alloys only show the peak of the α-Mg matrix, and no peak of the Gd-riched second phase is detected. The results show that the solution treatment of VK21 and VK41 with low Gd content is very good with the Gd completely dissolved into the matrix. The Gd-riched second phase in VK61 with relatively high Gd content cannot be detected due to an extremely low amount. As the lattice structure of Zr and Mg is the same, which are both hexagonal close-packed structure, and the peak of Zr coincides with Mg peak in XRD results. At the same time, the content of Zr after solution treatment in the alloy is less, which is not obvious in XRD results.

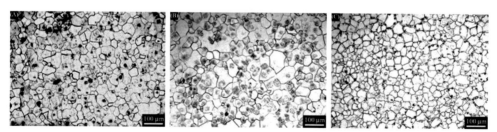

Figure 3.8 Microstructure of solution treated Mg$-$Gd$-$Zr alloy (A) VK21; (B) VK41; (C) VK61.

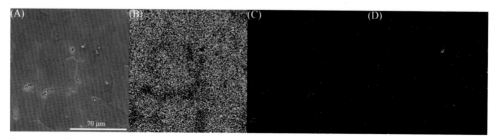

Figure 3.9 The solid solution treated VK61 alloy (A) scanning microstructure and the distribution of (B) Mg, (C) Gd, (D) Zr element.

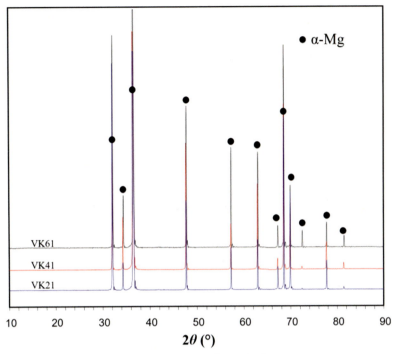

Figure 3.10 XRD pattern of solution treated Mg–xGd–0.6Zr alloy.

It can be seen from Fig. 3.11 that the peak of $(11\bar{2}0)$ plane shifts to a lower angle with the increase of Gd content. This is because a large number of Gd dissolved into the matrix, resulting in an increase of the plane spacing and the decrease of the diffraction angle of the corresponding plane. In order to further quantify the peak shift caused by the solid solution of the Gd element, Vegard's empirical formula is proposed to calculate the change of lattice parameters caused by Gd addition.

Vegard's rule is an empirical relation between two single substances and lattice parameters of their solid solution. It is usually used to describe alloy or two mixtures. Vegard firstly proposed to express KBr and KCl. He thought that the lattice parameters of KBr and KCl could be calculated by the specific gravity of their lattice parameters. After that, many researchers also applied such a proposal, which has been verified by many experiments. The simplest mathematical expression of Vegard's law is:

$$a(\text{mixture}) = (1 - x)a1 + xa2$$

where a_n ($n = 1, 2$) represents the lattice parameters of single-phase and solid solution, respectively, and x is the molar percentage of one of the single-phase substances. The modified formula through a lot of experimental modification is as follows:

$$a3(\text{mixture}) = (1 - x)a13 + xa23$$

In this chapter, the changed percentage of lattice parameters caused by the content of Gd in the matrix is quantified by the above-mentioned formula. Before the calculation, the factors that cause the change of matrix lattice parameters are divided into two categories: the addition of Gd, and other factors that can cause peak shift, such as the addition of Zr, residual stress and testing equipment.

Based on the XRD results of the solution-treated alloy in Fig. 3.10, the lattice parameters of the alloy matrix are calculated after solid solution treatment, as shown in Table 3.2. Compared with the lattice parameters (Table 3.1) of as-cast Mg–Gd–Zr alloy obtained from XRD results (Fig. 3.2), both lattice constants a and c increase after solid solution treatment due to the sufficient solid solution of Gd. However, the c/a axis ratio of the VK21 alloy decreases, while that of VK41 and VK61 increases, especially VK61 increases more.

The solid solution molar percentage of Gd in directions of a and c axis is calculated by the Vegard's empirical formula using the lattice data $a = 3.20890$, $c = 5.21020$ in Mg and the lattice parameter $a = 5.40000$ in Gd, as shown in Table 3.3. Comparing

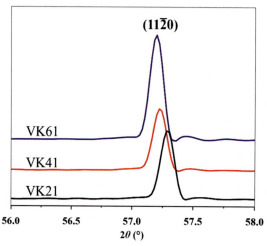

Figure 3.11 The peak shift of XRD pattern of solution treated Mg–xGd–0.6Zr alloy caused by the addition of Gd.

Table 3.2 Lattice parameters a, c(Å) of solid solution α–Mg matrix in the alloy.

Alloy	a/Å	c/Å
VK61	3.21798	5.21813
VK41	3.21610	5.21789
VK21	3.21346	5.21567

the calculated soluted quantity of Gd in a and c directions, it is found that the soluted quantity of Gd in the c direction is far greater than that in a direction. Therefore with increasing the Gd content, the c value in VK41 and VK61 increase more than a value. As a result, the c/a value of solution treated alloy will be larger than that of the as-cast alloy. The soluted quantity of Gd in the c direction is selected to calibrate the solution quantity of Gd in the matrix, the results are shown in (Table 3.4). The lattice parameter c of the matrix after solid solution of Gd, which is calculated by Vegard's empirical formula, is smaller than that calculated by XRD diffraction (Table 3.2). This may be due to neglecting the influence of Zr solution in alloy and the peak shift caused by residual stress. According to the shifting percentage, the shift degree increases with the increase of Gd content, which is in consistent with the peak shift of XRD result (Fig. 3.11). This method also proves the good solid solution effect of Gd in the magnesium matrix to some extent.

To further study the effect of solid solution of Gd on the properties of the alloy, the microhardness distribution of the solution-treated Mg—xGd—0.6Zr alloy are shown in Fig. 3.12. The microhardness of VK21 and VK41 alloys with low Gd content after solution treatment has almost no change. The solution of Gd element does not obviously affect the change in strength. The microhardness of VK61 alloy with relatively high Gd content slightly decreases, which may be due to the significant dissolution of the second phase which surpasses the solid solution strengthening effect. Thus, the microhardness of the alloy declines to some extent. At the same time, the homogeneity of alloys with low Gd content becomes better after solution treatment.

Table 3.3 Molar percentage of solid solution Gd calculated by Vegard's empirical formula.

Alloy	x1(a/Å)/mole percent	x2(c/Å)/mole percent
VK61	0.41440	4.17808
VK41	0.32860	4.05163
VK21	0.20811	2.88198

Table 3.4 Effect of Gd content calculated on lattice parameters of alloy matrix based on Vegard's rule.

Alloy	Actual addition of Gd/mole percent	Lattice parameters of matrix after Gd solution c/Å	Percentage of Gd solid solution result in lattice change %
VK61	1.291	5.21230	26.48
VK41	0.720	5.21140	15.60
VK21	0.320	5.21072	9.51

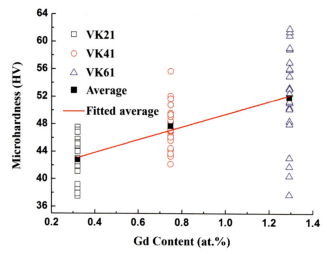

Figure 3.12 Microhardness distribution of solution treated Mg−Gd−Zr alloy.

3.3 Microstructures and mechanical properties of extruded Mg−Gd−Zr alloy

The previous results show that the solid solution of Gd in the magnesium alloy increases the matrix lattice constant. The c/a axis ratio decreases compared to pure magnesium, and the microhardness of all the alloys decreases and the microstructure becomes uniform after solid solution treatment. Therefore the solid solution treatment can provide a good ingot and billet for the extrusion process. In this section, the ingot and billet with different pretreatment will be extruded to study its microstructure and properties. The extrusion process is as follows: preheated at 480°C for 20 min, followed by extrusion at 450°C with an extrusion ratio of 28 and an extrusion rate of 2.1 m min^{-1}.

3.3.1 Effect of pretreatment on microstructure and mechanical properties of extruded Mg−xGd−0.6Zr alloy

Fig. 3.13 shows the microstructure of extruded Mg−xGd−0.6Zr (x = 2, 4, 6, wt.%) alloy without pretreatment. The alloy consists of fine equiaxed grains and a small number of coarse grains. The dynamic recrystallization during the extrusion process is close to complete. The second phase particles in the alloy and dissolved segregated Gd atoms provide the nucleation sites for the recrystallization. Compared with the as-cast alloy, the microstructure after extrusion becomes finer and more uniform, and all the average grain sizes are below 10 µm, as shown in Table 3.5. Unlike the microstructure of extruded alloy without pretreatment, the microstructure of extruded alloy after solid solution pretreatment consists of more coarse grains and a small number of fine grains, as shown in Fig. 3.14. The solid solution at high temperature makes a large amount of Gd atoms dissolve into the magnesium matrix and becomes an obstacle to dislocation movement during the extrusion process. The deformation during the extrusion process is mainly affected by the solid solution effect. As the addition of Gd for all the three alloys is within the range of solid solubility, the solid solution effect is not much different. Hence, the increase of Gd content has almost no effect on the microstructure of the alloy, and their average grain size is about 10 μm (Table 3.5). However, the grain size of the extruded alloy with pretreatment is slightly higher than that without pretreatment. The solid solution of microalloying element can achieve high addition amount of alloying element. In order to obtain a fine precipitated phase as the recrystallization nuclei based on solid solution treatment, a pretreatment of solid solution+aging is proposed. The microstructure of the extruded alloy is slightly coarser than that without pretreatment, but it is finer than that of the alloy before solution

Table 3.5 Average grain size of each alloy (μm).

Processing status	VK21	VK41	VK61
F–Ext	8.1	7.0	6.9
T4–Ext	10.9	9.4	11.4
T6–Ext	9.0	7.6	8.0

Figure 3.13 Microstructure of as-extruded Mg–xGd–0.6Zr alloy: (A) VK21, (B) VK41, (C) VK61.

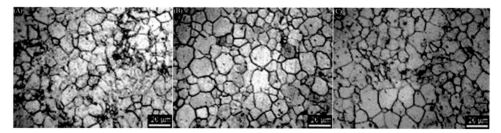

Figure 3.14 Microstructure of Mg–xGd–0.6Zr alloy in solution-extruded state: (A) VK21, (B) VK41, (C) VK61.

treatment (Fig. 3.15). The extrusion process of the alloy is affected by both the solid solution effect and the second phase. In low Gd containing alloy, the deformation is mainly affected by the former, while deformation of high containing Gd alloy is mainly affected by the latter.

The mechanical properties of the corresponding alloys are shown in Table 3.6. The tensile strength and yield strength of VK21 and VK41 alloys with relatively low Gd content are very close, and VK61 exhibits the highest tensile strength and yield strength. Only the solid solution pretreatment at a high temperature slightly reduces the strength of the alloy, the solid solution+aging pretreatment retains the strength of the extruded alloy to a certain extent. However, the strength of extruded alloys with pretreatment is still lower than the extruded alloy without pretreatment. It is considered that after the addition of the Gd element, a part of Gd dissolves into the matrix to generate solid solution strengthening, and the Gd exceeding the solid solution portion can react with Mg to form a stable compound to generate a second phase strengthening. Therefore the strength of the alloy with low Gd content mainly comes from solid solution strengthening. While the strength of the alloy with high Gd content comes from both solid solution strengthening and second phase strengthening,

Figure 3.15 Microstructure of solid solution aging-extruded Mg−xGd−0.6Zr alloy: (A) VK21, (B) VK41, (C) VK61

Table 3.6 Tensile strength, yield strength, and elongation of Mg−Gd−Zr alloy in different states.

Alloy	Tensile strength/Mpa	Yield strength/Mpa	Elongation/%
VK21-EXT	207	150	36.8
VK41-EXT	206	145	43.4
VK61-EXT	237	168	33.4
VK21-T4-EXT	192	125	33.9
VK41-T4-EXT	189	112	41.8
VK61-T4-EXT	216	137	30.9
VK21-T6/10h-EXT	204	137	45.6
VK41-T6/10h-EXT	206	136	44.0
VK61-T6/10h-EXT	228	159	37.9

which makes VK61 shows the highest strength. Solid solution pretreatment weakens the effect of second phase strengthening and simultaneously increases the contribution of solid solution strengthening. The solid solution+aging increases the second phase strengthening by sacrificing part of the solution strengthening, but both pretreatments decline the strength of the alloy in comparison with that of the alloy without pretreatment. The VK alloy processed by direct extrusion using as-cast ingot has good strength.

As an important index to evaluate the plastic deformation ability of the alloy, the elongation increases with the increase of Gd content within the composition range of alloy which mainly exhibits solution strengthening. When it exceeds the composition range, the elongation decreases with the increase of Gd content. Among the extruded alloys without pretreatment, the elongation of VK41 alloy is the highest of 43.3%, while the elongation of VK21 and VK61 alloy is 36.8% and 33.4%, respectively. Although the elongation of VK21 and VK61 decreased, they still have good plasticity. According to the calculation by Moitra et al. on the effect of solid solution on the slip of Mg-based binary alloys, it can be seen that the solid soluted Gd atom can promote the nucleation and slip of $<c+a>$ dislocation and improve the forming coefficient of the alloy by reducing the ratio of the unstable stacking fault energy between the basal plane and the pyramidal plane. At the same time, the Gd atom will generate solid solution strengthening in magnesium alloy. Therefore the solid soluted Gd plays an important role in the strength and plasticity of magnesium alloy. Especially the contribution to the plasticity of the alloy has a good effect both in theory and practice.

3.3.2 Effect of Gd addition on lattice parameters of extruded alloys

The XRD results of as-cast Mg–xGd–0.6Zr alloy after extrusion are shown in Fig. 3.16. There is only α-Mg phase observed in all the three alloys and no second phase is detected. Such results indicate that even if there is second phase in the alloy, its content is very low. The disappearance of the second phase in VK61 alloy is due to the dispersive distribution of the fragmented second phase. Besides, the fragmented dispersive phase is partially dissolved into the matrix by diffusion and other actions during extrusion process. As shown in Fig. 3.17, the peak of $(11\bar{2}0)$ plane shifts to a low angle with the increase of Gd content. The results show that the spacing between crystal planes decreases after extrusion, and thus the diffraction angle increases and the peak shifts to a high angle. The lattice parameters of the alloy matrix are calculated by XRD and shown in Table 3.7. Compared with the as-cast alloy, both lattice parameters a and c increase after extrusion. Only the c/a axis ratio of VK21 is reduced, and the c/a axis ratio of the other two alloys is increased. This means that VK21 after extrusion has better plastic deformability.

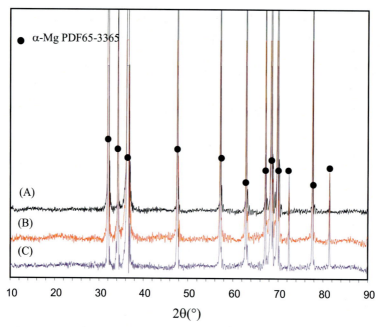

Figure 3.16 XRD patterns of cast-extruded Mg−xGd−0.6Zr alloy (A) VK61; (B) VK41; (C) VK21.

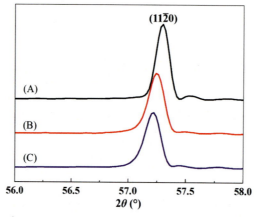

Figure 3.17 Peak shift of cast extruded Mg−xGd−0.6Zr alloy due to Gd addition: (A) VK61, (B) VK41, (C) VK21.

Table 3.7 The lattice parameters of extruded Mg−xGd−0.6Zr matrix obtained from Fig. 3.17.

Alloy	a/Å	c/Å	c/Å
VK21–Ext	3.21818	5.22045	1.62217
VK41–Ext	3.21528	5.21728	1.62265
VK61–Ext	3.21266	5.21638	1.62370

3.4 High plasticity mechanism of Mg—Gd—Zr alloy

The change of plasticity in Mg—Gd alloy can be analyzed using the change of solid solubility and the theory of solid solution strengthening and plasticizing. The Gd atoms in the solid solution can significantly increase the CRSS of the basal plane and reduce the CRSS of the pyramidal plane. Such effects help the Mg—Gd solid solution to activate more slip systems during the deformation, and thus improve its plasticity. Before reaching the maximum solubility, the plasticity of magnesium alloy increases with the increase of solid solution quantity of Gd. When the content of Gd exceeds the solid solubility, the plasticity of magnesium alloy begins to decrease due to the formation of second phases. Therefore for high plasticity Gd containing magnesium alloy, the content of Gd should not exceed the maximum solid solubility. In these alloys, 6 wt.% has exceeded the maximum solubility, so the plasticity begins to decrease.

It can be seen from the previous research that Gd in VK21 can completely dissolve into α-Mg, which has a very good solution strengthening and plasticizing effect. VK41 contains a small amount of the second phase with a higher Gd content, which benefits from both solution strengthening and second phase strengthening. The strength of VK61 alloy increases with the increase of the content of the second phase, but the effect of solution plasticization is counteracted.

The elongation δ of the alloy can be simplified to be affected by two factors: solution plasticization and precipitation reducing the plasticity.

$$\delta = k_1 c_1 + k_2 c_2$$

where c_1 is the atomic ratio of solid solution Gd, c_2 is the atomic ratio of precipitated Gd, and k_1 and k_2 are the corresponding coefficients. The results show that $k_1 \approx 106$, $k_2 \approx -2500$. It can be seen that the precipitation of the second phase largely reduces the elongation.

In order to further improve the comprehensive mechanical properties of the alloy, the corresponding relationship between process parameters, microstructure, mechanical properties based on conventional rapid extrusion is established. For ultrahigh plasticity Mg—Gd—Zr magnesium alloy, VK21 alloy with completely Gd solid soluted is the best option for further investigation.

3.4.1 Microstructure and plasticity of as-cast and extruded VK21 alloys

The microstructure of as-cast and extruded VK21 magnesium alloy is compared in Fig. 3.18. It can be seen that the microstructure of as cast VK21 alloy is uniform, showing obvious equiaxed crystal structure without dendrite structure. The average grain size of as-cast VK21 alloy measured by the linear intercept method is about 26 μm. The grain size of the alloy is refined after extrusion at 420°C. The microstructure of the extruded bar is characterized by double grain size distribution, which is composed of relatively large conventional grains and a large number of fine grains distributed along the extrusion direction.

It can be seen from the metallographic photograph that the fine grain structure of the VK21 alloy bar formed by conventional extrusion is uniform.

Fig. 3.19 shows the inverse pole figure (IPF) map and grain size distribution of as-cast and extruded VK21 magnesium alloy. It can be seen from Fig. 3.19A that the

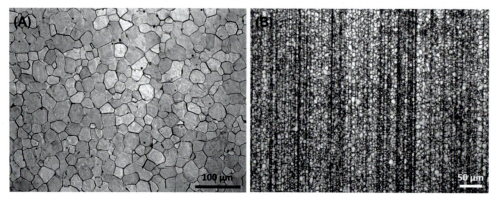

Figure 3.18 Microstructure of VK21 magnesium alloy: (A) as-cast, (B) as-extruded.

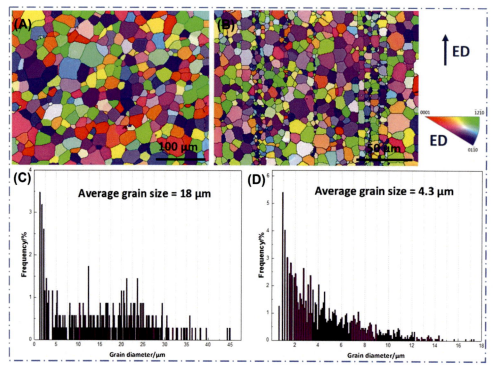

Figure 3.19 Inverse pole figure map and grain size distribution of VK21 alloy: (A), (C) as-cast, (B), (D) as-extruded.

grains of as-cast alloy are evenly distributed without texture characteristics. The average grain size measured by EBSD is about 18 μm, which is slightly smaller than the data measured by OM. This is because twins and deformation substructures are inevitably introduced by EBSD sampling, which is misconsidered as grain in orientation calibration. Fig. 3.19B shows the IPF map of the longitudinal section of VK21 extruded bar. The fine grain zone is clearly visible, and the grains in it are obviously smaller than the conventional grains, showing a banded distribution along (ED.) The average grain size calculated by EBSD is 4.3 μm, while the average grain size in the fine grain band is about 1.9 μm.

The XRD results of as-cast and extruded VK21 magnesium alloy are shown in Fig. 3.20. The results show that both as-cast and extruded alloys are mainly composed of α - Mg solid solution phase. A weak diffraction peak also exists at $2\theta = 22.3$ degrees. In addition, the peak value of (0002) plane after extrusion is low, which indicates a weak basal texture. In order to further determine the phase of the alloy, SEM and TEM are used to characterize the alloy. Fig. 3.21(B, C) is the amplification of A and B regions in (A), respectively. It can be seen that there are very few nanobulk phase and spherical particle phase in extruded VK21. The results of EDS under TEM show that the bulk phase is Mg_3Gd and the spherical phase is α-Zr. The length of bulk Mg_3Gd is about 500 nm, and the radius of the spherical α-Zr is about 300 nm. By comparing the standard XRD patterns of Mg_3Gd phase, the diffraction peak of which is at 22.3 degrees. There is no corresponding diffraction peak in the XRD

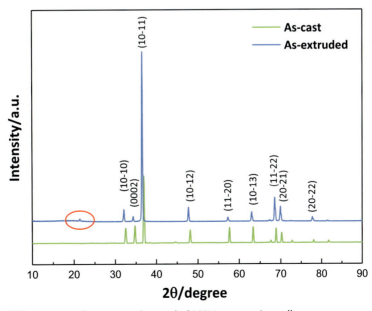

Figure 3.20 XRD patterns of as-cast and extruded VK21 magnesium alloy.

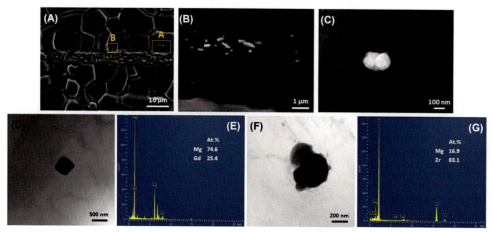

Figure 3.21 SEM, TEM and EDS of extruded VK21 alloy.

results mainly because of its extremely low amount. Due to the low amount and small size of Mg_3Gd and α-Zr phases, the dynamic recrystallization cannot be promoted by the PSN mechanism, and the dislocation motion cannot be effectively pinned during deformation. Therefore VK21 alloy can be treated as a single-phase solid solution when studying its deformation mechanism.

The (0001) pole figure and IPF of the extruded VK21 magnesium alloy are shown in Fig. 3.22(A, B). Meanwhile, the macrotexture of pure magnesium extruded at 320°C is presented for comparison (Fig. 3.22C, D). In the figures, RD represents the radial direction of the alloy bar, and ED refers to the extrusion direction. The results show that the texture type of extruded VK21 alloy bar is obviously different from that of pure magnesium after extrusion. Namely, the basal texture of VK21 is effectively weakened. The c-axis of grains deviates from the ED direction by about 45 degrees and the texture composition is $<11\bar{2}2>\|ED$, which is a typical RE texture and similar to that of Mg−Gd alloy studied by Stanford [1−5]. However, the pure magnesium extruded at 320°C exhibits the fiber texture with its crystallographic direction of $<11\bar{2}0>-<10\bar{1}0>\|(ED.)$

Fig. 3.23 shows the typical engineering stress-strain curves of as-cast and extruded VK21 alloy. Mechanical properties of pure magnesium extruded at 320°C are also presented for comparison. The yield strength and ultimate tensile strength of as-cast VK21 alloy are 83 and 170 MPa, respectively. The fracture elongation of as-cast VK21 alloy is as high as 43%, which is higher than that of all the existing cast magnesium alloys. The high elongation is due to the fine and uniform equiaxed grains, the scattered orientation without sharp texture feature. There is an obvious yield point on the tensile stress-strain curve of VK21 alloy formed by conventional rapid extrusion. The yield strength is 145 MPa, the ultimate tensile strength is 225 MPa, and the

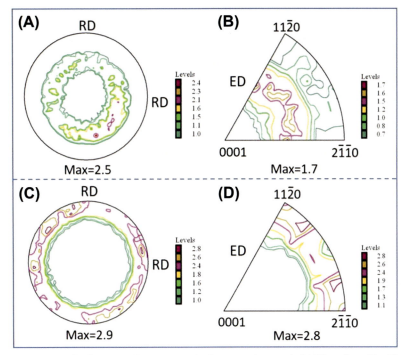

Figure 3.22 (0001) pole figure and inverse pole figures of extruded VK21 alloy (A), (B) and pure magnesium (C), (D).

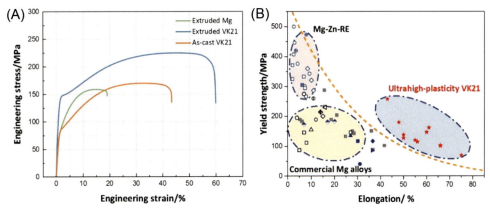

Figure 3.23 Stress-strain curve (A) and mechanical properties comparison (B) between VK21 and pure magnesium in as-cast and extruded state [6–15].

fracture elongation at room temperature is as high as 58%, which is identified as ultrahigh plasticity. In contrast, the strength and elongation of extruded pure magnesium are much lower, the yield strength is only 76 MPa, the ultimate tensile strength is about 170 MPa, and the fracture elongation is only 19%. The tensile mechanical properties of Mg—Zn—RE alloy and commercial wrought magnesium alloy at room temperature are compared in Fig. 3.23B. It can be seen that the yield strength of Mg—Zn—RE alloy can be as high as nearly 500 MPa. However, the fracture elongation is less than 5%. The yield strength and fracture elongation of commercial wrought magnesium alloy are both at a medium level, and the comprehensive mechanical properties are stable. The mechanical properties of VK21 magnesium alloy are superior after microstructural regulation, and the yield strength and fracture elongation of VK21 magnesium alloy are significantly higher than the C-curve of strength and plasticity tradeoff.

As mentioned above, VK21 wrought magnesium alloy prepared by conventional rapid extrusion exhibits ultrahigh plasticity, high yield strength, and ultimate tensile strength. The following study focuses on improving the comprehensive mechanical properties of extruded VK21 alloy through the optimization of process parameters and heat treatment.

3.4.2 Effect of extrusion process on microstructure and plasticity of VK21 alloy

3.4.2.1 Effect of cooling after extrusion on microstructure and plasticity of extruded Vk21 alloy

Fig. 3.24 shows the IPF map of extruded VK21 alloy with different cooling rates. Fig. 3.24A is the microstructure of the alloy obtained by conventional air cooling after extrusion with a surface cooling rate of $50°Cs^{-1}$. It can be seen that there are a large number of fine grain bands uniformly distributed along the (ED.) According to statistics,

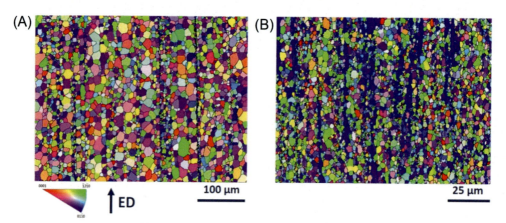

Figure 3.24 Inverse pole figure map of VK21 alloy with different cooling rates: (A) $50°Cs^{-1}$, (B) $200°Cs^{-1}$.

the recrystallization fraction of the alloy reaches 97.8%, and the average grain size is 5.8 μm. The surface cooling rate of the alloy bar treated by water cooling is $200°Cs^{-1}$. There are a lot of nonrecrystallized areas in the alloy along ED, and the nonrecrystallized grains show a hard orientation characteristic, namely $(10\bar{1}0)\perp$ED、$<01\bar{1}0>\parallel$ED, and the percentage of nonrecrystallized area is 24.1%. The average grain size of the alloy prepared by a water cooling of $200°Cs^{-1}$ is 1.5 μm, which becomes more refined than that obtained by a cooling rate of $50°Cs^{-1}$.

Fig. 3.25 shows the tensile stress–strain curves of the alloys prepared at two cooling rates. It can be seen that the yield strength of the alloy with a surface cooling rate of $200°Cs^{-1}$ is significantly improved, which is higher than that of the alloy treated with a cooling rate of $50°Cs^{-1}$. The yield strength reaches 260 MPa, and the fracture elongation remains at 42%. There is an obvious yield plateau on the stress–strain curve. The stress of the alloy decreases firstly and then remains unchanged after yield. There is a 24% (volume fraction) nonrecrystallized zone in the alloy prepared by rapid water cooling. Therefore when the alloy is under tension along ED, the hard orientation of the nonrecrystallized grains makes the basal slip and extension twins hard to be activated, hence the alloy shows a high yield strength. The tensile yield strength and fracture elongation of the alloy with a cooling rate of $50°Cs^{-1}$ are 133 MPa and 52%, respectively. Compared with the alloy with a surface cooling rate of $200°Cs^{-1}$, the alloy exhibits better deformability at room temperature.

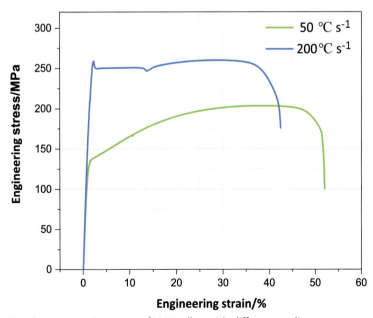

Figure 3.25 Tensile stress-strain curves of VK21 alloy with different cooling rates.

3.4.2.2 Effect of extrusion speed on microstructure and plasticity of extruded Vk21 alloy

The IPF map of the longitudinal section of VK21 alloy bar prepared at different extrusion speeds is shown in Fig. 3.26. Fig. 3.26A is the IPF map of the alloy with an extrusion speed of 1 m min^{-1}. It can be seen that the grains are uniform and there are few fine grains. The average grain size is about 3.1 μm. However, the grain size of the alloy bar extruded at 20 m min^{-1} is larger than that of the alloy prepared by 1 m min^{-1}. The average grain size is about 3.7 μm which is obviously larger than that of the alloy prepared by slow extrusion. The alloy obtained by the two extrusion speeds both exhibit recrystallization microstructure. Compared with the slow extrusion, the alloy prepared by 20 m min^{-1} rapid extrusion has more fine grain bands, and there is an obvious grain size difference between the grain within bands and the conventional grain.

Fig. 3.27 shows the tensile stress-strain curves of the alloys prepared at two extrusion speeds. When the extrusion speed is 1 m min^{-1}, the tensile yield strength is 187 MPa, the ultimate tensile strength is 223 MPa, and the fracture elongation is 52%. However, when the extrusion speed is 20 m min^{-1}, the yield strength is only 105 MPa, the ultimate tensile strength is 182 MPa, and the fracture elongation is 63%. The results show that when the extrusion speed is high, and a lot of heat is generated during the extrusion process, which leads to the rapid increase of temperature in the extrusion cylinder and thus promotes the grain growth after dynamic recrystallization. Therefore the average grain size of VK21 alloy prepared by 20 m min^{-1} rapid extrusion is larger, and its tensile yield strength at room temperature becomes lower.

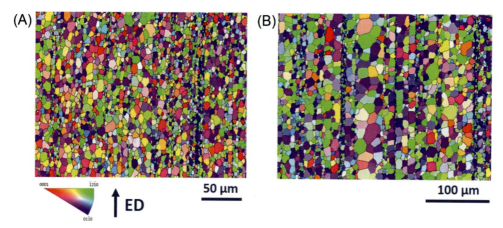

Figure 3.26 Inverse pole figure map of VK21 alloy with different extrusion speeds: (A) 1 m min^{-1}, (B) 20 m min^{-1}.

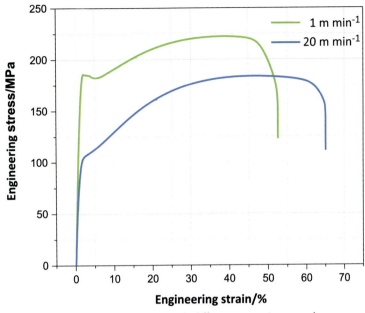

Figure 3.27 Stress-strain curves of VK21 alloy with different extrusion speeds.

Fig. 3.28 shows the grain size figure and inverse pole figure of the typical microstructure of the ultrahigh plasticity VK21 magnesium alloy prepared by conventional rapid extrusion. According to the grain size and the fine grain band distribution, the grains in the fine-grained band and the conventional grain are selected separately. It can be seen that the fine grain bands distribute along (ED.) Both grains with different sizes are recrystallized grains, which is obviously different from the nonrecrystallized area obtained by cooled with a high cooling rate. Compared to the IPF of fine grain bands and conventional grains, it can be seen that although the grain size is different, all of them exhibit RE texture. In order to clarify the grain structure characteristics of fine grains and conventional grains, TEM was used to characterize the grains with different sizes. Fig. 3.29 shows the TEM bright field images of fine grains and conventional grains in VK21 alloy. The blue arrow indicates the fine grains in the fine-grained bands, and the red arrow indicates the conventional grains. It can be seen that there are no dislocations in both types of grains, indicating that the fine grains and conventional grains are both recrystallized grains.

3.4.2.3 Effect of preupsetting treatment on microstructure and plasticity of extruded Vk21 alloy

Fig. 3.30 is the IPF map of VK21 alloy prepared by different preupsetting deformations. Fig. 3.30A is the IPF map of the alloy with preupsetting deformation of 63%. It

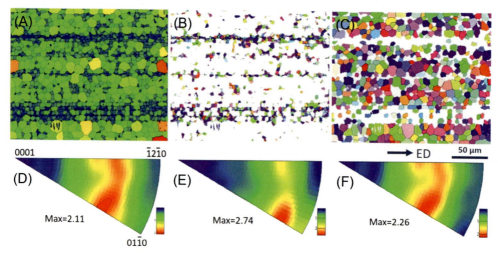

Figure 3.28 Fine grains and conventional grains of VK21 alloy formed by conventional rapid extrusion (A) grain size figure, (B) the inverse pole figure (IPF) map of fine grains, (C) the IPFmap of conventional grains, (D) the inverse pole figure of the whole area, (E) the inverse pole figure of fine grains, and (F) the inverse pole figure of conventional grains.

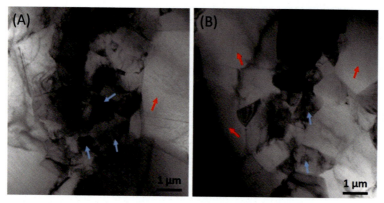

Figure 3.29 TEM bright field images of fine grains and conventional grains in VK21 alloy.

can be seen that the grain size of the alloy is relatively uniform, and the difference between the fine grain size and the conventional grain size is not obvious. The average grain size is 2.3 μm. However, the sample with 7% preupsetting deformation has larger grain size and uniform distribution of fine grain bands, and the grain size in the fine grain bands is different from that of conventional grains. The average grain size of the alloy is 3.9 μm, which is significantly larger than that of the alloy with 63% preupsetting deformation. It can be seen that the alloys prepared by the two kinds of preupsetting deformation present both full recrystallized microstructure.

Fig. 3.31 shows the stress–strain curves of the alloy prepared by extrusion after 63% and 7% preupsetting deformation. The results show that the tensile yield strength of the alloy prepared with 63% preupsetting deformation is 180 MPa, and there is a yield plateau. The ultimate tensile strength is 223 MPa and the fracture elongation is 47%. However, the yield strength of the alloy prepared by preupsetting deformation of 7%

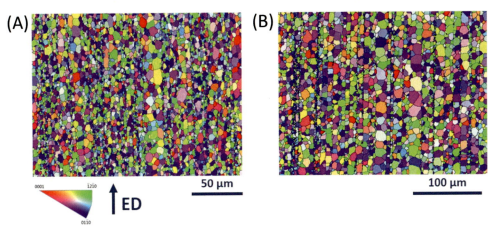

Figure 3.30 Inverse pole figure map of VK21 alloy prepared by different preupsetting deformation: (A) 63%, (B) 7%.

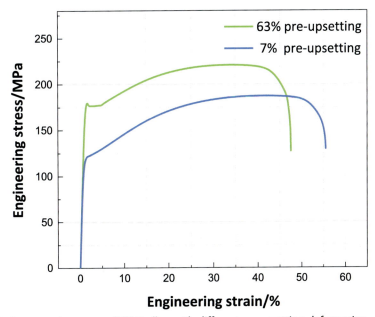

Figure 3.31 Stress-strain curves of VK21 alloy with different preupsetting deformation.

is only 120 MPa, and its ultimate tensile strength is 188 MPa. The fracture elongation reaches 55%, which is significantly higher than that of 63% preupsetting deformation. Large preupsetting deformation before extrusion can effectively crush the initial grains, refine the grains after dynamic recrystallization, and improve the tensile yield strength of the alloy.

3.4.2.4 Effect of extrusion ratio on microstructure and plasticity of extruded Vk21 alloy

Fig. 3.32 shows the IPF map of alloys prepared with different extrusion ratios. The alloys with extrusion ratios of 85 and 27 are obtained by extrusion at 420°C. In order to ensure the small grain size of alloys, the extrusion temperature of small extrusion ratio is reduced accordingly. The extrusion temperature of the alloy with an extrusion ratio of 8 is set as 350°C. The results show that there are obvious fine grain bands in the alloy with extrusion ratios of 85 and 27, and the average grain size is 3.8 and 2.7 μm, respectively. A large number of nonrecrystallized regions appear in the alloy with an extrusion ratio of 8 at low temperature, and the average grain size is 1.6 μm. The nonrecrystallization zone exhibits the characteristic of the hard orientation of $(10\bar{1}0)\perp$ED, $<01\bar{1}0>$∥ED, and the percentage of the nonrecrystallized zone is 25.6%.

The tensile stress–strain curves of VK21 alloy prepared by different extrusion ratios are shown in Fig. 3.33. The tensile yield strength and ultimate tensile strength of the alloy with an extrusion ratio of 8 are significantly higher than those of the alloy with an extrusion ratio of 85 and 27. The yield strength and ultimate tensile strength of the alloy with an extrusion ratio of 8 are 193 and 227 MPa, respectively. And the fracture elongation is 46%. The mechanical properties of the alloy with an extrusion ratio of 85 and 27 have little difference at room temperature. The corresponding yield strength of the two alloys is 140 and 137 MPa, respectively. And the ultimate tensile strength is 197 and 204 MPa, respectively. The fracture elongation is 51% and 52%, respectively.

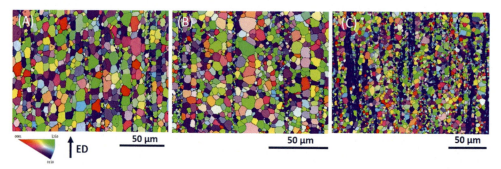

Figure 3.32 Inverse pole figure map of VK21 alloy with different extrusion ratios: (A) 85, (B) 27, (C) 8.

3.4.3 Effect of heat treatment on microstructure and plasticity of Vk21 alloy

3.4.3.1 Effect of preheat treatment on microstructure and plasticity of Vk21 alloy

IPF map of longitudinal section in VK21 magnesium alloy bar prepared by different heat treatments before extrusion is shown in Fig. 3.34. Fig. 3.34A is the IPF map of the alloy prepared by extrusion after homogenization treatment. The grain size of the

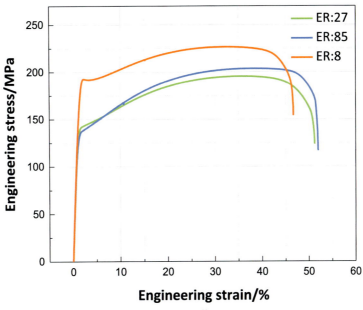

Figure 3.33 Stress-strain curves of VK21 alloy with different extrusion ratios.

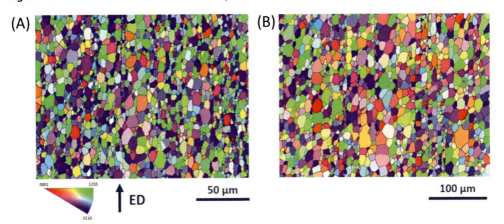

Figure 3.34 Inverse pole figure map of VK21 alloy with different heat treatment processes before extrusion: (A) homogenization, (B) solution treatment.

alloy is relatively uniform and fine grain bands are few. The average grain size calculated by EBSD is 3.6 μm. Fig. 3.34B is the IPF map of the alloy prepared by extrusion after solid solution treatment. The grain size of the alloy is larger than that of the alloy treated by homogenization. The average grain size is 7.5 μm, and there are still fine grain bands. In addition, it can be seen that the microstructure of the alloy prepared by the two heat treatment processes are both recrystallized grains.

Fig. 3.35 shows the stress-strain curves of VK21 magnesium alloy prepared for the two heat treatment processes. The results show that the strength and elongation of the extruded alloy after homogenization treatment are better than those of extruded alloy after solid solution treatment. The tensile yield strength, ultimate tensile strength, and fracture elongation of the alloy with homogenization treatment are 183, 220 MPa, and 46%, respectively. The tensile yield strength, ultimate tensile strength and fracture elongation of the alloy with solid solution are 153, 212 MPa and 44%, respectively. It should be noted that the VK21 alloy prepared by extrusion after solid solution still presents a yield plateau on the tensile stress-strain curve at room temperature.

3.4.3.2 Effect of annealing on microstructure and plasticity of extruded Vk21 alloy

Fig. 3.36 shows the IPF map of VK21 alloy with 63% preupset deformed, 420°C extruded and subsequently annealed at 420°C for different time. The results show that

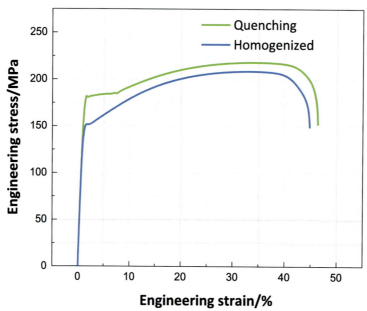

Figure 3.35 Stress-strain curves of VK21 alloy with different heat treatment processes before extrusion.

with the increase of annealing time, the grains of VK21 alloy gradually coarsen. The average grain size of VK21 alloy is 6.2, 7.6 and 8.5 μm after annealing for 90 min, 180 min and 270 min, respectively. Fine grains still exist in the alloy annealed for 270 min, indicating that the grains of the ultrahigh plasticity VK21 alloy with fine grains grow uniformly during annealing without preferential growth. The results of EBSD show that the texture of the alloy does not change obviously with the increase of annealing time.

Fig. 3.37 shows the stress-strain curves of the alloy annealed at 420°C for different time at room temperature. It can be seen that with the increase of annealing time, the

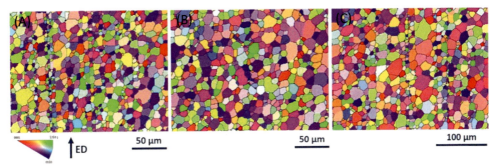

Figure 3.36 Inverse pole figure map of VK21 alloy with different annealing time: (A) 90 min, (B) 180 min, (C) 270 min.

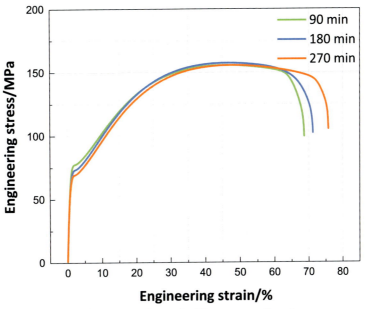

Figure 3.37 Stress-strain curves of VK21 alloy with different annealing time.

tensile yield strength of the alloy decreases slightly, while the fracture elongation increases correspondingly, which is mainly due to the grain growth during annealing. After annealing for 90 min, 180 min and 270 min, the tensile yield strength of those three alloys is 80, 72 and 64 MPa; the ultimate tensile strength is 161, 162 and 164 MPa; and the fracture elongation is 68%, 71% and 75%, respectively.

3.6 Summary

1. The microstructure of as-cast Mg−xGd−0.6Zr (x = 2, 4, 6, wt.%) alloy is refined with the increase of Gd content in a certain range of solid solubility and then coarsened with the increase of Gd content after exceeding this range. In VK21 and VK41 with low Gd content, the Gd is mainly dissolved into the matrix, while the Gd in VK61 has exceeded the maximum solid solubility. Besides the solid solution, $Mg_{5.05}Gd$, Mg_3Gd, and Mg_2Gd alloy phases are formed due to severe segregation. The results show that the lattice parameters c and a of Mg matrix increase with the solid solution of Gd, and the c/a ratio decreases with the increase of Gd content. Before reaching the maximum solid solution content, with the increase of Gd content, the plasticity of magnesium alloy increases greatly. After exceeding the maximum solution limit, the plasticity of magnesium alloy begins to decrease.

2. The results show that different pretreatments have no obvious effect on the microstructure of VK alloy, which are all typical dynamic recrystallization microstructures. The microstructure of the alloy with different Gd content has little difference. After solid solution pretreatment, the plasticity of the alloy increases with the increase of Gd content in the range of solution limit, and decreases with the increase of Gd content after exceeding the solution limit. The elongation of VK21 alloy reaches 45%.

3. The grains of as-cast VK21 alloy are fine, uniform, and equiaxed. The microstructure of VK21 alloy contains a large number of fine grain bands along the extrusion direction. The texture of VK21 is a typical rare earth texture with a crystallographic direction $<11\bar{2}2>\|ED$, which is different from traditional extruded fiber texture. The extruded alloy shows ultrahigh plasticity, and the elongation at room temperature is more than 60%. After proper annealing treatment, the elongation of extruded VK21 alloy reaches 75% and has ultrahigh plasticity.

References

[1] Yasi JA, Hector JLG, Trinkle DR. First-principles data for solid-solution strengthening of magnesium: from geometry and chemistry to properties. Acta Materialia 2010;58:5704−13.
[2] Wen L, Chen P, Tong ZF, et al. A systematic investigation of stacking faults in magnesium via first-principles calculation. European Physical Journal B 2009;72:397−403.

[3] Wang Y, Chen LQ, Liu ZK, et al. First-principles calculations of twin-boundary and stacking-fault energies in magnesium. Scripta Materialia 2010;62:646—9.

[4] Shi DK. Fundamentals of materials science. China Machine Press; 2003.

[5] Hirth JP, Lothe J. Theory of dislocations. John Wiley and Sons, Inc; 1982. p. 857.

[6] Chetty N, Weinert M. Stacking faults in magnesium. Physical Review B 1997;56(17):10844.

[7] Tabache MG, Bourret ED, Elliot AG. Measurements of the critical resolved shear stress for indium-doped and undoped GaAs single crystals. Applied Physics Letters 1986;49:289—91.

[8] Wu DH, Hu GX, et al. Relationship between critical shear stress (CRSS) and crystal orientation of <011] superlattice dislocation slip in TiAl single crystal. Journal of metals 1999;35:337—106.

[9] Zhu M, Wang LY, Song GF, et al. Preparation and properties of single crystal Cu al Ni be shape memory alloy. Functional materials 2007;38:1474—7.

[10] Chen YJ, Chen Q, Wang ZD, et al. Preparation and microstructure analysis of Al-1% Si single crystal. Journal of Beijing University of Science and Technology 2005;27:50—4.

[11] Monnet G, Pouchon MA. Determination of the critical resolved shear stress and the friction stress in austenitic stainless steels by compression of pillars extracted from single grains. Materials Letters 2013;98:128—30.

[12] Note R, Kimura S, Inoue A. Preparation of decagonal Al-Ni-Co single quasicrystal by Czochralski method. Materials Transactions JIM 1997;38:943—9.

[13] Huntington H. Modification of the Peierls-Nabarro model for edge dislocation core. In: Proceedings of the Royal Society of London Series B, 1955, 68: 1043.

[14] Joos B, Duesbery M. The Peierls stress of dislocations: an analytic formula. Physical Review Letters 1997;78(2):266—9.

[15] Ando S, Tonda H. Non-basal slip in magnesium-lithium alloy single crystals. Materials Transactions JIM 2000;41:1188—91.

Further reading

Agnew S, Horton J, Yoo M. Transmission electron microscopy investigation of <c + a> dislocations in Mg and α-solid solution Mg-Li alloys. Metallurgical and Materials Transactions A 2002;33:851—8.

Akhtar A, Teghtsoonian E. Solid solution strengthening of magnesium single crystals—I alloying behaviour in basal slip. Acta Metallurgica 1969;17:1339—49.

Akhtar A, Teghtsoonian E. Solid solution strengthening of magnesium single crystals—II the effect of solute on the ease of prismatic slip. Acta Metallurgica 1969;17:1351—6.

Akhtar A, Teghtsoonian E. Supplement to trans. JIM 1968;9:692—7.

Arrabal R, Mingo B, Pardo A, et al. Role of alloyed Nd in the microstructure and atmospheric corrosion of as-cast magnesium alloy AZ91. Corrosion Science 2015;97:38—48.

Chino Y, Kadob M, Mabuchi M. Enhancement of tensile ductility and stretch formability of magnesium by addition of 0.2 wt.% (0.035 at%) Ce. Materials Science and Engineering A 2008;494:343—9.

Chino Y, Kadob M, Mabuchi M. Texture and stretch formability of a rolled Mg-Zn alloy containing dilute content of Y. Materials Science and Engineering A 2009;513-514:394—400.

Ding WJ, Li DQ, Wang QD, et al. Microstructure and mechanical properties of hot-rolled Mg-Zn-Nd-Zr alloys. Materials Science and Engineering A 2008;483-484:228—30.

Fang C, Zhang J, Pan FS. First-principles study on solute-basal dislocation interaction in Mg alloys. Journal of Alloys and Compounds 2019;785:911—17.

Hantzsche K, Bohlen J, Wendt J, et al. Effect of rare earth additions on microstructure and texture development of magnesium alloy sheets. Scripta Materials 2010;63:725—30.

He SM, Zeng XQ, Peng LM, et al. Microstructure and strengthening mechanism of high strength Mg-10Gd-2Y-0.5Zr alloy. Journal of Alloys and Compounds 2007;427(1-2):316—23.

Huang H, Miao HW, Yuan GY, et al. Fabrication of ultra-high strength magnesium alloys over 540, MPa with low alloying concentration by double continuously extrusion. Journal of Magnesium and Alloys 2018;6(2):107—13.

Huang H, Yuan GY, Chu ZH, et al. Microstructure and mechanical properties of double continuously extruded Mg-Zn-Gd-based magnesium alloys. Materials Science and Engineering A 2013;560:241−8.

Jiang B, Liu WJ, Qiu D, et al. Grain refinement of Ca addition in a twin-roll-cast Mg-3Al-1Zn alloy. Materials Chemistry and Physics 2012;133(2-3):611−16.

Jiang B, Yin HM, Yang QS, et al. Effect of stannum addition on microstructure of as-cast and as-extruded Mg-5Li alloys. Transactions of Nonferrous Metals Society of China 2011;21(11):2378−83.

Jiang B, Zeng Y, Yin HM, et al. Effect of Sr on microstructure and aging behavior of Mg-14Li alloys. Progress in Natural Science: Materials International 2012;22(2):160−8.

Jiang B, Zeng Y, Zhang MX, et al. Effects of Sn on microstructure of as-cast and as-extruded Mg−9Li alloys. Transactions of Nonferrous Metals Society of China 2013;23(4):904−8.

Jiang B, Zhou GY, Dai JH, et al. Effect of second phases on microstructure and mechanical properties of as-cast Mg-Ca-Sn magnesium alloy. Rare Metal Materials and Engineering 2014;43(10):2445−9.

Kang F, Li Z, Wang JT, et al. The activation of <c+a> non-basal slip in Magnesium alloys. Journal of Materials Science 2012;47:7854−9.

Koizumi T, Egami M, Yamashita K, et al. Platelet precipitate in an age-hardening Mg-Zn-Gd alloy. Journal of Alloys and Compounds 2018;752:407−11.

Kresse G, Furthmüller J. Efficiency of ab-initio total energy calculations for metals and semiconductors using a plane-wave basis set. Computational Materials Science 1996;6:15−50.

Kresse G, Hafner J. Ab initio molecular dynamics for open-shell transition metals. Physical Review B 1993;48:13115.

Li JC, He ZL, Fu PH, et al. Heat treatment and mechanical properties of a high-strength cast Mg−Gd−Zn alloy. Materials Science and Engineering A 2016;651:745−52.

Li RH, Pan FS, Jiang B, et al. Effect of Li addition on the mechanical behavior and texture of the as-extruded AZ31 magnesium alloy. Materials Science and Engineering A 2013;562:33−8.

Li RH, Pan FS, Jiang B, et al. Effects of yttrium and strontium additions on as-cast microstructure of Mg-14Li-1Al alloys. Transactions of Nonferrous Metals Society of China 2011;21(4):778−83.

Liu GB, Zhang J, Dou YC. First-principles study of solute−solute binding in magnesium alloys. Computational Materials Science 2015;103:97−104.

Liu SJ, Wang K, Wang JF, et al. Ageing behavior and mechanisms of strengthening and toughening of ultrahigh-strength Mg-Gd-Y-Zn-Mn alloy. Materials Science and Engineering A 2019;758:96−8.

Liu TT, Pan FS. Development and application of "solution strengthening plasticization" theory for magnesium alloys. Chinese Journal of nonferrous metals 2019;29(9):2050−63.

Liu TT, Pan FS, Zhang XY. Effect of Sc addition on the work-hardening behavior of ZK60 magnesium alloy. Materials & Design 2013;43:572−7.

Miao JS, Sun WH, Klarner AD, et al. Interphase boundary segregation of silver and enhanced precipitation of Mg17Al12 Phase in a Mg-Al-Sn-Ag alloy. Scripta Materialia 2018;154:192−6.

Moitra A, Kim SG, Horstemeyer MF. Solute effect on the <a+c> dislocation nucleation mechanism in magnesium. Acta Materialia 2014;75:106−12.

Momma K, Izumi F. VESTA 3 for three-dimensional visualization of crystal, volumetric and morphology data. Journal of Applied Crystallography 2011;44:1272−6.

Nie JF. Precipitation and hardening in magnesium alloys. Metallurgical and Materials Transactions A 2012;43(11):3891−939.

Pan FS, Mao JJ, Zhang G, et al. Development of high-strength, low-cost wrought Mg−2.0mass% Zn alloy with high Mn content. Progress in Natural Science: Materials International 2016;26(6):630−5.

She J, Pan FS, Guo W, et al. Effect of high Mn content on development of ultra-fine grain extruded magnesium alloy. Materials & Design 2016;90:7−12.

Sun G, Wang SH, Zhang XF, Lu Z, Feng ZH, Ma ZF. Effect of solution treatment and pre-stretching deformation on microstructure and properties of 2197 Al Li alloy. Metal heat treatment 2011;36(10):75−8.

Tan J, Sun YH, Xie HB, et al. Atomic-resolution investigation of Y-rich solid solution with an invariable orientation in Mg-Y binary alloy. Journal of Alloys and Compounds 2018;766:716−20.

Tan YX, Hu ZZ. Material research methods. Beijing: China Machine Press; 2013.

Tong X, You GQ, Wang YC, et al. Effect of ultrasonic treatment on segregation and mechanical properties of as-cast Mg-Gd binary alloys. Materials Science and Engineering A 2018;731:44−53.

Vítek V. Intrinsic stacking faults in body-centred cubic crystals. Philosophical Magazine 1968;18:773—86.

Wang C, Zhang HY, Wang HY, et al. Effects of doping atoms on the generalized stacking-fault energies of Mg alloys from first-principles calculations. Scripta Materiars 2013;69:445—8.

Wang F, Hu T, Zhang YT, et al. Effects of Al and Zn contents on the microstructure and mechanical properties of Mg-Al-Zn-Ca magnesium alloys. Materials Science and Engineering A 2017;704:57—65.

Wang JF, Song PF, Huang S, et al. High-strength and good-ductility Mg—RE—Zn—Mn magnesium alloy with long-period stacking ordered phase. Materials Letters 2013;93:415—18.

Wang JF, Wang K, Hou F, et al. Enhanced strength and ductility of Mg-RE-Zn alloy simultaneously by trace Ag addition. Materials Science and Engineering A 2018;728:10—19.

Wang LF, Huang GS, Quan Q, et al. The effect of twinning and detwinning on the mechanical property of AZ31 extruded magnesium alloy during strain-path changes. Materials & Design 2014;63:177—84.

Wang QH, Shen YQ, Jiang B, et al. A good balance between ductility and stretch formability of dilute Mg-Sn-Y sheet at room temperature. Materials Science and Engineering A 2018;736:404—16.

Wang YY. Calculation and study of stacking fault energy of magnesium alloy based on molecular dynamics [D]. Chongqing: Chongqing University; 2019. p. 51—61.

Wang ZJ, Jia WP, Cui JZ. Study on the deformation behavior of Mg-3.6% Er magnesium alloy. Journal of Rare Earths 2007;25(6):744—8.

Wu D, Chen RS, Han EH. Excellent room-temperature ductility and formability of rolled Mg-Gd-Zn alloy sheets. Journal of Alloys and Compounds 2011;509:2856—63.

Wu L, Pan FS, Yang MB, et al. As-cast microstructure and Sr-containing phases of AZ31 magnesium alloys with high Sr contents. Transactions of Nonferrous Metals Society of China 2011;21(4):784—9.

Wu ZX, Ahmad R, Yin BL, Sandloebes S, Curtin WA. Mechanistic origin and prediction of enhanced ductility in magnesium alloys. Science 2018;359:447—52.

Wu ZX, Curtin WA. The origins of high hardening and low ductility in magnesium[J]. Nature 2015;526:62—7.

Xua J, Jiang B, Song JF, et al. Unusual texture formation in Mg-3Al-1Zn alloy sheets processed by slope extrusion. Materials Science and Engineering A 2018;732:1—5.

Yan H, Chen RS, Han EH. A comparative study of texture and ductility of Mg-1.2Zn-0.8Gd alloy fabricated by rolling and equal channel angular extrusion. Materials Characterization 2011;62:321—6.

Yang QS, Jiang B, Jiang W, et al. Evolution of microstructure and mechanical properties of Mg-Mn-Ce alloys under hot extrusion. Materials Science and Engineering A 2015;628:143—8.

Yin HM, Jiang B, Huang XY, et al. Effect of Ce addition on microstructure of Mg-9Li alloy. Transactions of Nonferrous Metals Society of China 2013;23(7):1936—41.

Yu ZJ, Huang YD, Qiu X, et al. Fabrication of a high strength Mg—11Gd—4.5Y—1Nd—1.5Zn—0.5Zr (wt. %) alloy by thermomechanical treatments. Materials Science and Engineering A 2015;622:121—30.

Yu ZP, Yan YH, Yao J, et al. Effect of tensile direction on mechanical properties and microstructural evolutions of rolled Mg-Al-Zn-Sn magnesium alloy sheets at room and elevated temperatures. Journal of Alloys and Compounds 2018;744:211—19.

Yuasa M, Miyazawa N, Hayashi M, et al. Effects of group II elements on the cold stretch formability of Mg-Zn alloys. Acta Materials 2015;83:294—303.

Yuasa M, Hayashi M, Mabuchi M, et al. Improved plastic anisotropy of Mg-Zn-Ca alloys exhibiting high-stretch formability: a first-principles study. Acta Materials 2014;65:207—14.

Zeng Y, Jiang B, Li RH, et al. Effect of Li content on microstructure, texture and mechanical properties of cold rolled Mg-3Al-1Zn alloy. Materials Science and Engineering A 2015;631:189—95.

Zhang HY, Wang HY, Wang C, et al. First-principles calculations of generalized stacking fault energy in Mg alloys with Sn, Pb and Sn+Pb dopings. Materials Science and Engineering A 2013;584:82—7.

Zhang J, Dou Y, Dong H. Intrinsic ductility of Mg-based binary alloys: a first-principles study. Scripta Materials 2014;89:13—16.

Zhang J, Dou YC, Liu GB, et al. First-principles study of stacking fault energies in Mg-based binary alloys. Computational Materials Science 2013;79:564—9.

Zhang J, Li WG, Zhang BX, et al. Influence of Er addition and extrusion temperature on the microstructure and mechanical properties of a Mg-Zn-Zr magnesium alloy. Materials Science and Engineering A 2011;528(13-14):4740—6.

Zhang J, Liu M, Dou YC, et al. Role of alloying elements in the mechanical behaviors of an Mg-Zn-Zr-Er Alloy. Metallurgical and Materials Transactions A 2014;45(12):5499–507.

Zhang J, Zhang XF, Li WG, et al. Partition of Er among the constituent phases and the yield phenomenon in a semi-continuously cast Mg-Zn-Zr alloy. Scripta Materialia 2010;63(4):367–70.

Zhang JM, Yu CM. Effect of aging on microstructure and mechanical properties of Nb containing low alloy high strength steel. Journal of Materials Heat Treatment 2019;40(6):123–9.

Zhang J, Qi M, Fusheng P. Effects of trace Er addition on the microstructure and mechanical properties of Mg−Zn−Zr alloy. Materials & Design 2010;31(9):4043–9.

Zhao YZ, Pan FS, Peng J, et al. Effect of neodymium on the as-extruded ZK20 magnesium alloy. Journal of Rare Earths 2010;28:631–5.

Zhao ZL, Sun ZW, Liang W, et al. Influence of Al and Si additions on the microstructure and mechanical properties of Mg-4Li alloys. Materials Science and Engineering A 2017;702:206–17.

CHAPTER 4

Medium-strength and high-plasticity Mg−Mn-based alloys

Mg−Gd-based RE containing Mg alloys exhibit excellent properties. However, the high price and relatively high density of Gd largely increase the cost and density of the alloy, which is not beneficial for its wide commercialization. Mn is a cheap alloying element, its addition in magnesium alloy can reduce the impurity content (Fe) to improve the corrosion resistance, and its precipitates can refine the recrystallized grains. Besides, according to the previous theoretical calculation, it is found that Mn is one of the most effective solid solution strengthening and plasticizing elements. However, Mn has a low solid solubility in Mg matrix, which makes it difficult to effectively achieve solid solution strengthening and plasticization compared to that of Gd element. In addition, the influence of Mn precipitate on recrystallization has a great impact on the development of ultrafine grain magnesium alloy.

4.1 Microstructures and properties of Mg−Mn-based alloy

4.1.1 Effect of Mn on microstructures of Mg−Mn-based alloy

According to the Mg−Mn binary phase diagram (Fig. 4.1), the solid solubility of Mn in Mg is relatively low (about 2.2wt.%), and Mn does not form any sort of intermetallic. With

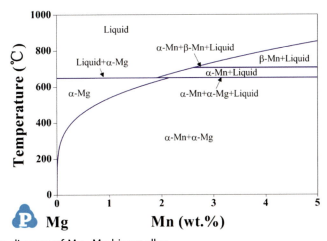

Figure 4.1 Phase diagram of Mg−Mn binary alloy.

High Plasticity Magnesium Alloys
DOI: https://doi.org/10.1016/B978-0-12-820110-7.00004-5

the decrease of temperature, a large number of Mn particles precipitate in the matrix. In order to ensure the uniform distribution of Mn in Mg—Mn alloy ingot, homogenization treatment is usually adopted with a temperature of 550°C and a holding time of 24 h.

Fig. 4.2 is the optical microstructures of as-cast Mg—Mn binary alloys. It can be seen that the microstructures of as-cast alloys are mainly composed of coarse equiaxed grains. With the increase of Mn content, the microstructures of the Mg—Mn alloys are refined. The average grain size is shown in Table 4.1. The average grain size of pure magnesium (without Mn addition) is about 1378.6 µm (as shown in Fig. 4.2A). With the addition of small amount of Mn (<2 wt.%), the microstructure of cast alloy is seldom changed (>1 mm), indicating that the refining effect in magnesium alloy with a small amount of Mn addition is not obvious. When the addition of Mn is high (≥2 wt.%), the microstructure of cast alloy is largely refined. Especially after adding 3 wt.% of Mn, the average grain size of the alloy is about 766.4 µm (as shown in Fig. 4.2E), which is about half of that of pure magnesium.

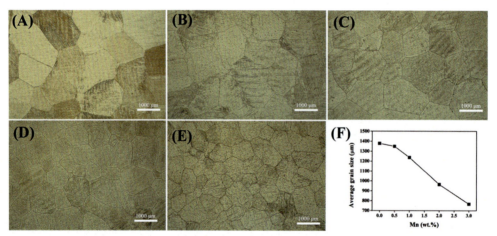

Figure 4.2 Microstructures of as-cast Mg—Mn alloys: (A) Mg, (B) Mg—0.5Mn, (C) Mg—1.0Mn, (D) Mg—2.0Mn, (E) Mg—3.0Mn and (F) average grain size as a function of Mn content.

Table 4.1 Grain size of Mg—Mn binary alloys.

Alloy	Average grain size (µm)	
	As cast	As extruded
Mg	1378.6	15.8
Mg—0.5Mn	1348.8	4.6
Mg—1.0Mn	1236.8	3.1
Mg—2.0Mn	964.9	1.7
Mg—3.0Mn	766.4	1.1

Alloying elements have great influences on the microstructures of as-cast magnesium alloys. Elements (e.g., Ti, Zr, Ca, and Sr) can remarkably change the microstructures and refine the grains of magnesium alloy. However, the additions of Mn, Sb, Pb have little impact on the microstructure and their grain refining effects are not obvious. Thus they are rarely added as the main alloying elements in magnesium. Nevertheless, with the increase of Mn content, the microstructure of the alloy becomes finer, which is attributed to a large number of Mn particles precipitate prior to α-Mg matrix in the solidification process. With the gradual decrease of temperature, Mn particles concentrate in front of the solid—liquid interface, which hinder the further growth of grains and result in a refined microstructure.

Hot extrusion is beneficial to eliminate some defects in magnesium alloy ingots, modify the microstructures, refine the grains and improve the mechanical properties of alloys, such as strength and ductility at room temperature. Fig. 4.3 shows the optical microstructures of the alloys after hot extrusion at 250°C. It can be seen that the alloys undergo dynamic recrystallization (DRX) during hot extrusion and their grains are significantly refined. Their average grain size decreases significantly with the increase of Mn content (Fig. 4.3F). Pure Mg has an average grain size of about 15.8 μm after hot extrusion (Fig. 4.3A). By adding Mn element in Mg, the average grain size is

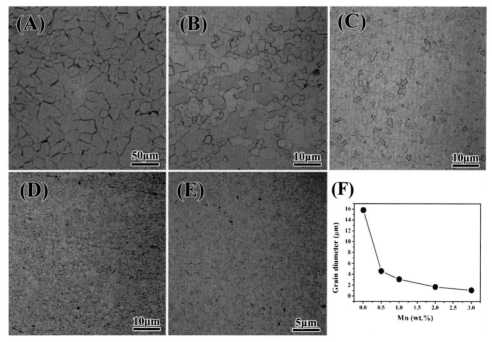

Figure 4.3 Microstructures of extruded Mg—Mn bar: (A) Mg, (B) Mg—0.5Mn, (C) Mg—1.0Mn, (D) Mg—2.0Mn, (E) Mg—3.0Mn and (F) average grain size as a function of Mn content.

significantly reduced. Specifically, the average grain size is only about 1.1 μm after adding 3.0 wt.% Mn (Fig. 4.3E). The main reason for the refinement can be attributed to the extremely low solid solubility of Mn in Mg matrix at low temperature. During hot extrusion process, the fine and dispersed Mn particles are precipitated, which hinder the movement of the grain boundaries and thereby inhibit the growth of DRX grains. With the increase of Mn content, the amounts of the precipitated Mn particles increase significantly, leading to a stronger pinning effect and a more obvious refining effect.

Fig. 4.4 shows the microstructures of the as-cast Mg−Mn alloy revealed by scanning electron microscopy (SEM). Fig. 4.4D, E, and F are enlarged views of Fig. 4.4A, B, and C, respectively. It can be seen from the figures that with the increase of Mn content, the amount of precipitated particles increases significantly. To further determine the phase composition of the precipitates in the alloy, energy dispersive spectrometer (EDS) analysis is performed on the second phase particles in Mg−0.5Mn alloy (as shown in Fig. 4.5). The precipitates are thus confirmed as Mn particles.

Fig. 4.6 shows the microstructures of Mg−Mn alloy after extrusion at 250°C. With the increase of Mn content, the amount of Mn particles that precipitates from the alloy also increases accordingly after hot extrusion. A large number of fine Mn

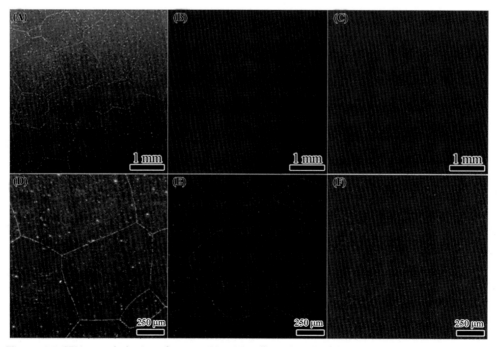

Figure 4.4 SEM morphologies of as-cast Mg−Mn alloys: (A, D) Mg−0.5Mn; (B, E) Mg−1.0Mn and (C, F) Mg−3.0Mn.

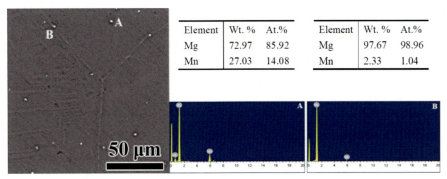

Element	Wt. %	At.%	Element	Wt. %	At.%
Mg	72.97	85.92	Mg	97.67	98.96
Mn	27.03	14.08	Mn	2.33	1.04

Figure 4.5 EDS spectrum of the second phase in Mg—0.5Mn alloy.

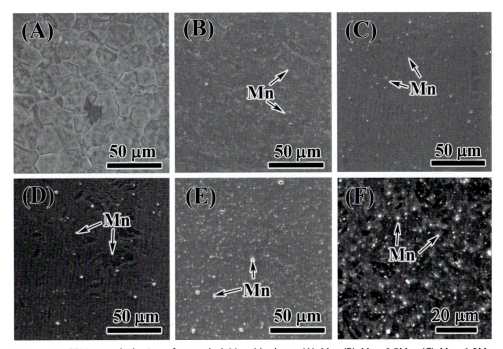

Figure 4.6 SEM morphologies of extruded Mg—Mn bars: (A) Mg; (B) Mg—0.5Mn; (C) Mg—1.0Mn; (D) Mg—2.0Mn and (E, F) Mg—3.0Mn.

particles (Fig. 4.6F) and a small number of coarse ones (Fig. 4.6E) are both precipitated in the Mg—3.0Mn alloy. Qi Fugang et al. studied the microstructure of wrought ZM61 magnesium alloy, they also found that a large number of Mn particles precipitated dispersively in the alloy.

Transmission electron microscopy analysis was carried out on Mg—1.0Mn alloy and the results are shown in Fig. 4.7. It can be seen from Fig. 4.7A—B that a large number of fine particles are dispersively precipitated in the alloy. According to EDS

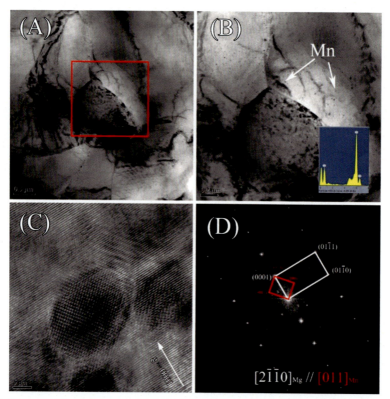

Figure 4.7 Transmission electron microscopy (TEM) image of Mg−1.0Mn alloy, (A) and (B) bright field images and corresponding EDS spectrum results, (C) high-resolution image of Mn precipitate, (D) Fourier conversion chart of Mn and Mg matrix.

results (Fig. 4.7B), the precipitated phase is identified as Mn particle. Fig. 4.7C shows a high-resolution image of the second phase in the alloy. It can be clearly seen that the precipitated phase is coherent with the matrix, which is inconsistent with the non-coherent relationship reported in previous literature. Mn phase reported in previous literature is the primary precipitate, while the Mn particles in this chapter are precipitated dynamically during hot extrusion. The diffraction spots of the second phase and the matrix are clearly marked in Fig. 4.7D. The orientation between Mn particle and Mg matrix is (0001) Mg//(111) Mn, $[2\bar{1}\bar{1}0]_{Mg}//[011]_{Mn}$ by comparing the standard PDF (32−0637) card. In summary, the dynamically precipitated Mn particles in Mg−1.0Mn alloy have a coherent relationship with the matrix, which are beneficial for hindering the growth of recrystallized grains during hot extrusion process and for refining the microstructures of the alloy, which thusly improving the mechanical properties of the alloys at room temperature.

The morphologies of the secondary phases in Mg−3.0Mn alloy are shown in Figs. 4.8 and 4.9. A large number of fine Mn particles are dispersively precipitated in Mg−3.0Mn

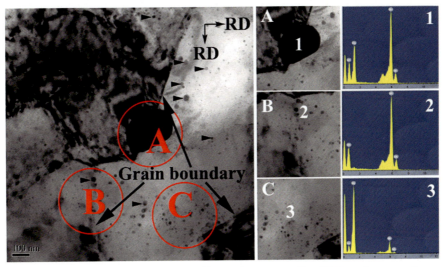

Figure 4.8 Bright field image of Mg—3.0Mn alloy and corresponding EDS spectrum results.

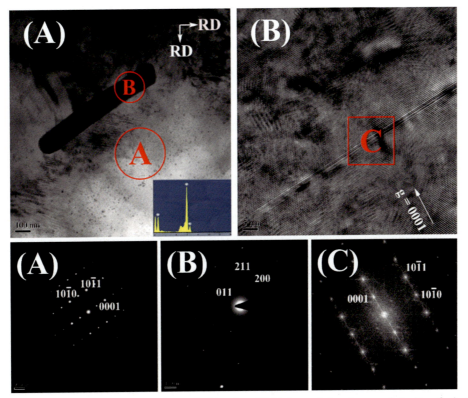

Figure 4.9 (A) Bright field image of Mg—3.0Mn alloy, (B) high-resolution image of stacking faults in the alloy, An and B are diffraction spots of the magnesium matrix and coarse second phase in figure (A), and C is Fourier transformation diagram of stacking fault and matrix in the figure (B).

alloy (Fig. 4.8), while a small number of coarse Mn particles can also be observed (Figs. 4.8A and 4.9A) which show a noncoherent relationship with the matrix (Fig. 4.8A). It is found that the long rod-shaped Mn particles are parallel to the *c*-axis of magnesium matrix by analyzing the orientation between the coarse Mn particles and the matrix. According to the results of Nie et al., the coarse Mn particle is beneficial to hinder the basal slip of Mg alloy and thus improve the room temperature strength. Meanwhile, a large number of fine Mn particles precipitated in Mg—3.0Mn alloy are also coherent with Mg matrix, which are similar to that in Mg—1.0Mn alloy. The fine Mn precipitates are not only beneficial to refine the microstructures, but also favorable for improving the mechanical properties of the alloys. Furthermore, a large number of basal stacking faults are found in Mg—3.0Mn alloy (Fig. 4.9A and B). These are helpful for hindering basal slip, which in turn improves its yield strength at room temperature. In conclusion, there are not only a small amount of coarse Mn particles in Mg—3.0Mn alloy, but also a large number of fine and dispersed Mn particles which can refine the microstructures and improve the mechanical properties.

Full recrystallization occurs during extrusion in all Mg—Mn alloys. With the increase of Mn content, the microstructures of the alloys are obviously refined. When the content of Mn in the alloy is ≤ 1.0%, a large number of fine Mn particles are dispersively precipitated, which are coherent with the Mg matrix. These precipitates are beneficial to prevent the growth of recrystallized grains and refine the microstructures of alloys during hot extrusion. When the Mn content increases (about 3.0%), a large number of fine Mn particles together with a small amount of coarse Mn particles are precipitated in the alloy. Similar to that in Mg—1.0Mn alloy, these fine Mn particles have a coherent relationship with Mg matrix, which are helpful for refining the microstructures and improving the mechanical properties of the alloy.

4.1.2 Effect of Mn on mechanical properties of Mg—Mn alloys

1. Mechanical properties of as-cast Mg—Mn alloy at room temperature

Fig. 4.10 shows the engineering stress—strain curve, the yield strength and fracture elongation in different Mg—Mn alloys. The corresponding yield strength, tensile strength and fracture elongation are listed in Table 4.2. From Fig. 4.10 and Table 4.2, it can be seen that the yield strength of Mg—Mn alloy at room temperature increases significantly with the increase

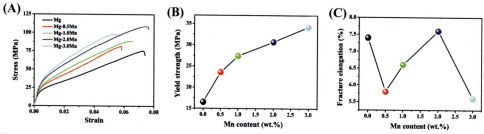

Figure 4.10 (A) Stress—strain curves of as-cast Mg—Mn alloys, (B) yield strength as the function of Mn contents in alloys, (C) elongation as a function of Mn contents in alloys.

of Mn content. When the content of Mn increases to 3.0 wt.%, the tensile yield strength of the alloy reaches its maximum value of 34 MPa, which is 107.3% higher than that of pure magnesium. The ultimate tensile strength of the alloy increases firstly and then decreases with the increase content of Mn. When the content of Mn increases to 2.0 wt.%, the ultimate tensile strength of the alloy reaches the maximum value of 109 MPa, which is 44.9% higher than that of pure magnesium. When the content of Mn continues to increase to 3.0 wt.%, the ultimate tensile strength of the alloy decreases to 98 MPa. In addition, the elongation does not change much with the increase of Mn content in the alloy at room temperature. When the content of Mn increases to 2.0 wt.% in the alloy, the elongation reaches its maximum value of 7.6% at room temperature, which is equivalent to that of pure Mg.

2. Mechanical properties of extruded Mg—Mn alloy bars

Fig. 4.11 shows the engineering stress—strain curves of extruded Mg—Mn alloy bars, and the corresponding mechanical properties at room temperature are listed in Table 4.3. In order to study the effect of Mn content on the mechanical properties of the alloys at room

Table 4.2 Mechanical properties of as-cast Mg—Mn alloys.

Samples	YS (MPa)	UTS (MPa)	ε%
Mg	17	75	7.4
Mg—0.5Mn	24	81	5.8
Mg—1.0Mn	27	88	6.6
Mg—2.0Mn	31	109	7.6
Mg—3.0Mn	34	98	5.6

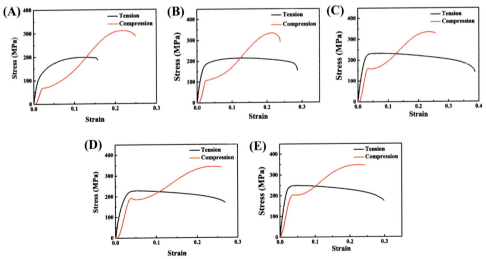

Figure 4.11 Mechanical properties of extruded Mg—Mn alloy bars at room temperature: (A) Mg; (B) Mg—0.5Mn; (C) Mg—1.0Mn; (D) Mg—2.0Mn and (E) Mg—3.0 Mn.

temperature, the relationships between the strength, elongation, tensile asymmetry, compressive asymmetry of the alloys and the variation of Mn content are shown in Fig. 4.12. The tensile yield strength, ultimate tensile strength, compressive yield strength and ultimate compressive strength of the extruded Mg—Mn alloy bars are significantly increased with the increase of Mn content at room temperature. Mg—3.0Mn alloy shows a high strength at

Table 4.3 Mechanical properties of extruded Mg—Mn bars at room temperature.

Alloy	Tensile property			Compression performance			Tensile/compressive yield strength ratio
	Yield strength (MPa)	Tensile strength (MPa)	Elongation (%)	Yield strength (MPa)	Tensile strength (MPa)	Elongation (%)	
Mg	98	199	15.7	60	312	24.9	0.62
Mg—0.5Mn	159	215	28.7	105	335	23.8	0.66
Mg—1.0Mn	204	234	38.8	158	336	25.6	0.77
Mg—2.0Mn	208	230	26.9	189	345	25.9	0.90
Mg—3.0Mn	213	248	29.9	193	349	24.3	0.91

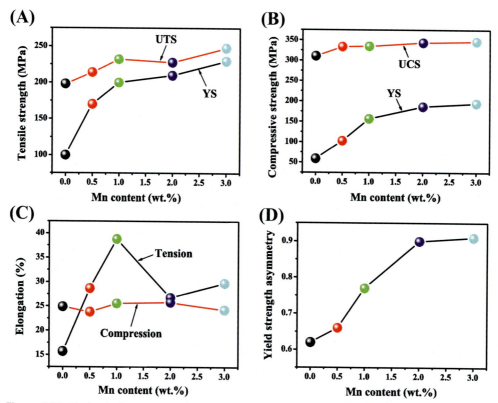

Figure 4.12 Mechanical properties of extruded Mg—Mn alloys (A) tensile strength, (B) compressive strength, (C) fracture elongation and (D) tensile-compressive yield strength ratio.

room temperature, with the tensile yield strength, ultimate tensile strength, compressive yield strength and ultimate compressive strength being 213, 248, 193 and 349 MPa, respectively. According to Fig. 4.12C, the tensile fracture elongation of the alloy increases firstly and then decreases with the increase of Mn content. When the content of Mn reaches to 1.0 wt.%, the fracture elongation of Mg—1Mn alloy is 38.8%, which is 147.1% higher than that of pure magnesium, indicating excellent room temperature plasticity. It can also be seen from Fig. 4.12D that the tension and compression asymmetry of the alloy also increases with the increase of Mn content.

In conclusion, Mg—Mn alloys show excellent mechanical properties at room temperature. When the content of Mn in the alloy is 1.0 wt.%, it shows relatively high plasticity and yield strength at room temperature. Its tensile yield strength, compressive yield strength, tensile fracture elongation and compressive fracture elongation are 204, 158 MPa, 38.8% and 25.6%, respectively. When the content of Mn is higher (about 3.0 wt.%), the alloy exhibits relatively high yield strength and excellent elongation at room temperature. Its tensile yield strength, compressive yield strength, tensile fracture elongation and compressive fracture elongation are 213, 193 MPa, 29.9% and 24.3%, respectively. In addition, Mg—3.0Mn alloy also has excellent tensile and compressive yield strength symmetry properties. Its ratio of tensile and compressive yield strength is about 0.91, which is 46.8% higher than that of pure magnesium.

4.1.3 Effect of Mn on fracture surfaces of Mg—Mn alloys

Fig. 4.13 shows the tensile fracture surfaces of extruded Mg—Mn alloy bars. The fracture morphologies of the alloy are related to the content of Mn. The fracture of pure Mg has a typical cleavage fracture feature (Fig. 4.13A). A large number of cleavage steps as well as dimples appear on the fracture surface. The fracture surface indicates that local plastic deformation occurs before the final fracture, which reveals a low strength and elongation of the alloy. When a small amount of Mn is added, a large number of dimples are accompanied by a few cleavage steps (Fig. 4.13B) appear in the alloy. Such fracture feature indicates that local plastic deformation occurs to a large extent before the final fracture. The fracture modes are both ductile fracture and brittle fracture, which reveals that the alloy has a certain strength and undergoes a certain plastic deformation. With further increase in Mn content, the fracture surface of the alloy shows typical ductile fracture characteristics. There are a lot of small dimples in the alloy (Fig. 4.13C—E), indicating that the alloy undergoes an extremely large extent of plastic deformation before the final fracture. Hence, the alloy shows excellent plastic characteristics and high tensile strength.

4.2 Effect of Al on the microstructure and mechanical properties of Mg—1Mn—Al alloy

The addition of Mn in Mg—Al alloys has been widely studied and used. For example, both AZ and AM magnesium alloys contain Mn with a content of less than 0.6 wt.%.

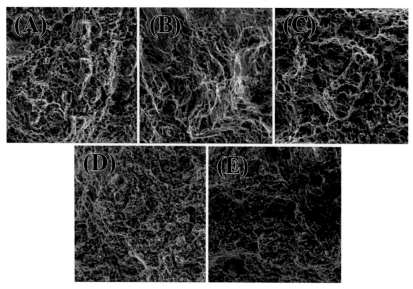

Figure 4.13 Fracture surfaces of extruded Mg−Mn alloy bars: (A) Mg; (B) Mg−0.5Mn; (C) Mg−1.0Mn; (D) Mg−2.0Mn and (E) Mg−3.0Mn.

Shang et al. showed that Mn existed in the form of Al_8Mn_5 in AZ31 alloy with the calculation of FactSage software. In addition, Mg−Al−Sn−Mn alloy exhibits a good microstructure and mechanical properties with a minor addition of Mn. The newly developed Mg−Al−Ca−Mn/Mg−Al−Ca−Mn−Zn alloy show good bake hardening effect due to the presence of Al−Mn particles. However, Mn is usually added as a trace element in these studies and the content of Mn is usually less than 0.6 wt.%. Investigations show that when Mn content is 1.4 wt.% in the alloy containing high Al content, the sizes of the second phase particles become very large. Further increasing the Mn content will coarsen these second phase particles. Therefore Mn content should not be too high in magnesium alloys containing high Al content.

In this chapter, based on the excellent comprehensive mechanical properties of Mg−1Mn binary alloy, the influence of Al content on the type of second phase is studied in Mg−1Mn alloy. Effect of second phase on the DRX behavior is investigated by modifying their types and morphologies. The evolution of microstructures and mechanical properties of the extruded alloy are studied as well.

4.2.1 Effect of Al on the microstructure of Mg−1Mn−Al alloy

Fig. 4.14 shows the microstructures of as-cast Mg−xAl−1Mn alloys as indicated by optical microscopy. After the addition of Al, the grains in the alloy are refined to equiaxed grains. With the increase of Al content, the average grain size of the alloy is significantly reduced, which is shown in Table 4.4. Without Al addition, the grain of

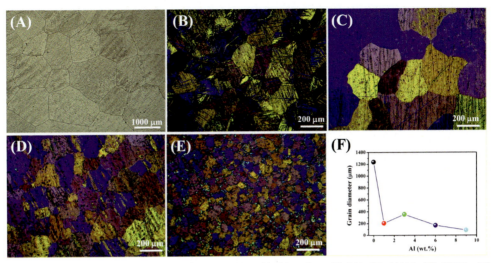

Figure 4.14 Optical microstructure of as-cast Mg—Al—Mn alloy, (A) M1, (B) AM11, (C) AM31, (D) AM61, (E) AM91, (F) average grain size as a function of Al content.

Table 4.4 Grain size of Mg—Al—Mn alloy.

ALLOY	Average grain size (μm)	
	As cast	As extruded
M1	1236.8	5.8
AM11	206.3	2.4
AM31	355.7	22.7
AM61	192.4	16.8
AM91	89.6	14.2

M1 alloy is relatively coarse with an average grain size of about 1236.8 μm (Fig. 4.14A). When 1.0 wt.% Al is added, the microstructure of AM11 alloy is obviously refined with an average grain size of about 206.3 μm (Fig. 4.14B). However, when the content of Al increases continuously to 3.0 wt.%, the grain size of AM31 inversely becomes large with an average grain size of 355.7 μm (Fig. 4.14C). The coarse grain in Mg—3Al—1Mn alloy is due to the changed morphology, type and quantity of the second phase. When the content of Al continues to increase, the microstructure of the alloy is refined continuously. Specifically, for AM91 alloy, its microstructure is significantly refined with an average grain size of about 89.6 μm (Fig. 4.14E). This is ascribed to a large number of $Mg_{17}Al_{12}$ eutectic phase precipitate at the grain boundary in AM91 alloy, which hinder the growth of the grains.

Fig. 4.15 shows the microstructures of wrought Mg—xAl—1Mn alloys along the extrusion direction (ED). It can be seen from the figure that with the addition of Al,

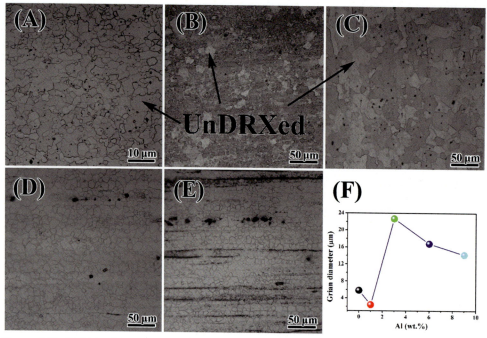

Figure 4.15 Optical microstructures along the extrusion direction of (A) M1, (B) AM11, (C) AM31, (D) AM61, (E) AM91, (F) the average grain size as a function of Al content.

the grains of the alloy are larger than those of M1 alloy. The average grain sizes of the recrystallized grains are listed in Table 4.4. Recrystallization in AM11 and AM31 alloy is not fully happened during hot extrusion. There are still a small amount of non-recrystallized grains in the alloy (Fig. 4.15A−C). When the content of Al in the alloy further increases, the amount of second phase increases remarkably. A large number of precipitated second phases act as the heterogeneous nuclei for recrystallized grains in the hot extrusion process, inducing the nucleation and growth of the recrystallized grains and leading to the completion of recrystallization (Fig. 4.15D and E). The average grain sizes of AM61 and AM91 are 16.5 and 15.1 μm, respectively, which are decreased significantly in comparison to that of AM31 alloy.

In order to further analyze the microstructure of as-cast Mg−xAl−1Mn alloy, SEM observations are also carried out (Fig. 4.16). When Al is not added, Mn particles only precipitate in the grains and grain boundaries in M1 alloy (Fig. 4.16A). When Al is added, Mn particles react with Al to form various Al−Mn precipitates. According to the energy dispersive spectrum analyses (Table 4.5), it is found that the second phase in AM11 alloy is mainly Al_8Mn_5 phase precipitated at grain boundaries and a small amount of $Al_{11}Mn_4$ phase precipitated in the grains (Fig. 4.16B). With further increase of Al content, the amount of $Mg_{17}Al_{12}$ eutectic precipitated at the grain boundaries increases obviously and a small amount of Al_4Mn phase is also observed. However, a small amount of blocky $Al_{11}Mn_4$ phase (Fig. 4.16C) is also observed in AM31 alloy. Part of $Al_{11}Mn_4$ phase could not be transformed into Al_4Mn phase due to the fast cooling rate during the non-equilibrium solidification process of the alloy.

Fig. 4.17 shows the SEM and EDS line scan results of as-cast AM91 alloy. According to the EDS spectrum results, $Mg_{17}Al_{12}$ eutectic phase mainly precipitates along the grain boundaries while a small amount of Al—Mn phase precipitates inside the grains. Combined with the results shown in Fig. 4.16, it is believed that the small amount of Al—Mn phase is Al_4Mn phase. Additionally, the addition of Mn does not change the morphology of $Mg_{17}Al_{12}$ eutectic phase.

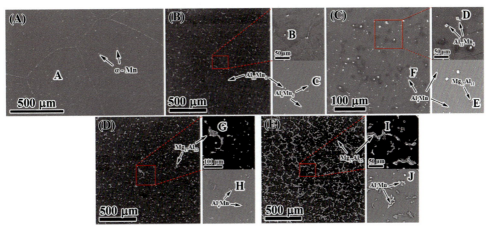

Figure 4.16 SEM microstructures of the as-cast Mg—Al—Mn alloy, (A) M1, (B) AM11, (C) AM31, (D) AM61, (E) AM91.

Table 4.5 Energy spectrum results of each point in Fig. 4.16.

Position	Element (at.%)			Existing compounds
	Mg	**Al**	**Mn**	
A	99.47	—	0.53	α-Mn
B	8.62	35.86	55.53	$Al_{11}Mn_4$
C	41.34	29.75	28.90	Al_8Mn_5
D	2.51	37.71	59.78	$Al_{11}Mn_4$
E	67.20	32.80	—	$Mg_{17}Al_{12}$
F	44.93	46.9ik0	8.17	Al_4Mn
G	84.50	15.50	—	$Mg_{17}Al_{12}$
H	25.21	43.93	30.86	Al_4Mn
I	68.10	31.90	—	$Mg_{17}Al_{12}$
J	2.86	55.82	41.32	Al_4Mn

Fig. 4.18 shows the SEM microstructures and EDS results of the second phase in wrought Mg—xAl—1Mn alloy. When Al is not added, a large number of second phase Mn particles are dispersively precipitated in the alloy (Fig. 4.18A). With the addition of Al, Mn particles react with Al to form Al—Mn intermetallics in the alloy. The

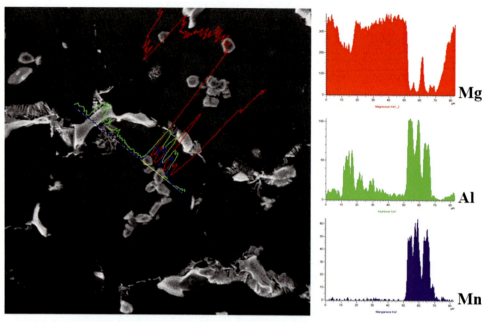

Figure 4.17 EDS line scanning energy spectrum of as-cast AM91 alloy.

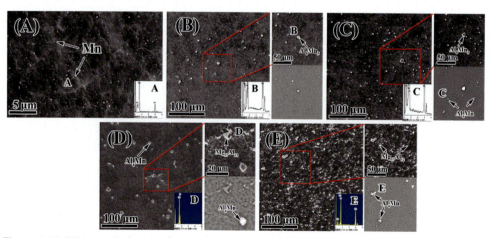

Figure 4.18 SEM morphologies of wrought Mg—xAl—1.0Mn alloy, (A) M1, (B) AM11, (C) AM31, (D) AM61, (E) AM91.

second phase is mainly Al_8Mn_5 in AM11 alloy (Fig. 4.18B) and mainly Al_8Mn_5 and Al_4Mn in AM31 alloy (Fig. 4.18C). Unlike as-cast AM31 alloy, no $Mg_{17}Al_{12}$ phase is found in the extruded counterpart. When the alloy is extruded at $300°C$, no dynamic precipitation occurs after the $Mg_{17}Al_{12}$ phase dissolves into the magnesium matrix during the extrusion process, resulting in partial solid solution of Al and Mn in α-Mg matrix. Fig. 4.18D and E show the SEM microstructures of AM61 and AM91 alloys, respectively. A small amount of typical $Mg_{17}Al_{12}$ phases are observed in AM61. While a large number of $Mg_{17}Al_{12}$ phases are uniformly distributed in the matrix and at grain boundaries in AM91 alloy after being fragmented under extrusion, which is conducive to strengthen the properties at room temperature.

4.2.2 Effect of Al on the mechanical properties of Mg–1Mn–Al alloy at room temperature

Fig. 4.19 and Table 4.6 show the tensile properties of as-cast Mg–xAl–1Mn alloy at room temperature. It can be seen from the figure that the yield strength of the alloy increases obviously with the increase of Al content. The yield strength of AM91 alloy is the highest with a value of 123 MPa, which is about 3.5 times of M1 alloy

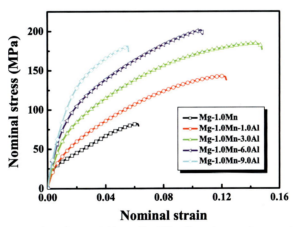

Figure 4.19 Tensile properties of as-cast Mg–xAl–1Mn alloy at room temperature.

Table 4.6 Tensile properties of as-cast Mg–xAl–1Mn alloy at room temperature.

Alloy	Yield strength (MPa)	Tensile strength (MPa)	Elongation (%)
M1	27	88	6.6
AM11	42	143	12.3
AM31	60	185	14.8
AM61	88	201	10.7
AM91	123	181	5.6

(27 MPa). Moreover, the addition of Al can improve the elongation of the alloy at room temperature. When Al is not added, the elongation of the alloy is only 6.6%. With the increase of Al content, the elongation of the alloy increases firstly and then decreases obviously. The elongation of AM31 alloy reaches the maximum value of approximately 14.8%, which is about 124.2% higher than that of M1 alloy.

Fig. 4.20 and Table 4.7 show the mechanical properties of wrought Mg−xAl−1Mn alloy at room temperature. The yield strength, tensile fracture elongation and tensile/compressive yield ratio of M1 alloy are 191 MPa, 32.4%, and 0.79, respectively. After adding 1.0 wt.% Al, the yield strength and tensile/compressive yield ratio of AM11 alloy is obviously improved, whereas the tensile fracture elongation of the alloy is obviously reduced. The yield strength, tensile fracture elongation and tensile/compressive yield ratio of AM11 alloy are 250 MPa, 21.4%, and 0.87, respectively. With further increase of Al content to 3.0 wt.%, the yield strength, tensile fracture elongation and tensile/compressive yield ratio of the alloy are

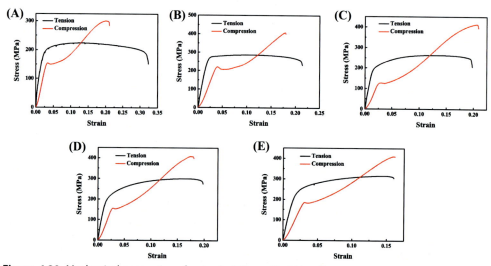

Figure 4.20 Mechanical properties of extruded Mg−xAl−1Mn alloy at room temperature, (A) M1, (B) AM11, (C) AM31, (D) AM61, (E) AM91.

Table 4.7 Mechanical properties of extruded Mg−xAl−1Mn alloy bars at room temperature.

Alloy	Tensile property			Compression property			Tensile/ compressive yield ratio
	Yield strength (MPa)	Tensile strength (MPa)	Elongation (%)	Yield strength (MPa)	Tensile strength (MPa)	Elongation (%)	
M1	191	224	32.4	150	301	21.1	0.79
AM11	250	287	21.4	217	408	18.1	0.87
AM31	179	264	19.8	119	415	20.9	0.67
AM61	189	301	19.8	152	410	18.0	0.81
AM91	205	317	16.1	182	412	16.2	0.89

179 MPa, 19.8% and 0.67, respectively, which are significantly lower than those of M1 and AM11. With further increase of Al content to 6.0 and 9.0 wt.%, the strength and tensile/compressive yield ratio of the alloy increase obviously, but the tensile fracture elongation decreases. The yield strength, tensile fracture elongation and tensile/compressive yield ratio of AM91 alloy are 205 MPa, 16.1% and 0.89, respectively. Compared with M1 alloy, the yield strength and yield strength ratio of AM91 alloy increase by 7.8% and 12.6%, respectively, while the tensile fracture elongation of the alloy decreases by 50.3%.

4.2.3 Fracture surfaces

The fracture surface of M1 alloy presents typical ductile fracture feature (Fig. 4.21A) with a large number of small dimples on the fracture surface, indicating its high plasticity at room temperature. When the content of Al in the alloy is low (such as in AM11 and AM31 alloy), a large number of dimples are observed in the alloy (Figs. 4.21B and C), which shows typical ductile fracture feature. Therefore AM11 and AM31 alloys exhibit high plasticity at room temperature. However, when the content of Al increases continuously (such as AM61 alloy), a large number of dimples as well as cleavage steps appear on the fracture surface (Fig. 4.21D). It is indicated that local plastic deformation occurs to a certain extent before the final fracture, resulting in ductile fracture and brittle fracture features. Hence, AM61 alloy shows certain strength as well as plastic deformation characteristics. When the content of Al is relatively high (such as in AM91 alloy), a large

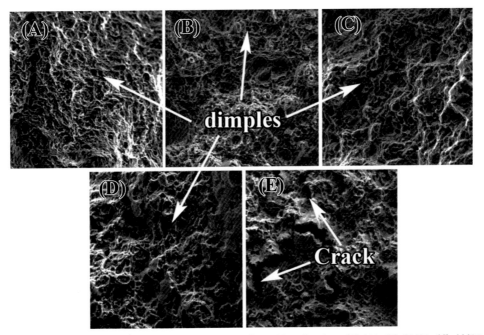

Figure 4.21 Fracture surfaces of extruded Mg—xAl—1.0Mn alloy bar, (A) M1, (B) AM11, (C) AM31, (D) AM61, (E) AM91.

number of cleavage steps and cracks (Fig. 4.21E) are observed in the alloy. The fracture of AM91 alloy is thus considered as typical brittle fracture and it shows a high yield strength and relatively low plasticity at room temperature.

4.2.4 Effect of minor Al on microstructures of Mg−1Mn−Al Alloy

Previous studies have shown that second phases in Mg−Al−Mn alloys include α-Mn, Al_8Mn_5, $Al_{11}Mn_4$. Mg−1Mn−1Al alloy exhibits a relatively high yield strength and elongation. According to the phase diagram of Mg−1Mn−Al system, the alloy contains α-Mn, Al_8Mn_5 and $Al_{11}Mn_4$ second phases when the content of Al is less than 1%. Therefore this section mainly focuses on further improve the microstructures and mechanical properties of AM11 alloy with addition of minor Al. The influence of alloy composition on the type of the second phase is investigated. The influences of Al−Mn phases on DRX behavior, microstructure and mechanical properties are also studied by adjusting the type and size of these second phases.

Fig. 4.22 is the vertical section of Mg−1.0Mn−xAl ($x \leq 1\%$) alloy phase diagram calculated by Pandat. It can be seen from the figure that with the increase of Al content, the type of the second phase in the alloy changes significantly. When Al content increases from 0 to 1 wt.%, phase combinations are α-Mn+Mg, Al_8Mn_5+α-Mn+Mg, Al_8Mn_5+Mg, Al_8Mn_5+$Al_{11}Mn_4$+Mg, respectively. Based on the phases in the alloy, four groups of alloys including Mg−1.0Mn, Mg−1.0Mn−0.3Al, Mg−1.0Mn−0.5Al and Mg−1.0Mn−1.0Al are designed. The four groups of alloys correspond to the four types of second phases as mentioned above. For description purposes, the four alloys are abbreviated as M1, MA103, MA105 and MA11. Subsequently, the effects of the combinations of these four different second phases on the microstructures and mechanical properties of the extruded Mg−1Mn−xAl ($x \leq 1\%$) alloys are studied.

Fig. 4.23 shows the X-ray diffraction (XRD) patterns of as-cast Mg−1.0Mn−xAl ($x \leq 1\%$) alloys. The second phases in the M1, MA103, MA105 and MA11 alloys are

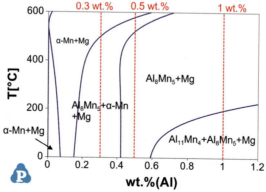

Figure 4.22 Vertical section of Mg−1.0Mn−xAl ($x \leq 1\%$) alloy phase diagram calculated by Pandat.

α–Mn, Al$_8$Mn$_5$+α–Mn, Al$_8$Mn$_5$, Al$_8$Mn$_5$+Al$_{11}$Mn$_4$, respectively. This is in consistent with the calculation by Pandat.

Fig. 4.24 shows SEM images and EDS mapping results of as–cast Mg—1.0Mn—xAl ($x \leq 1$%) alloys. The α–Mn phase with granular shape distributes both along the grain

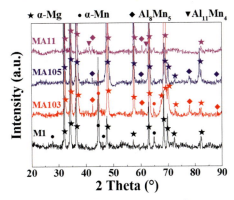

Figure 4.23 XRD pattern of as-cast Mg—1.0Mn—xAl ($x \leq 1$%) alloys.

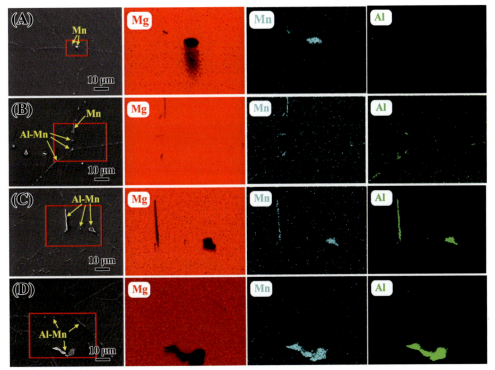

Figure 4.24 SEM pictures and area mapping results of as-cast Mg—1.0Mn—xAl ($x \leq 1$%) alloy, (A) M1, (B) MA103, (C) MA105, and (D) MA11.

boundaries and within the grains in M1 alloy. In MA103 alloy, α-Mn phase exhibits long-strip shape and distributes along the grain boundaries, while Al_8Mn_5 phase exhibits both granular and long-strip shapes and distributes both along the grain boundaries and within the grains. Similarly, Al_8Mn_5 presents granular and long-strip shapes and distributes both along the grain boundaries and within the grains in MA105 alloy. In MA11 alloy, the granular and long strip-like Al_8Mn_5 phase distributes both along the grain boundaries and within the grains. However, $Al_{11}Mn_4$ phase is not seen in the SEM images of MA11 alloy. This is because the $Al_{11}Mn_4$ phase is mainly a precipitated phase, which is difficult to be distinguished via SEM. Meanwhile, it can be seen that with the increase of Al content, the size of Al_8Mn_5 phase increases gradually. For example, the diameter of the largest particle is more than 10 μm in MA11 alloy. These coarse second phase particles may have negative impacts on the mechanical properties.

The OM of Mg−1.0Mn−xAl ($x \le 1\%$) alloys after hot extrusion with a speed of 50 mm s^{-1} at 250°C is shown in Fig. 4.25. The OM is the combined images from the

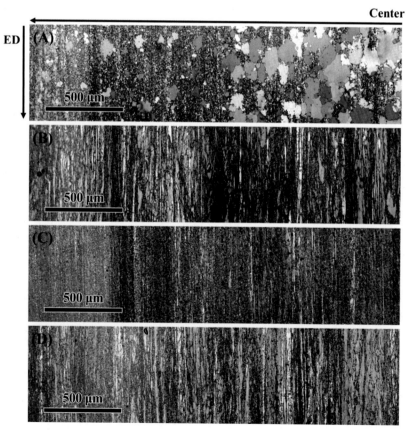

Figure 4.25 OM of extruded Mg−1.0Mn−xAl ($x \le 1\%$) alloys, (A) M1, (B) MA103, (C) MA105, and (D) MA11.

center to the edge of the sample so as to show a wider range of characteristics. With the increase of Al content, the grain size of the alloy has been significantly refined. The average grain size of M1, MA103, MA105 and MA11 is 8.6, 2.8, 1.9 and 3.6 μm, respectively. The M1 alloy exhibits a mixed microstructure with equiaxed coarse grains (~ 50 μm) and fine grains (~ 5 μm), as shown in Fig. 4.25A. The coarse grains mainly appear in the center of the bar, which is considered to be due to the abnormal grain growth during the extrusion process. With the addition of Al element, the grains with abnormal growth disappear and are replaced by some elongated grains which are generally considered as unrecrystallized microstructure. The MA103 and MA11 alloys have comparatively larger unrecrystallized regions (Fig. 4.25B and C). In contrast, MA105 alloy has fewer unrecrystallized regions and the characteristic is close to complete recrystallization (Fig. 4.25D). The changed microstructures result from the different second phases in the as-cast alloys. Therefore it is concluded that the optimum grain refinement effect is obtained if the extruded Mg—1.0Mn—xAl ($x \leq 1\%$) alloy contains only Al_8Mn_5 as second phase.

Fig. 4.26 is the SEM photographs of the extruded Mg—1.0Mn—xAl ($x \leq 1\%$) alloys. The second phase particles distribute along the ED. Second phase particles with long-strip shapes are rarely observed in the extruded alloy, indicating the long strip-like second phases are fragmented and distributed along the extrusion direction. With the increase of Al content, the size of Al—Mn phase particle increases gradually, which is in consistent with that of as-cast alloy.

Fig. 4.27 shows the EBSD results of extruded Mg—1.0Mn—xAl ($x \leq 1\%$) alloys. The microstructures are basically the same as that indicated from OM, showing mixed microstructures. All the samples show typical fiber textures, and the maximum polar

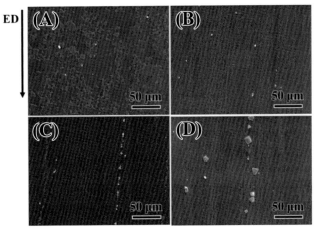

Figure 4.26 SEM photograph of extruded Mg—1.0Mn—xAl ($x \leq 1\%$) alloys, (A) M1, (B) MA103, (C) MA105, and (D) MA11.

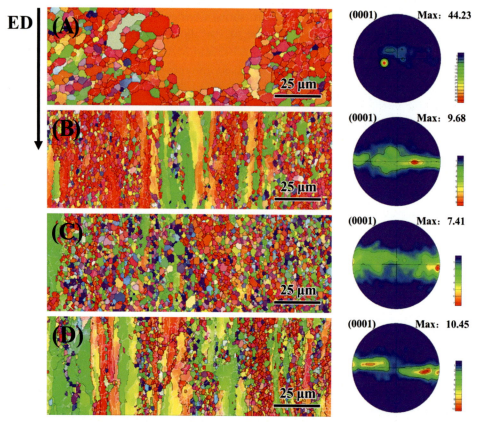

Figure 4.27 Inverse pole figure and pole figure of extruded Mg−1.0Mn−xAl ($x \leq 1\%$) alloys, (A) M1, (B) MA103, (C) MA105, and (D) MA11. The observation direction is parallel to the extrusion direction.

density decreases firstly and then increases with the increase of Al content. It has been reported in literature that the non-recrystallized region and the coarse grains tend to enhance the texture, while the fine dynamic recrystallized microstructure can weaken the texture. For the grains in M1 alloy, the major parts are coarse equiaxed grains with only one orientation, resulting in a very high texture intensity of 44.23. Meanwhile, MA103 and MA11 alloys also show strong textures due to a large number of non-recrystallized grains. The larger the proportion of the non-recrystallized region, the stronger the texture is. Otherwise the texture becomes weaker. Therefore MA105 alloy has the smallest non-recrystallized region and the weakest texture intensity of 7.41.

Fig. 4.28 is the bright field TEM photographs of extruded Mg−1.0Mn−xAl ($x \leq 1\%$) alloys. There are a large number of uniformly distributed second phases in all four samples. In M1 alloy, α-Mn particles are spherical with an average size of about 15 μm. With the addition of Al element, the block and strip-shaped second phases can

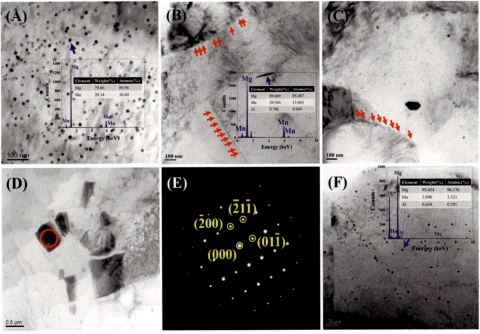

Figure 4.28 Bright field TEM photographs of extruded Mg—1.0Mn—xAl alloys, (A) M1, (B) MA103, (C) MA105, (D) MA11, (E) SAED of the red area marked in (D), and (F) magnified bright field phase of MA11 alloy.

be observed in MA103 alloy, among which the strip-like particle is α-Mn phase. Both the type and morphology of the second phases are basically in consistent with the SEM results. The second phase with block shape can be detected in both MA105 and MA11 alloys. The block phase is proved as Al_8Mn_5 phase (Fig. 4.28E), as indicated from the diffraction analysis (Fig. 4.28D). Meanwhile, numerous nanoparticles can be observed in the enlarged image of MA11 alloy, which is speculated as $Al_{11}Mn_4$ phase through the EDS results. The volume fraction (f) and average size (d) of the second phase in MA103, MA105 and MA11 are f_{MA103}=5.1%, d_{MA103}=7 nm, f_{M105}=5.3%, d_{M105}=11 nm, f_{MA11}=3.2%, d_{MA11}=21 nm, respectively. The average size of the second phase increases with the increase of Al element, while the volume fraction of the second phase increases firstly and then decreases. When the Al content reaches 0.5 wt. %, the volume fraction of the second phases is the largest. This shows that large amounts of Al and Mn are consumed by the formations of coarse Al_8Mn_5 particles, resulting in the decrease of the volume fraction of the second phases. In addition, the second phases distribute along the grain boundaries in MA103 and MA105 alloys, as indicated by the red arrow in the figures. Such phenomenon implies that these second phase particles can pin the grain boundaries of recrystallized grains and thus hinder the grain boundary migration, leading to grain refinement.

In order to further analyze the DRX behavior of the extruded alloy, the DRX distribution from EBSD is shown in Fig. 4.29. Among them, M1 alloy has the highest DRX degree of 58.3%. It is worth noting that the coarse grains in M1 alloy show the DRX feature, indicating that these coarse grains are resulted from abnormal growth. Combined with the distributions of coarse grains in OM, it is concluded that this is due to the gradient heat distribution during the extrusion process. The heat dissipation rate on the surface of bar is fast whereas that in the center is slow. Thus the driving force of recrystallization is high in the center of bar, promoting the growth of DRX grains. Therefore large grains are more distributed in the center rather than in the

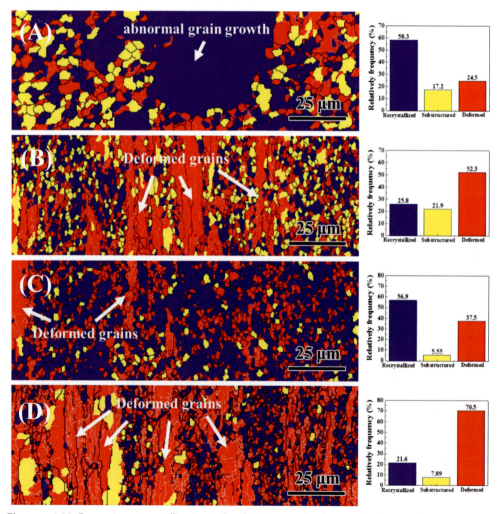

Figure 4.29 Dynamic recrystallization distribution and statistical data of as-extruded Mg−1.0Mn−xAl (x ≤ 1%) alloys, (A) M1, (B) MA103, (C) MA105, and (D) MA11.

edge. In addition, there are still some deformed grain regions and substructures in M1 alloy. The proportions of these deformed grain regions and substructures are much higher than the fully dynamic recrystallized areas in literature, such as that in Mg–Sn–Zn–Ca and Mg–Al–Sn–Zn alloys. Therefore some unrecrystallized areas remain in M1 alloy, indicating that the abnormal growth of recrystallized grains occur before the fully DRX in M1 alloy. In the three Mg–1Mn–xAl alloys, the elongated grains show the characteristics of the deformed microstructure (red area in the figure). It can be seen from the statistical data that the proportions of these deformation regions in the Mg–1Mn–xAl alloys are higher than that of the M1 alloy. As we all know, the DRX process includes the nucleation and growth of new grains. Such long strip-like deformed region shows that the nucleation process in DRX is hindered. That is to say, the addition of Al will hinder the DRX due to the formation of Al–Mn phase. In addition, it can be observed that numerous Al_8Mn_5 particles pin the grain boundaries from the TEM results (Fig. 4.28), indicating that these nano-scale particles can hinder the migration of the grain boundaries. Grain boundary migration is an important factor for the nucleation and growth of new grains during dynamic recrystallization. Therefore when the grain boundary migration is hindered, the degree of DRX decreases.

With the increase of Al content, the degree of DRX increases first and then decreases, indicating the constituent of the second phase also has an important influence on the DRX behavior. The results show that Al_8Mn_5 phase particles with relatively small size strongly hinder dynamic recrystallization, leading to the formation of long strip-like deformed region. With increase of Al content to 0.5 wt.%, Al_8Mn_5 phase is the only second phase in the alloy. Some Al_8Mn_5 particles with the size larger than 1 μm can effectively stimulate the DRX nucleation and induce particle-stimulated nucleation (PSN). Therefore the degree of DRX is improved. When the content of Al increases to 1.0 wt.%, the $Al_{11}Mn_4$ precipitates appear in the alloy. These nano-precipitates further hinder the recrystallization and weaken the PSN effect, thus reducing the degree of dynamic recrystallization. In conclusion, the constituent of the second phase has an important effect on the DRX behavior of the alloy. When Al_8Mn_5 phase is the only second phase in the alloy, the alloy shows the best DRX behavior and eliminates the abnormal growth of grains. In addition, the PSN effect by Al_8Mn_5 phase reduces deformed grain areas and results in good grain refinement.

The constituent of the second phase not only affects the grain size, but also significantly affects the texture. The PSN effect both accelerates the DRX process and optimizes the recrystallization texture. During hot deformation, both the recrystallization nucleation generated by the PSN and the shear band generated by the fine dispersive phases have the potentials to form random texture. However, unrecrystallized regions and DRX with abnormal growth are not conducive to texture weakening. In order to

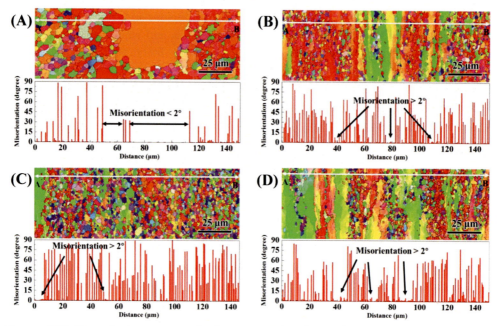

Figure 4.30 The misorientation angle between point An and point B in the extruded Mg−1.0Mn−xAl (x ≤ 1%) alloy, (A) M1, (B) MA103, (C) MA105, and (D) MA11.

further study the texture evolution of Mg−1.0Mn−xAl alloy through the addition of Al element, the point-to-point misorientation angle analysis of the EBSD results is shown in Fig. 4.30. In M1 alloy, misorientation angle greater than 2 degrees is hardly observed in the abnormal grown grains, implying the same orientation. Such a large number of regions with the same orientation results in strong texture with a very high maximum pole density of 44.23 in M1 alloy. In addition, there are some sub-grain boundaries with 5 degrees−10 degrees misorientation angles in the coarse unrecrystal-lized grains. The presences of these local sub-grain boundaries slightly deflect the deformed grains, which weaken the texture to a certain extent. Therefore in the three alloys containing Al, the maximum pole density is lower than that of M1 alloy. But in general, the large unrecrystallized area still has a similar orientation, which increases the maximum polar density and enhances the texture. Therefore the MA105 alloy with more dynamic recrystallized grains has the weakest texture.

4.2.5 Effect of minor Al on mechanical properties of Mg−1Mn−Al alloy

Fig. 4.31 compares the tensile and compressive curves and mechanical properties of Mg−1Mn−xAl (x ≤ 1%) alloys. The strength and plasticity values are listed in Table 4.8. The results show that the addition of Al has a significant effect on the mechanical properties of the extruded M1 alloy. With the addition of Al element, the

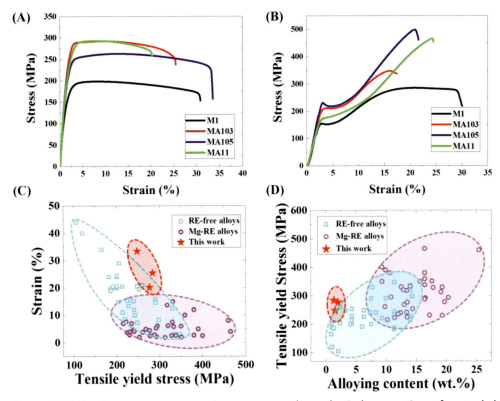

Figure 4.31 Tensile curves, compressive curves and mechanical properties of extruded Mg−1.0Mn−xAl ($x \leq 1.0$) alloy, (A) tensile curves, (B) compressive curves, (C) comparison of tensile yield strength and strain of Mg−1.0Mn−xAl ($x \leq 1.0$) alloys, rare earth-free alloys and rare earth-containing alloys, (D) comparison of tensile yield strength and alloying content of Mg−1.0Mn−xAl alloys, rare earth-free alloys and rare earth-containing alloys.

Table 4.8 Tensile yield strength, ultimate tensile strength, fracture elongation, compressive yield strength and tensile-compressive yield strength ratio of Mg−1Mn−xAl ($x \leq 1.0$) alloys with low Al content.

Alloy	Tensile yield strength (MPa)	Ultimate tensile strength (MPa)	Elongation (%)	Compressive yield strength (MPa)	Tensile/ compressive yield strength ratio
M1	186	200	30.8	154	0.83
MA103	284	292	25.5	212	0.75
MA105	248	263	33.4	232	0.94
MA11	277	292	20.2	173	0.62

tensile yield strength and ultimate tensile strength of Mg−1Mn−xAl alloy improve significantly. However, the increase in fracture elongation is only observed in MA105. The tensile yield strength (284 MPa) and ultimate tensile strength (292 MPa) of MA103 alloy are the highest. MA105 alloy exhibits simultaneously improved yield strength and plasticity, and the best tensile/compressive yield asymmetry. The tensile yield strength, ultimate tensile strength, fracture elongation and the tensile/compressive yield asymmetry are 248, 263 MPa, 33.4% and 0.94, respectively.

The Mg−1Mn−xAl alloys described in this chapter have good comprehensive mechanical properties and low alloying contents (Al ≤ 1 wt.%). Therefore typical alloys (rare-earth alloys such as Mg−Gd and Mg−Y alloy; and nonrare-earth alloys such as Mg−Al, Mg−Ca, Mg−Sn and Mg−Zn system alloy) in the literature are compared with Mg−1Mn−xAl alloys. The yield strength and elongation are compared in Fig. 4.31C, while the yield strength and alloying content are compared in Fig. 4.31D. The comprehensive mechanical properties of Mg−1Mn−xAl alloys are better than those of most nonrare-earth magnesium alloys (Fig. 4.31C). Compared with the rare earth-containing alloy, Mg−1Mn−xAl alloys have better plasticity than the Mg−RE alloys and have strength comparable to most Mg−RE alloys. Additionally, Mg−1Mn−xAl alloys have low alloying contents and exhibit superior yield strengths in comparison with the alloys with equivalent alloying contents (Fig. 4.31D). Besides, both alloying elements Al and Mn are comparatively cheap. Therefore Mg−1Mn−xAl (x ≤ 1.0) are magnesium alloys with potentially low costs and high performances.

Extruded Mg−1Mn−xAl (x ≤ 1.0) alloys have good comprehensive mechanical properties. In order to further reveal the evolution of microstructures and mechanical properties, grain refinement and texture evolutions of Mg−1Mn−xAl alloys during DRX will be further discussed. The influences of grain refinement and texture evolution on the mechanical properties will also be discussed.

It is previously shown that the addition of Al has an important impact on the microstructures and mechanical properties of Mg−1Mn−xAl alloy. The evolution of mechanical properties is closely related to the change of microstructure. The main factors influencing yield strength and fracture elongation are grain size, texture and second phase particles. The effect of grain size on yield strength is usually illustrated by Hall−Petch relation. Yield strength is inversely proportional to grain size. Compared with M1 alloy, the tensile yield strength and compressive yield strength increase significantly with the addition of Al. There is a linear relationship between the compressive yield strength and the grain size. The compressive yield strength increases with the grain refinement. However, the tensile yield strength presents an inverse Hall−Petch relationship, which shows nonlinear relationship with grain size. The nonlinear relationship is mainly found in MA105 alloy, which has the smallest average grain size and yet lower tensile yield strength than those of MA103 and MA11 alloys. Such phenomenon indicates that there are other factors to the yield strength.

Generally speaking, the abovementioned inverse Hall—Petch relationship can be understood through the evolution of deformation mechanism. The {0001} <11−20> basal slip is considered to be the main deformation mechanism in magnesium alloys, especially in the initial stage of tensile test performed along (ED.) Thus the inverse Hall—Petch relationship can be explained by the Schmid factor (SF) of the basal slip,

$$\Delta\sigma_y = \frac{\tau_{CRSS}}{M} \tag{4.1}$$

where τ_{CRSS} is the critical resolved shear stress, M is the Schmid factor.

The SF distribution of basal $<a>$ slip with loading tensile stress along ED is calculated and shown in Fig. 4.32. The larger the SF, the easier the basal slip operates. As a result, the yield strength is easily decreased and the plasticity is improved at a high SF.

The contribution of the second phase particles to the mechanical properties is very important. However, the contribution is relatively complex, which is related to the shape, size and distribution of the second phase particle. In general, the dispersive

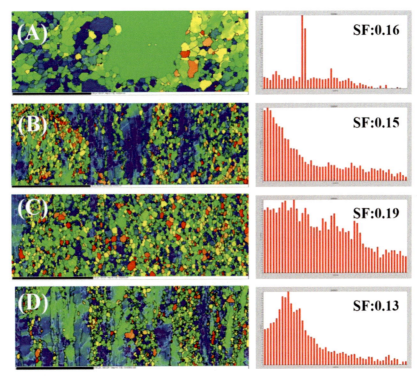

Figure 4.32 Schmid factor distribution of basal $<a>$ slip and statistical data of the extruded Mg−1.0Mn−xAl ($x \leq 1.0$) alloy, (A) M1, (B) MA103, (C) MA105 and (D) MA11.

distribution of second phase particles in the magnesium matrix can greatly improve the mechanical properties. According to Orowan formula, the increase of tensile yield strength is related to the size (d) and volume fraction (f) of the second phase particles

$$YS \propto f^{1/2} d^{-1} \ln d \qquad (4.2)$$

Obviously, increasing the volume fractions and decreasing the average sizes of the second phases can effectively improve the yield strength of the alloy. According to the TEM results, the average size and volume fraction of the nano second phases are $d_{M1} = 15$ nm, $f_{M1} = 4.6\%$, $d_{MA103} = 7$ nm, $f_{MA103} = 5.1\%$, $d_{MA105} = 10$ nm, $f_{MA105} = 5.3\%$, $d_{MA11} = 11$ nm, $f_{MA11} = 3.2\%$. It is seen that with the increase of the Al content, the fraction of the nanosecond phases decreases and average size increases, leading to a weakening of Orowan strengthening effect. The decrease in volume fraction of nanoparticles is mainly due to the formation of coarse second phase particles. Especially when the Al content increases to 1.0 wt.%, the size of Al_8Mn_5 phase particles reaches $\sim 10\ \mu m$, as shown in Fig. 4.26. These coarse particles cannot effectively hinder the movements of dislocations and can easily generate microcracks. Therefore the plasticity of MA11 alloy is poor.

Magnesium alloys with typical fiber textures usually exhibit strong tensile/compressive yield asymmetry. This is mainly due to the low operating stress of {10−12} <10−11> extension twins. The SF of {10−12} <10−11> extension twins of magnesium alloy with fiber texture during compressive test is higher than that in tensile test. Therefore extension twins are easily activated during compression. Grain refinement and texture weakening are important ways to improve the yield anisotropy of magnesium alloys. This asymmetry is mainly improved by suppressing the twin behavior during compression. It has been reported that when the grain size of magnesium alloy is less than 2.7 μm, the twinning behavior is completely suppressed and almost no twinning occurs during deformation. The addition of Al effectively refines the microstructure and weakens the texture, so the tensile/compressive yield asymmetry of the three alloys containing Al is better than that of M1 alloy. In particular, MA105 alloy has the smallest grain size and the weakest texture, and the tensile/compressive yield asymmetry is greatly improved with a $\sigma_{CYS}/\sigma_{TYS} = 0.94$.

As mentioned above, when the content of Al is 0.5 wt.%, the mechanical properties of Mg−1Mn−xAl alloy are optimized with high yield strength, fracture elongation and good tensile/compressive yield asymmetry. The good mechanical properties are due to the uniform and fine microstructure, weakened texture and fine second phase particles. Replacing α-Mn particles in M1 alloy with Al_8Mn_5 particles can improve its DRX behavior and obtain good microstructure and mechanical properties.

4.3 Effect of Y on microstructures and properties of Mg−1Mn−Y alloy

As an important rare earth-alloying element, Y is widely used in magnesium alloys. The maximum solid solubility of Y in Mg matrix is 12.47 wt.% and mainly exists in the form of solid solution in α-Mg matrix, which can effectively refine the grain size of the alloy. Recent studies have shown that Y can not only increase the strength, but also improve the plasticity of magnesium alloy. Y is a solid solution strengthening and plasticizing element in magnesium alloy. Therefore the effect of Y addition on the microstructures and properties of Mg−1Mn alloy will be studied in this section. The mechanical properties of Mg−1Mn alloy will be further optimized by using the solution strengthening and plasticizing effect of Mn and Y elements.

4.3.1 Effect of Y on the microstructure of Mg−1Mn−Y alloy

Fig. 4.33 shows the OM of the extruded alloy. With the increase of extrusion temperature, the microstructure of Mg−1Mn alloy becomes obviously coarse and the average grain size also increases significantly (Fig. 4.33A−C). The microstructure of Mg−1Mn alloy is obviously refined with the Y addition. The average grain size of the alloy decreases significantly with the increase in Y content (Fig. 4.33D−F). When 0.2 wt.% Y is added, the average grain size of the alloy is 8 μm, which is larger than that of Mg−1Mn alloy at the same extrusion temperature. When the content of Y in the alloy is as high as 1.0 wt.%, the microstructure of the alloy is obviously refined and the average grain size is about 1.9 μm. The grain size is obviously finer than that of Mg−1Mn−0.2Y alloy and that of Mg−1Mn alloy under the same extrusion

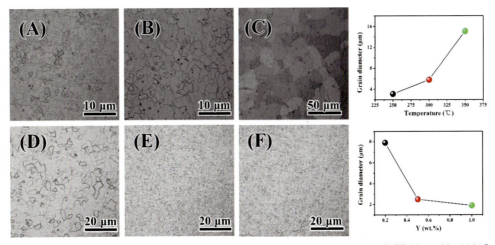

Figure 4.33 The OM of the extruded Mg−1Mn−Y alloys, (A) Mg−1Mn-250°C, (B) Mg−1Mn-300°C, (C) Mg−1Mn-350°C, (D) Mg−1Mn−0.2Y, (E) Mg−1Mn−0.5Y and (F) Mg−1Mn−1Y.

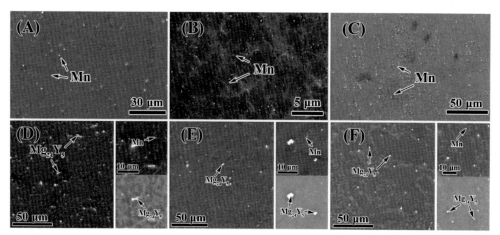

Figure 4.34 SEM pictures of extruded Mg−1Mn−Y alloy: (A) Mg−1Mn-250°C; (B) Mg−1Mn-300°C; (C) Mg−1Mn-350°C; (D) Mg−1Mn−0.2Y; (E) Mg−1Mn−0.5Y and (F) Mg−1Mn−1Y.

temperature. Therefore the addition of rare-earth element Y can improve the microstructure and thus improve the mechanical properties.

Fig. 4.34 is the SEM morphologies of the Mg−1Mn−Y alloys. Among them, Fig. 4.34A−C are the morphologies of the Mg−1Mn alloy extruded at different temperatures. It can be seen that as the extrusion temperature increases, a large amount of Mn particles are precipitated in the alloy. When Y is added to the alloy, the $Mg_{24}Y_5$ phase is precipitated in the alloy and the total amounts of precipitates significantly increase as well. Meanwhile, the amounts of Mn precipitates in Y containing alloy are significantly decreased in comparison with that in Mg−1Mn alloy extruded under the same temperature (Fig. 4.34D−F).

Fig. 4.35 shows the TEM bright field image of the alloy and the corresponding EDS analysis results. Similar to the TEM microstructure of extruded Mg−1Mn alloy, fine spherical particles are participated dispersively in Mg−1Mn−1Y alloy. Combined with the EDS and XRD results, the participated particles are identified as α-Mn particles (Fig. 4.35C). In addition, a small amount of coarse second phase particles (Fig. 4.35B) are also observed in the alloy, which is identified as $Mg_{24}Y_5$ phase according to EDS results.

4.3.2 Effect of Y on the mechanical properties of Mg−1Mn−Y alloy

Fig. 4.36 and Table 4.9 show the mechanical properties of extruded Mg−1Mn and Mg−1Mn−xY alloys at room temperature. The yield strength and plasticity of Mg−1Mn alloy decrease significantly with the increase of extrusion temperature (Fig. 4.36A−C). The Mg−1Mn alloy extruded at 250°C exhibits excellent mechanical properties at room temperature. The tensile yield strength and fracture elongation of

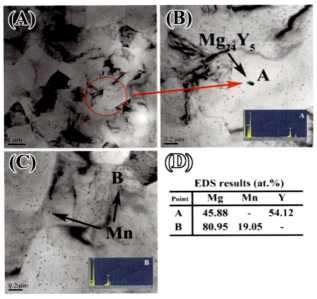

Figure 4.35 TEM of the extruded Mg−1Mn−1Y alloy and the corresponding EDS results, (A) and (C) TEM bright field images of the alloy, (B) enlarged view of the red area in (A), (D) EDS results.

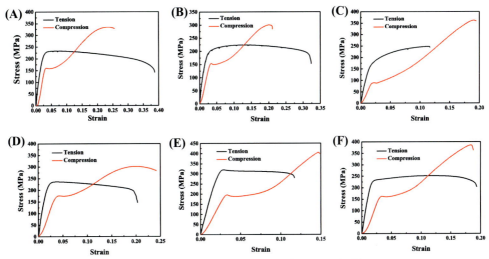

Figure 4.36 The mechanical properties of Mg−1Mn−xY alloy extruded bars at room temperature: (A) Mg−1Mn-250°C; (B) Mg−1Mn-300°C; (C) Mg−1Mn-350°C; (D) Mg−1Mn−0.2Y; (E) Mg−1Mn−0.5Y and (F) Mg−1Mn−1Y.

Table 4.9 The mechanical properties at room temperature of Mg−1Mn and Mg−1Mn−xY alloys extruded bars.

Alloy	Tensile property			Compression property			Tensile/compressive yield strength ratio
	Yield strength (MPa)	Tensile strength (MPa)	Elongation (%)	Yield strength (MPa)	Tensile strength (MPa)	Elongation (%)	
Mg−1Mn-250°C	204	234	38.8	158	336	25.6	0.77
Mg−1Mn-300°C	191	224	32.4	150	301	21.1	0.79
Mg−1Mn-350°C	154	248	11.7	86	363	19.4	0.56
Mg−1Mn−0.2Y	186	236	20.4	157	303	24.3	0.84
Mg−1Mn−0.5Y	311	321	11.7	193	406	14.8	0.62
Mg−1Mn−1Y	209	253	19.4	154	387	18.8	0.74

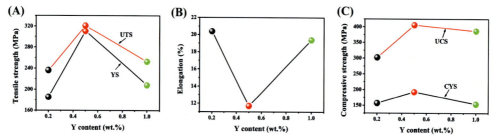

Figure 4.37 The mechanical properties of Mg−1Mn−xY extruded alloy as a function of Y content (A) tensile strength, (B) fracture elongation and (C) compressive strength.

Mg−1Mn are 204 MPa and 38.8%, respectively. After adding Y element, the yield strength and elongation of the alloy change significantly (Fig. 4.36D−F). The relationships between the strength and plasticity of the alloy and Y content are shown in Fig. 4.37.

It can be seen from Fig. 4.37 that the strength of the alloy firstly increases and then decreases with the increase of the Y content. With addition of minor Y of 0.2 wt.%, the yield strength and ultimate tensile strength of the alloy are 186 and 236 MPa, respectively. With the increase of Y content to 0.5 wt.%, the yield strength and ultimate tensile strength are significantly promoted to 311 and 321 MPa, respectively, which are 67.5% and 35.9% higher than those of Mg−1Mn−0.2Y alloy. Meanwhile, the plasticity at room temperature of Mg−1Mn−0.2Y and Mg−1Mn−0.5Y alloy decreases by about 11.7% and 42.6%, respectively. When the content of Y increases to 1.0 wt.%, the strength of the alloy decreases obviously. The yield strength and ultimate tensile strength of the alloy are 209 and 253 MPa, respectively, which are equivalent to those of Mg−1Mn−0.2Y alloy. Compared with Mg−1Mn−0.5Y alloy, the yield strength and tensile strength decrease by about 32.9% and 21.3% respectively. Therefore the addition of a small amount of Y (up to 0.5 wt.%) can significantly improve the strength but reduce the plasticity of the alloy at room temperature.

4.3.3 Fracture surface

Fig. 4.38 shows the fracture surfaces of Mg—1Mn—xY alloy bars after tensile tests along (ED). With the increase of extrusion temperature, the fracture surface of Mg—1Mn alloy exhibits typical ductile fracture characteristic (Fig. 4.38A—C), indicating that the alloy has undergone large plastic deformation before final fracture. When Y is added to the alloy, the fracture surface of the alloy changes obviously. After adding minor Y (Fig. 4.38D), a large amount of dimples as well as cleavage steps appear on the fracture surface, which indicates that local plastic deformation occurs before fracture. Thus the fracture mode is between brittle fracture and ductile fracture, resulting in the alloy with a certain strength and high plasticity at room temperature. When the Y content increases to 0.5 wt.%, large areas of cleavage steps appear on the fracture surface of the alloy (Fig. 4.38E). The fracture of the alloy presents typical brittle fracture characteristic. With the further increase of Y content to 1.0 wt.% (Fig. 4.38F), the fracture surface is similar to that of Mg—1Mn—0.2Y alloy, which includes a large number of dimples and a certain amount of cleavage steps. Similarly, it is indicated that local plastic deformation occurs before fracture and the fracture mode is between ductile fracture and brittle fracture, resulting in the alloy with a certain strength and plasticity at room temperature.

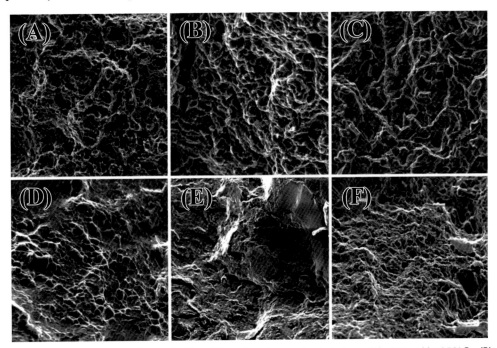

Figure 4.38 Fracture surface of Mg—1Mn—xY alloy extruded bar, (A) Mg—1Mn-250°C; (B) Mg—1Mn-300°C; (C) Mg—1Mn-350°C; (D) Mg—1Mn—0.2Y; (E) Mg—1Mn—0.5Y and (F) Mg—1Mn—1Y.

In conclusion, the fracture mechanism of Mg−1Mn−xY alloy is mainly a mixture of ductile fracture and brittle fracture. The coarse $Mg_{24}Y_5$ precipitates after extrusion easily become the crack source, which promotes the crack initiation and propagation, and thus reduces the strength and plasticity of the alloy. Therefore the content of Y addition to Mg−1Mn alloy should be controlled within 0.5 wt.%.

4.4 The mechanism of high plasticity of Mg−Mn alloy and the influence of aluminum and yttrium

4.4.1 Analysis on the formation of microstructures of Mg−Mn alloys

The relationship between the grain size of DRX and Zener−Holloman parameters during hot working is as follows:

$$Z = \dot{\varepsilon}\exp(Q/RT) \tag{1}$$

$$D = CZ^{-n_D} \tag{2}$$

where $\dot{\varepsilon}$ is the strain rate during the deformation, R is the Avogadro constant, Q is the relative activation energy, T is the deformation temperature, D is the recrystallized grain size, C and n_D are the experimental constants.

It can be seen from formula (1) and (2) that the lower the deformation temperature, the finer the grain size.

From the relationship between the size of the recrystallized new grain, the nucleation rate and the grain growth rate (Johnson−Mehl formula):

$$d = K \times \left(\dot{G}/\dot{N}\right)^{-4} \tag{3}$$

where K is a constant, $\dot{G}$ is the grain growth rate, $\dot{N}$ is the nucleation rate.

High nucleation rate and low grain growth rate are needed to obtain fine recrystallized grains. Therefore in order to obtain finer microstructure, Mg−Mn alloy needs to meet the following requirements: a high deformability at low temperature to avoid rapid grain growth in deformation at high temperature; plenty of heterogeneous nucleation sites can be generated during the DRX process in hot deformation, and the boundary of new recrystallized grains can be dragged to prevent the grain growth.

1. High deformability at low temperature

Recent studies have shown that the additions of small amounts of alloying elements (including Mn) can regulate the activation energy differences between type I and type II pyramidal $<c+a>$ dislocation slips, and promote the cross-slip rates of $<c+a>$ dislocations, which greatly improve the plastic deformability of magnesium alloys. The addition of alloying elements to activate pyramidal $<c+a>$ slip is an important method to improve the deformability of Mg alloys at low temperature. Mn can improve the resistance of basal slip and reduce the CRSS difference between the basal slip and non-basal slip. Therefore the Mn-containing magnesium alloy has good uniform deformability at low temperature. During the hot working process, the dissolved Mn in the matrix improves the plasticity and formability of the Mg matrix. The Mg–Mn alloy can be plastically deformed at a low temperature (less than 250°C), which also ensures that recrystallized grains do not grow too much.

2. Sufficient stimulating nucleation particles

The nucleation mechanism of DRX of magnesium alloy is generally divided into continuous DRX nucleation and discontinuous DRX nucleation. The mechanism of continuous DRX is a sub-grain nucleation, which is generated by the low angle grain boundary (LAGB) due to the dislocation accumulation during deformation. Then the LAGB gradually tilts into the high angle grain boundary, thus forms new dynamic recrystallized grains. The discontinuous DRX nucleation mechanism is the classical grain boundary bowing out, which usually occurs at high temperature during the deformation. The grain boundary migrates under the effect of strain and forms zigzag shape grain boundary. These zigzag grain boundaries further diffuse and migrate to form new dynamic recrystallized grains. In addition, twining DRX is also an important deformation mechanism at low temperature. During the deformation at low temperature, twins can easily form due to its low CRSS, which is only slightly higher than that of basal slip. Therefore the DRX nucleation on the twin boundary can be also considered as the twining recrystallization. In addition to the above three recrystallization nucleation mechanisms, PSN is also an important mechanism. In this mechanism the dispersed and fine primary second phase can induce recrystallization nucleation and reduce the non-recrystallization zone, which improve the degree of recrystallization and refine the recrystallization microstructure. The proper size of second phase to induce nucleation is generally about 1 μm. The coarse second phase can easily initiate crack in the loading process, which is harmful to the strength and plasticity of the alloy.

Among the abovementioned recrystallization nucleation mechanisms, the PSN nucleation mechanism is mostly affected by alloying elements. The type, size and distribution of the second phase in the alloy can be controlled by alloy composition, and thus to regulate the recrystallization nucleation sites. Meanwhile, these second phases can not only increase the number of nucleation sites, but also increase the randomness of the recrystallization grain orientation, thereby weakening the texture and improving the anisotropy of the wrought magnesium alloy. The PSN effect can be observed in AZ, ZM, Mg–RE and Mg–Mn-based wrought magnesium alloys, which both refines the grains and weakens the texture of the alloy. In the Mg–Mn binary ultrafine-grained magnesium alloy with high

Mn content, it is found that the PSN effect induced by plenty of fine primary α-Mn particles is one of the key factors to obtain a uniform ultrafine-grained microstructure.

3. Pinning effect of precipitates on grain boundary

After solving the shortage of recrystallized nucleation sites during the deformation at low temperature, the degree of grain growth is the key to finally obtain ultrafine microstructure. Therefore it is particularly important to regulate the grain boundary migration of recrystallized grains. Grain boundary migration is due to the movement of the grain boundary under thermal activation, which causes the small grains to merge with each other and thus the grain grows. Therefore to hinder the migration of the grain boundary, it is necessary to reduce the driving force and hinder the migration rate of the grain boundary. Reducing the deformation temperature can decrease the driving force of the grain boundary migration. Hindering the grain boundary migration requires second phases to pin the grain boundary and to increase the resistance of the grain boundary migration, and thereby reduce the migration rate of grain boundary. Generally, second phases with smaller diameters are believed to be capable of preventing the grain boundary migrations (Smith–Zener Pinning, SZP). Most of the second phases are nano-sized aging precipitates, which can prevent grain boundary migration and hinder the secondary growth of the recrystallized grains, leading to refined grains. It is common to obtain massive dispersively precipitated second phases through aging heat treatment before plastic processing. Yu et al. used the same method to obtain a large amount of Mg–Zn and Mn phases in ZM61 ingot before extrusion. The tensile yield strength of ZM61 alloy has been improved (~ 70 MPa), and the tensile and compression asymmetry of the alloy has also been improved. Jung et al. refined the recrystallized grains of the alloy by $\sim 2\,\mu m$ through aging before extrusion in AZ80 alloy and improved the yield strength by ~ 20 MPa. These studies have fully demonstrated that the SZP effect can effectively suppress the grain boundary migration, and achieve grain refinement and performance improvement in the wrought magnesium alloys. It can be seen from the Mg–Mn phase diagram that the solid solubility of Mn in Mg matrix decreases sharply with the temperature decreasing. Therefore during hot deformation, fine and dispersed α-Mn will precipitate out of the Mg matrix. The dynamically precipitated fine Mn particles during extrusion can effectively pin the recrystallized grain boundaries and prevent the grain boundaries migration. Especially when deformed at low temperatures, such precipitates can preserve the fine recrystallized grains.

4.4.2 High-plasticity mechanism of Mg–Mn alloy

The factors to elongation are complicated for metallic materials, including grain size, grain orientation, content and type of second phase, and test conditions, etc. These factors have different effects on the plastic deformation mechanism and thus show different impacts on the elongation of the material. Therefore it is extremely important to analyze the plastic deformation mechanism during the deformation process.

Texture is an important factor to the plasticity of wrought magnesium alloy. The additions of alloying elements have significant effects on the deformation textures. Especially, the additions of rare-earth elements have great effects on weakening the textures. Meanwhile, the weakened texture resulted from adding rare earth is also considered as the rare-earth texture. For nonrare-earth elements, their abilities to weaken texture are limited. However, it is reported in some literatures that nonrare-earth elements can also weaken the texture. Fang et al. found that adding Al to Mg—8Gd—5Y—2Zn alloy can form a large amount of Al—Gd second phases. This type of second phase is not eutectic compound, and most of them locate in the grain. During deformation, a large number of PSN mechanisms are activated, which can greatly reduce the basal texture. In addition, in the study of Ding et al., the effects of rare earth and Ca on the texture weakening of Mg—2Zn alloy were compared. It is found that the addition of Ca can effectively weaken the texture of Mg—2Zn alloy, and the weakening effect is comparable to that of rare earth. These studies have shown that as long as the contents of nonrare-earth elements are adjusted appropriately, good weakened texture effects can be achieved. From the macro-texture of Mg—Mn binary alloy (Fig. 4.39), it can be seen that although the basal texture is distributed in a ring shape, its distribution is scattered and the maximum polar density is small. And as the Mn content increases, the texture intensity shows a weakened tendency. This is mainly because with more Mn addition, it is easy to form coarse primary Mn particles, which can induce recrystallization nucleation and thus weaken the texture. Therefore the Mg—3Mn alloy has the weakest texture. It is concluded that the addition of Mn

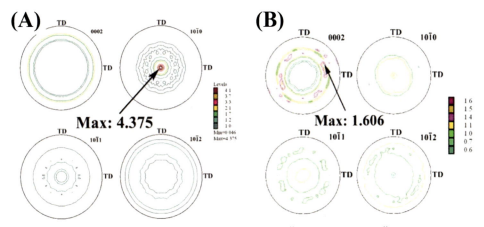

Figure 4.39 Macro-texture of Mg—Mn alloy (A) Mg—1Mn alloy, (B) Mg—3Mn alloy.

can refine the grains of Mg—Mn—Zn alloy as well as weaken the texture to some extent.

As we all know, dislocations will accumulate and release during the deformation process. High cumulative storage rate of dislocations results in more dislocations pile up, leading to strain hardening and strength improvement. However, when the dislocations are accumulated beyond the limit of the matrix, cracks will initiate to release the stress. Therefore with a higher dislocation storage rate, if the local stress cannot be effectively released, the stress concentration will result in the initiation and propagation of cracks which finally lead to material failure. In large grains, the dislocations can be easily stored but not easy to be released. Thus the strain hardening rate is high which leads to premature failure of the material. In contrast, the dislocation storage rate is low in fine grains and the dislocation can be easily annihilated at the grain boundary. Therefore the strain hardening rate tends to be stable in the fourth stage. With the increase of strain, the work hardening ability can be improved and thus promote the plastic deformability. In wrought magnesium alloy, the ultra-fine microstructure can increase the strength and induce the grain boundary slip to coordinate deformation, and hence improve the plasticity. Recent studies have shown that when the grain size of magnesium alloy reaches submicron level, $<c+a>$ slip can be activated and the amount of slips is increased, which both lead to the improvement the plasticity of magnesium alloy. When the grains are refined to below 2.7 μm, the nucleation of {10−12} twins can be suppressed. Such fine grain is beneficial to improve the room temperature formability and results in high strength and plasticity of magnesium alloy.

In this chapter, the high plasticity of Mn-containing magnesium alloys is mainly due to the improved plasticity of the matrix and grain refinement. The improved plasticity of the matrix is mainly derived from the effects of solid solution strengthening and plasticization caused by the alloying elements of Mn, Al, and Y. The refined grains can activate the non-basal slips and coordinate deformation, and thus improve the plasticity of the alloy.

4.4.3 Effect of alloying elements aluminum and yttrium

In the Mg—xAl—1Mn ($x = 1, 3, 6, 9$) alloys, the relatively high Mn content consumes part of the Al. Thus the solid solution Al in the Mg matrix is lower than that in traditional AZ and AM alloys, and the Al—Mn phase becomes the main second phase. During the hot deformation process, the alloying elements Mn and Al can improve the plastic formability of the Mg matrix, realizing deformation at a lower temperature. In addition, the Al—Mn phase can act as a heterogeneous nucleus to promote recrystallization nucleation, the nano-sized Al—Mn phase can effectively pin the recrystallized grain boundaries and hinder the grain growth. The fine grains improve both the strength and plasticity of the alloy. In general, Mg—xAl—1Mn ($x = 1, 3, 6, 9$)

alloys with high Mn content can be processed at lower temperatures than traditional AZ and AM alloys. The resulting microstructure is fine and the elongation is better. By tuning the addition of Al, Al_8Mn_5 is found to possess the best recrystallization effect among all Al–Mn compounds. As can be seen from Fig. 4.32, the Schmid factor of basal slip in MA105 alloy is higher than those of the other three alloys. Therefore although the MA105 alloy has the finest microstructure, numerous dynamically recrystallized grains are randomly oriented, which is beneficial to the activation of basal slip. Thus the yield strength is reduced and the plasticity is improved.

In the Mg–xY–Mn alloys, Y and some Mn are dissolved in the matrix during hot deformation which improve the plastic formability of the Mg matrix. In addition, the primary Mn particles become heterogeneous nuclei to promote the recrystallization nucleation. The solid solubility of Mn in the Mg matrix is limited at the hot deformation temperature. Part of Mn precipitates in the form of fine α-Mn phase, which can effectively pin the recrystallized grain boundaries and hinder the grain growth. Similarly, the dissolved Y and Mn in the matrix enhance the strength and plasticity of Mg alloy. Furthermore, the fine grains also improve the strength and plasticity of the alloy.

4.5 Summary

Currently, magnesium alloys exhibit high costs and poor comprehensive mechanical properties at room temperature. This chapter introduces Mg–Mn alloy with low cost and excellent comprehensive mechanical properties at room temperature. The microstructure evolution of the as-cast and extruded Mg–Mn alloys and the effect of grain orientation distribution on the properties of the alloys are systematically studied. The compositions of Mg–Mn alloys are optimized. Meanwhile, the influence of extrusion temperature on the microstructure and mechanical properties of the Mg–1Mn alloy is also studied, and the hot deformation parameters are optimized. The influences of Al and Y on the microstructures and mechanical properties of Mg–1Mn alloy are also studied, and the composition of this alloy is optimized. The main results are as follows:

1. The phases of as-cast Mg–Mn alloys mainly consist of α-Mg matrix and α-Mn precipitates. With the increase of Mn content, the intensity of diffraction peak of α-Mn obviously increases, which indicates that the amounts of the second phases increase sharply. Besides, the average grain size of the alloy decreases remarkably with the increase of Mn content. The main reason for the microstructure refinement is that Mn precipitates prior to α-Mg matrix during solidification and accumulates at the front of the solid–liquid interface, which effectively hinders grain growth and refines microstructure. The results show that the strength and microhardness of as-cast Mg–Mn alloy increase significantly with the increase of Mn

content. The improvement of strength and micro-hardness is mainly due to the fine grain strengthening and precipitation strengthening of Mn.

2. The microstructure of Mg−Mn alloy is significantly refined, and recrystallization is complete after extrusion at a low temperature. With the increase of Mn content, the recrystallization grain size in the alloy decreases significantly and the yield strength of the alloy increases significantly at room temperature. The results show that when the content of Mn in the alloy is 1 wt.%, the basal texture is strong, which is not conducive to the operation of the basal slip, resulting in a high yield strength of 204 MPa. When the content of Mn reaches 3 wt.%, the amount of the second phase Mn particles increases significantly, and the microstructure is refined obviously. The alloy has a high yield strength of 213 MPa and a maximum elongation of 29.9% at room temperature.

3. In the Mg−1Mn−xAl ($x < 1$) alloys with Mn as the compound phase, the amount of second phases changes with the increase of Al content. When the Al content is 0, 0.3, 0.5, and 1.0 wt.%, the corresponding phase constituents of the alloy is α-Mn+Mg, Al_8Mn_5+α-Mn+Mg, Al_8Mn_5+Mg, Al_8Mn_5+$Al_{11}Mn_4$+Mg, respectively. After extrusion at 250°C, the grain size of the alloy is significantly refined with the addition of Al. Mg−1.0Mn−0.5Al alloy presents the smallest grain size of 1.9 μm. The yield strength, ultimate tensile strength, fracture elongation and the tensile/compressive yield asymmetry are 248, 263 MPa, 33.4% and 0.94, respectively. The good comprehensive mechanical properties are attributed to the uniform and fine grains, weak texture and fine second phase particles.

4. In wrought Mg−xAl−1Mn ($x > 1$) alloys, the alloy is not fully recrystallized when Al content is less than 6 wt.%. Among them, AM11 alloy has the biggest unrecrystallized area and exhibits a strong basal texture, which are not conducive to the operation of the basal slip, leading to improved yield strength of the alloy. The yield strength, tensile strength and elongation of AM11 alloy at room temperature are 250 MPa, 287 MPa and 21.4%, respectively. When the Al content in the alloy is higher than 6 wt.%, the $Mg_{17}Al_{12}$ phase is precipitated in the alloy which can induce the recrystallization nucleation. Complete recrystallization occurs in the alloy during the extrusion process. The yield strength, tensile strength and elongation of AM91 alloy at room temperature are 205, 317 MPa and 16.1%, respectively.

5. After extrusion, the grain size of the Mg−1Mn−xAl ($x < 1$) alloy is significantly refined. The average grain sizes of M1, MA103, MA105, and MA11 alloys are 8.6, 2.8, 1.9, and 3.6 μm, respectively. MA103 and MA11 alloys exhibit microstructures with mixed grains, while MA105 alloy exhibits a relatively uniform and fine microstructure. The tensile yield strength, ultimate tensile strength, compressive yield strength, fracture elongation and tensile/compressive yield asymmetry of MA105 alloy are 248 MPa, 263 MPa, 232 MPa, 33%, and 0.94, respectively. The

good mechanical properties are attributed to the uniform and fine microstructure, weak texture and fine second phase particles.

6. In the extruded Mg—1Mn—xY alloy, the phase of the alloy is mainly α-Mg matrix, precipitated α-Mn phase and $Mg_{24}Y_5$ phase. The amounts of $Mg_{24}Y_5$ precipitates increase obviously with the increase of Y content. With an increased Y content, the microstructure is significantly refined and the average grain size is significantly reduced. The Mg—1Mn—0.5Y alloy has high strength and plasticity with yield strength, ultimate tensile strength and tensile elongation of 311 MPa, 321 MPa and 11.7%, respectively.

Further reading

Bahmani A, Arthanari S, Shin KS. Corrosion behavior of Mg—Mn—Ca alloy: influences of Al, Sn and Zn. Journal of Magnesium and Alloys 2019;7(1):38—46.

Bhattacharyya JJ, Nakata T, Kamado S, et al. Origins of high strength and ductility combination in a Guinier-Preston zone containing Mg-Al-Ca-Mn alloy. Scripta Materialia 2019;163:121—4.

Cepeda-Jiménez CM, Pérez-Prado MT. Microplasticity-based rationalization of the room temperature yield asymmetry in conventional polycrystalline Mg alloys. Acta Materialia 2016;108:304—16.

Chang CI, Lee CJ, Huang JC. Relationship between grain size and Zener—Holloman parameter during friction stir processing in AZ31 Mg alloys. Scripta Materialia 2004;51:509—14.

Chen B, Lin D, Zeng X, et al. Microstructure and mechanical properties of ultrafine grained Mg97-Y2-Zn1 alloy processed by equal channel angular pressing. Journal of Alloys and Compounds 2007;440 (1—2):94—100.

Ding H, Shi X, Wang Y, et al. Texture weakening and ductility variation of Mg—2Zn alloy with CA or RE addition. Materials Science and Engineering: A 2015;645:196—204.

Dou Y. A fundamental study on strengthening and toughening of magnesium alloys based on first-principle and molecular dynamics methods (Ph.D. thesis). 2016.

Elsayed FR, Sasaki TT, Ohkubo T, et al. Effect of extrusion conditions on microstructure and mechanical properties of microalloyed Mg—Sn—Al—Zn alloys. Materials Science and Engineering: A 2013;588:318—28.

Fang C, Liu G, Liu X, et al. Significant texture weakening of Mg-8Gd-5Y-2Zn alloy by Al addition. Materials Science and Engineering: A 2017;701:314—18.

Gao L, Chen RS, Han EH. Microstructure and strengthening mechanisms of a cast Mg—1.48 Gd—1.13 Y—0.16 Zr (at.%) alloy. Journal of Materials Science 2009;44(16):4443—54.

He SM, Zeng XQ, Peng LM, et al. Microstructure and strengthening mechanism of high strength Mg—10Gd—2Y—0.5 Zr alloy. Journal of Alloys and Compounds 2007;427(1—2):316—23.

He Y, Lei T, Wu H, Cheng D. Dependence of dynamically recrystallized grain size with Zener-Holloman parameter. Journal of East China Institute of Metallurgy 1995;02:139—45.

Hono K, Mendis CL, Sasaki TT, et al. Towards the development of heat-treatable high-strength wrought Mg alloys. Scripta Materialia 2010;63(7):710—15.

Hu Y, Deng J, Zhao C, et al. Research status and development of rare Mg-Gd series alloys. Materials Review 2010;24(12):95—9 +103.

Huang K, Marthinsen K, Zhao Q, et al. The double-edge effect of second-phase particles on the recrystallization behaviour and associated mechanical properties of metallic materials. Progress in Materials Science 2018;92:284—359.

Itoi T, Takahashi K, Moriyama H, et al. A high-strength Mg—Ni—Y alloy sheet with a long-period ordered phase prepared by hot-rolling. Scripta Materialia 2008;59(10):1155—8.

Jeong YS, Kim WJ. Enhancement of mechanical properties and corrosion resistance of Mg—Ca alloys through microstructural refinement by indirect extrusion. Corrosion Science 2014;82:392—403.

Jung JG, Park SH, Yu H, et al. Improved mechanical properties of Mg−7.6Al−0.4Zn alloy through aging prior to extrusion. Scripta Materialia 2014;93:8−11.

Lee Y, Dahle A, Stjohn D. The role of solute in grain refinement of magnesium. Metallurgical and Materials Transactions A 2000;31(11):2895−906.

Li ZT, Qiao XG, Xu C, et al. Ultrahigh strength Mg-Al-Ca-Mn extrusion alloys with various aluminum contents. Journal of Alloys and Compounds 2019;792:130−41.

Liu K, Rokhlin LL, Elkin FM, et al. Effect of ageing treatment on the microstructures and mechanical properties of the extruded Mg−7Y−4Gd−1.5 Zn−0.4 Zr alloy. Materials Science and Engineering: A 2010;527(3):828−34.

Liu XB, Chen RS, Han EH. Effects of ageing treatment on microstructures and properties of Mg−Gd−Y−Zr alloys with and without Zn additions. Journal of Alloys and Compounds 2008;465 (1−2):232−8.

Liu T, Pan F. Development and application of "solid solution strengthening and ductilizing" for magnesium alloys. Chinese Journal of Nonferrous Metals 2019;9:12.

Luo K, Zhang L, Wu G, et al. Effect of Y and Gd content on the microstructure and mechanical properties of Mg−Y−RE alloys. Journal of Magnesium and Alloys 2019;7(2):345−54.

Miao L, Zhang X, Zhang K, et al. Effect of Y on microstructure and properties of as-cast Mg-Y alloy. Spec Cast & Nonferrous Alloy 2015;35(6):636−40.

Nakata T, Mezaki T, Xu C, et al. Improving tensile properties of dilute Mg-0.27 Al-0.13 Ca-0.21 Mn (at.%) alloy by low temperature high speed extrusion. Journal of Alloys and Compounds 2015;648:428−37.

Nakata T, Xu C, Matsumoto Y, et al. Optimization of Mn content for high strengths in high-speed extruded Mg-0.3 Al-0.3 Ca (wt.%) dilute alloy. Materials Science and Engineering: A 2016;673:443−9.

Nie JF. Effects of precipitate shape and orientation on dispersion strengthening in magnesium alloys. Scripta Materialia 2003;48(8):1009−15.

Nie JF. Precipitation and hardening in magnesium alloys. Metallurgical and Materials Transactions A 2012;43(11):3891−939.

Pan F, Han E. High performance wrought magnesium alloy and its processing technology. Science Press, China, 2007.

Pan H, Qin G, Huang Y, et al. Development of low-alloyed and rare-earth-free magnesium alloys having ultra-high strength. Acta Materialia 2018;149:350−63.

Pan H, Qin G, Xu M, et al. Enhancing mechanical properties of Mg−Sn alloys by combining addition of Ca and Zn. Materials & Design 2015;83:736−44.

Peng P, He X, She J, et al. Novel low-cost magnesium alloys with high yield strength and plasticity. Materials Science and Engineering: A 2019;766:138332.

Peng Q, Dong H, Wang L, et al. Microstructure and mechanical property of Mg−8.31 Gd−1.12 Dy−0.38 Zr alloy. Materials Science and Engineering: A 2008;477(1−2):193−7.

Polina M, Guy BH, Yael T, et al. The relation between Mn additions, microstructure and corrosion behavior of new wrought Mg-5Al alloys. Materials Characterization 2018;145:101−15.

Qi FG, Zhang DF, Zhang X, et al. Effect of Sn addition on the microstructure and mechanical properties of Mg−6Zn−1Mn (wt.%) alloy. Journal of Alloys and Compounds 2014;585(5):656−66.

Qi FG, Zhang DF, Zhu Z, et al. Effect of heat treatment on microstructure and mechanical properties of extruded ZM61 magnesium alloy. Materials Science and Technology 2012;28(12):1426−33.

Rong W, Zhang Y, Wu Y, et al. The role of bimodal-grained structure in strengthening tensile strength and decreasing yield asymmetry of Mg-Gd-Zn-Zr alloys. Materials Science and Engineering: A 2019;740:262−73.

Sasaki TT, Elsayed FR, Nakata T, et al. Strong and ductile heat-treatable Mg−Sn−Zn−Al wrought alloys. Acta Materialia 2015;99:176−86.

Sasaki TT, Yamamoto K, Honma T, et al. A high-strength Mg−Sn−Zn−Al alloy extruded at low temperature. Scripta Materialia 2008;59(10):1111−14.

She J, Pan F, Zhang J, et al. Microstructure and mechanical properties of Mg-Al-Sn extruded alloys. Journal of Alloys and Compounds 2015;657:893−905.

She J, Pan F, Peng P, et al. Microstructure and mechanical properties of as-extruded Mg—x Al—5Sn—0.3Mn alloys (x=1, 3, 6 and 9). Materials Science and Technology 2015;31(3):344—8.

She J, Pan F, Zhang J, et al. Microstructure and mechanical properties of Mg—Al—Sn extruded alloy. Journal of Alloys and Compounds 2016;657:893—905.

She J, Pan FS, Guo W, et al. Effect of high Mn content on development of ultra-fine grain extruded magnesium alloy. Materials & Design 2016;90:7—12.

She J, Peng P, Xiao L, et al. Development of high strength and ductility in Mg—2Zn extruded alloy by high content Mn-alloying. Materials Science and Engineering: A 2019;765:138203.

Somekawa H, Basha DA, Singh A. Deformation behavior at room temperature ranges of fine-grained Mg—Mn system alloys. Materials Science and Engineering: A 2019;766:138384.

Sun M, Wu G, Wang W, et al. Research progress of Mg-Gd alloy. Materials Review 2009;23 (6A):98—103.

Wang Q, Chen J, Zhao Z, et al. Microstructure and super high strength of cast Mg-8.5 Gd-2.3 Y-1.8 Ag-0.4 Zr alloy. Materials Science and Engineering: A 2010;528(1):323—8.

Wu HJ, Wang TZ, Wu RZ, Hou LG, Zhang JH, Li XL, et al. Effects of annealing process on the interface of alternate α/β Mg-Li composite sheets prepared by accumulative roll bonding. Journal of Materials Processing Technology 2018;254:265—76.

Wu WX, Jin L, Zhang ZY, et al. Grain growth and texture evolution during annealing in an indirect-extruded Mg—1Gd alloy. Journal of Alloys and Compounds 2014;585:111—19.

Xiao L, Pan F. Research progress of Mg—Mn alloys. Nonferrous Metals Engineering 2019;9:1.

Yamada K, Hoshikawa H, Maki S. 等. Enhanced age-hardening and formation of plate precipitates in Mg—Gd—Ag alloys. Scripta Materialia 2009;61(6):636—9.

Yin SM, Wang CH, Diao YD, Wu SD, Li SX. Influence of grain size and texture on the yield asymmetry of Mg-3Al-1Zn alloy. Journal of Materials Science & Technology 2011;27(1):29—34.

Yu D, Zhang D, Sun J, et al. Improving mechanical properties of ZM61 magnesium alloy by aging before extrusion. Journal of Alloys and Compounds 2017;690:553—60.

Zhang C, Wu L, Huang G, et al. Influence of microalloying with Ca and Ce on the corrosion behavior of extruded Mg-3Al-1Zn. Journal of the Electrochemical Society 2019;166(13):C445—53.

Zhang J, Liu S, Wu R, et al. Recent developments in high-strength Mg-RE-based alloys: focusing on Mg-Gd and Mg-Y systems. Journal of Magnesium and Alloys 2018;6(3):277—91.

CHAPTER 5

Medium-strength and high-plasticity Mg−Sn-based alloys

The maximum solid solubility of Sn in Mg is 14.48%, possessing relatively high solid solution strengthening capability. The solid solubility of Sn in Mg decreases with decreasing temperature, and the formed Mg_2Sn has relatively good precipitation hardening effect. In addition, Mg_2Sn possesses good high-temperature stability, which is beneficial to improve the high-temperature creep resistance of Mg alloys. Therefore the Mg−Sn alloy is considered as a very promising alloy system, and it is attracting more and more attention. The computational simulation results indicate that the Sn atom is preferentially dissolved at the $\{11\bar{2}0\}$ crystallographic plane of Mg crystals, reducing the unstable stacking fault energy of the pyramidal surface of $\{11\bar{2}2\}$ $\langle11\bar{2}3\rangle$, and consequently decreasing the ratio of the critical shear stress between the non-basal surface and the basal surface. Therefore in principle, the addition of Sn can not only improve the strength of Mg alloy but also enhance its formability. The previous studies indicate the comprehensive mechanical properties of Mg-5 wt.%Sn alloy are relatively good, and the creep resistance of Mg-10 wt.%Sn alloy is good as well, which is even superior to that of AE42 (Mg−4Al−2RE) alloy. In addition, according to the simulation result of the crystal structure of the Mg-3wt.%Al-3wt.%Sn (AT33) Mg alloy in completely solid solution state, the unstable stacking fault energy for both cylindrical plane $<a>$ and pyramidal plane $<c+a>$ of the Mg crystal doped with alloying elements are decreased. Thus the formability of the AT33 Mg alloy in solid solution state is relatively good. This chapter intends to proceed research on the medium-strength and high-plasticity Mg alloys based on Mg−Sn system.

5.1 Microstructure and property of extruded Mg−Sn alloy

(1) The microstructure of as-extruded sheet of Mg−Sn alloy

Fig. 5.1 shows the extruded sheet microstructure evolution of Mg−Sn alloy with different Sn contents. The most grains of the Mg−Sn extruded sheet are equiaxed, and their sizes are not uniform, indicating the alloy has subjected the incomplete dynamic recrystallization during the extrusion deformation. For Mg−0.5Sn and Mg−1Sn alloys, because of the low Sn content, there are few second phases in the alloys, and the second phases distributed along the extrusion direction cannot be

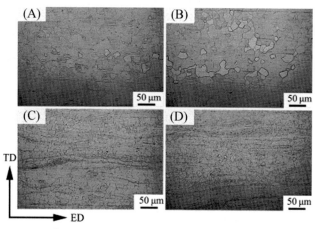

Figure 5.1 Microstructure of the extruded sheet of Mg—Sn alloys: (A) Mg—0.5Sn, (B) Mg—1Sn, (C) Mg—2Sn, and (D) Mg—2.5Sn.

observed by a metallurgical microscope; while the one with strip shape can be observed along the extrusion direction in Mg—2Sn and Mg—2.5Sn alloys.

The extruded sheet of Mg—Sn alloy was studied by using the backscattered electron, as shown in Fig. 5.2. It can be seen that both Mg—0.5Sn and Mg—1Sn have few second phases. When the Sn content is 2 wt.%, the second phases appear, distributing along the extrusion direction in the form of particles. According to the Mg—Sn binary alloy phase diagram, the second phase is Mg_2Sn.

(2) The properties of as-extruded sheet of Mg—Sn alloy

Fig. 5.3 shows the true stress-strain curves of Mg—Sn extruded sheets with different Sn contents. Their mechanical properties are presented in Table 5.1. Both tensile strength and yield strength increase with increasing the angle between the tensile direction and the extrusion direction, while the value of work hardening index (n) decreases with increasing the angle. The elongation of Mg—0.5Sn and Mg—1Sn alloys decrease with the increase of the angle between the tensile direction and the extrusion direction, while the elongation of Mg—2Sn and Mg—2.5Sn reach the maximum when the angle is 45 degrees.

Both Mg—0.5Sn and Mg—1Sn alloys do not have second phases, and their strength is almost the same. The second phases increase with increasing Sn content, and the tensile strength improves along the three directions. When the Sn content increases from 0.5 wt.% to 2.5 wt.%, the tensile strength along the transverse direction (TD) increases from 266 to 331 MPa. The overall trend of elongation and the value of n increase with increasing the Sn content.

Fig. 5.4 shows the macro texture of Mg—Sn as-extruded sheets with different Sn contents. It shows that the polar axes of the (0002) basal plane pole figures of the four

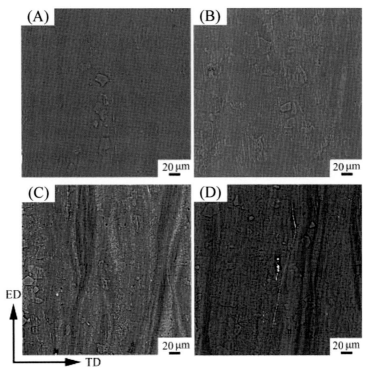

Figure 5.2 Backscattered electron scanning micrographs of the extruded sheet of Mg—Sn alloys: (A) Mg—0.5Sn, (B) Mg—1Sn, (C) Mg—2Sn, and (D) Mg—2.5Sn.

Mg—Sn extruded sheets are all in the center, which indicates that the grains in the alloy sheets are mainly oriented on the basal plane. The minimum texture level is uniformly set as 1.0 mrd to compare the changes in the weakly oriented textures of the sheets with different Sn contents. From Fig. 5.4, the texture of the (0002) basal plane pole figure becomes more and more divergent with increasing Sn content, while the range of weak orientation is wider and wider. Correspondingly, the texture intensity of the base plane becomes weaker and weaker.

Based on the results from the first principle calculations, the solid solution of Sn atoms can reduce the critical resolved shear stress (CRSS) ratio between the non-basal plane and the basal plane of the Mg alloys. The higher the Sn content in the solid solution, the larger the decreased CRSS ratio between the non-basal plane and the basal plane, the greater the contribution from slipping of the non-basal plane during plastic deformation, and the more grains oriented toward the non-basal plane in the deformed sheets. The second phases increase with increasing the Sn content. The brittle and hard second phases would split the matrix, so that the elongation of the material decreases. However, the actual results show that the improved strength of the

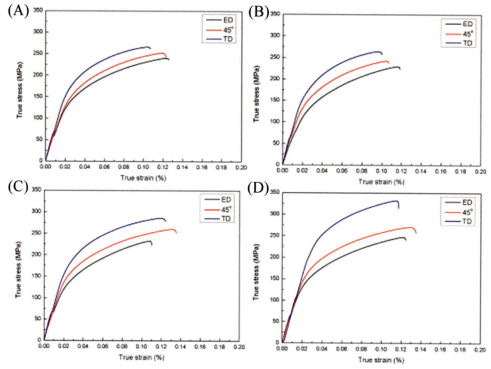

Figure 5.3 The room-temperature true stress-strain curves of the as-extruded sheet of Mg—Sn alloys: (A) Mg—0.5Sn, (B) Mg—1Sn, (C) Mg—2Sn, and (D) Mg—2.5Sn. The ED represents the extrusion direction, and the TD is the transverse direction.

extruded sheet does not sacrifice its elongation, instead, it is slightly increased. Therefore the solid solution of Sn can significantly improve the plasticity of Mg alloys.

Fig. 5.5 presents the fracture morphology of the experimental alloy after stretching along the extrusion direction (ED) 45 degrees, and TD at room temperature. When the Sn content is low, many dimples at different sizes exist in the fracture surface, and there are also a lot of cleavage lines. Therefore the fracture characteristics of the materialsare between cleavage fracture and ductile fracture when the Sn content is low. As the increase in the Sn content, the dimples distributed in the fracture become more uniform and finer, and there is only a small amount of cleavage lines, presenting ductile fracture.

5.2 Phase diagram and alloy design of Mg—Al—Sn alloy

The aforementioned researches indicate the strength and elongation of the binary Mg—Sn alloy are not good enough, which need to be further improved. Therefore the multi-component Sn-containing Mg alloys with various strengthening methods

Table 5.1 The tensile results at room temperature for the extruded sheets along the ED, 45 degrees, and TD.

Alloy	Ultimate tensile strength (UTS) (MPa)			Yield strength (YS) (MPa)			Fracture elongation (FE) (%)			Work hardening index n		
	ED	45 degrees	TD	ED	45 degrees	TD	ED	45 degrees	TD	ED	45 degrees	TD
Mg–0.5Sn–True	239	251	266	130	137	157	9.4	9.0	7.5	0.33	0.32	0.32
Nominal	212	224	240	123	137	155	10.4	10.7	8.1			
Mg–1Sn–True	233	242	264	132	142	156	9.8	8.4	7.9	0.38	0.34	0.31
Nominal	207	218	240	129	135	155	10.2	9.0	8.3			
Mg–2Sn–True	232	260	286	127	145	161	9.0	11.5	10.4	0.37	0.33	0.31
Nominal	208	227	253	121	127	153	11.5	14.2	13.3			
Mg–2.5Sn–True	247	270	331	134	156	218	10.3	11.4	9.0	0.34	0.3	0.25
Nominal	218	238	296	133	148	212	11.2	12.2	9.4			

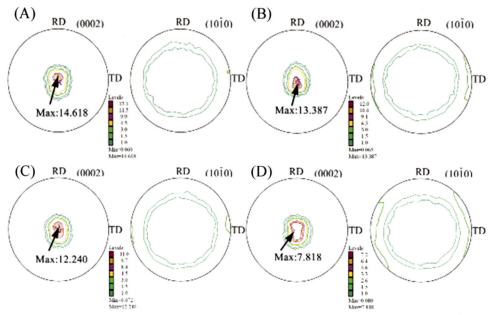

Figure 5.4 Macro texture of the extruded sheets of Mg—Sn alloys: (A) Mg—0.5Sn, (B) Mg—1Sn, (C) Mg—2Sn, and (D) Mg—2.5Sn.

need to be developed. Al is widely used in Mg alloys as the main alloying element due to its strong solid solution strengthening in Mg matrix. Currently, the commercial Mg alloys taking Al as the main alloying element include Mg—Al—Zn (AZ) and Mg—Al—Mn (AM), where the Al content has a very large composition range from 3% to 9%. Because no compound can be formed by Al and Sn, the Mg—Al—Sn Mg alloy possesses both the solid solution strengthening from Al element and the dispersion strengthening from Mg_2Sn, which is being developed to a new type of Mg alloy with excellent performances.

The phase diagram is one of the main foundations for the alloy design. Regarding the research on the phase diagram of Mg—Al—Sn system, Doernberg et al. [1] confirmed the phase diagram of Mg—Al—Sn system through experiments and developed the phase diagram of this system; Kang et al. [2] optimized the ternary system with Factsage. Following the researches, this chapter will develop the Mg—Al—Sn phase diagram with different phase composition based on the reported thermodynamic data of Mg—Al—Sn system, and verify the Mg—Al—Sn phase diagram with experiments. The applied software is Thermo-calc and Pandant.

The three binary phase diagrams of the Mg—Al—Sn system are shown in Fig. 5.6. From the binary phase diagrams, Mg—Al system contains three compounds including $Al_{30}Mg_{23}$, Al_3Mg_2, and $Mg_{17}Al_{12}$; Mg—Sn system has only one compound of Mg_2Sn;

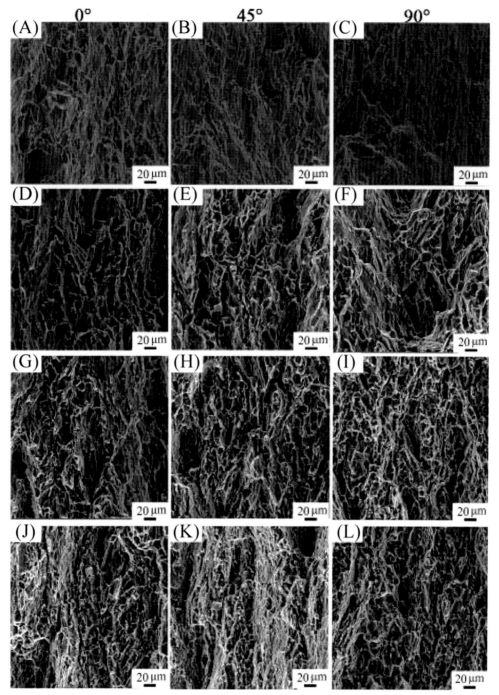

Figure 5.5 The fracture morphology of experimental alloys after strengthening toward the directions of ED, 45 degrees, and TD: (A), (B), and (C) for Mg−0.5Sn; (D), (E), and (F) for Mg−1Sn; (G), (H), and (I) for Mg−2Sn; (J), (K), and (L) for Mg−2.5Sn.

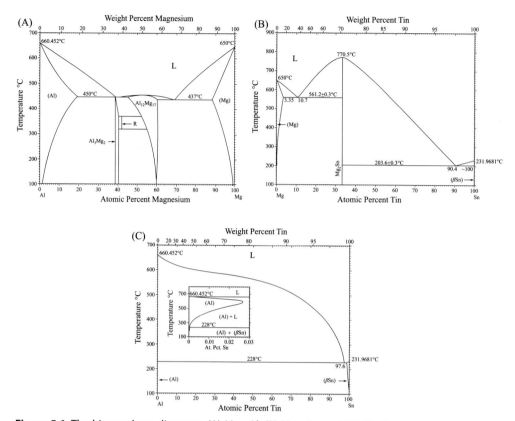

Figure 5.6 The binary phase diagrams: (A) Mg−Al, (B) Mg−Sn, and (C) Al−Sn.

and there is no compound in Al−Sn system, which cannot be found from references. According to Doernberg's report, the isothermal section of Mg−Al−Sn system at 400°C is presented in Fig. 5.7. The detailed parameters of the space group and unit cell for all compounds, Al, Mg, and Sn are shown in Table 5.2.

5.2.1 Construction and verification of phase diagram

(1) Construction of thermodynamic phase diagram

Because there is no ternary compound in this system, the thermodynamic data of the Al−Mg and Al−Sn binary systems are from COST507; Al−Sn system uses modified 0LfccAl, Snoo = 43410.66 + 11.76812*T by Din; Mg−Sn binary thermodynamic data follows the optimized data of Mg−Sn binary system by S. Fries. The description of ternary liquid phase is merely needed because the solid solubility of the ternary solid phase is very small. Fig. 5.8 shows the calculated liquid phase projection of Mg−Al−Sn. Fig. 5.9 presents the derived entire reaction diagram of Mg−Al−Sn

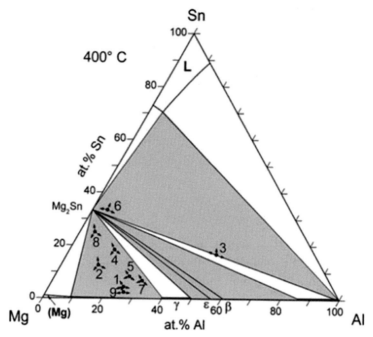

Figure 5.7 The isothermal section at 400°C for Mg—Al—Sn system.

Table 5.2 The phase and crystal structure of Mg—Al—Sn system.

Samples	Space group	Lattice parameters(Å)		
		a	*b*	*c*
Al(FCC)	Fm$\bar{3}$m	4.0488	—	—
Mg(Hcp)	P63/mmc	3.2089	—	5.2101
Sn(Bct)	I41/amd	5.8318	—	3.1818
Al$_3$Mg$_2$(β)	Fd$\bar{3}$m	28.16—28.24	—	—
Mg$_{17}$Al$_{12}$(γ)	I$\bar{4}$3m	10.5438	—	—
Mg$_2$Sn	Fm$\bar{3}$m	6.765	—	—
Mg$_{23}$Al$_{30}$(ε)	R$\bar{3}$	12.8254	—	—

ternary system in this work. The thermodynamic data related to the Mg—Al—Sn system are shown in Table 5.3.

(2) Verification of phase diagram

In order to verify the Mg—Al—Sn phase diagram, three samples were prepared at 300°C, referring to the isothermal section of Mg—Al—Sn at 300°C. The alloy distribution in the ternary phase diagram is shown in Fig. 5.10. All samples were smelted under the protection of mixed gas of sulfur hexafluoride and carbon dioxide, annealed

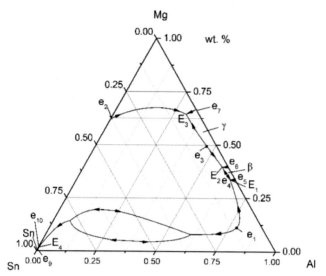

Figure 5.8 The liquid phase projection of Mg−Al−Sn.

in vacuum at 430°C for 30 days, reduced to 300°C with the cooling rate of 50°C per day, maintained at 300°C for 7 days, and quenched in an ice-water mixture to get Mg−Al−Sn alloy balance sample.

The composition of 1# sample is $Mg_{70}Al_{15}Sn_{15}$ (mass fraction, all refer to mass fraction unless noted in this chapter). The XRD pattern is presented in Fig. 5.11, where the sample has three phases, namely, α-Mg, Mg_2Sn, and $Mg_{17}Al_{12}$. Besides, the backscattered electron spectrometer (BSE) and energy-dispersive spectrometer (EDS) results (Fig. 5.12) of this sample further verify that this sample contains three phases of Mg, Mg_2Sn, and $Mg_{17}Al_{12}$, thereby determining the phase region of 1# sample presented in Fig. 5.10.

The composition of 3# sample is $Mg_{25}Al_{55}Sn_{20}$, with the XRD pattern as shown in Fig. 5.13. This sample contains three phases, including Al, Al_3Mg_2, and Mg_2Sn. In addition, its SEM-BSE image (Fig. 5.14) indicates that this sample contains three phases, which are bright, gray convex, and gray concave phases. They are determined to be Mg_2Sn, Al_3Mg_2, and Al in combination with the EDS result. It is difficult to directly distinguish from the EDS results because the compositions of Al_3Mg_2 and $Al_{30}Mg_{23}$ are similar. Therefore the phase composition of this sample is mainly evidenced by SEM and EDS and determined by XRD results. The XRD, SEM+EDS results confirm the phase region of the 3# sample in the Mg−Al−Sn phase diagram in Fig. 5.10.

Through the above two samples, the phase relationship of isothermal section of Mg−Al−Sn system at 300°C can be determined by the contact rules for phase regions. The correctness of this result can be further verified by testing the phase

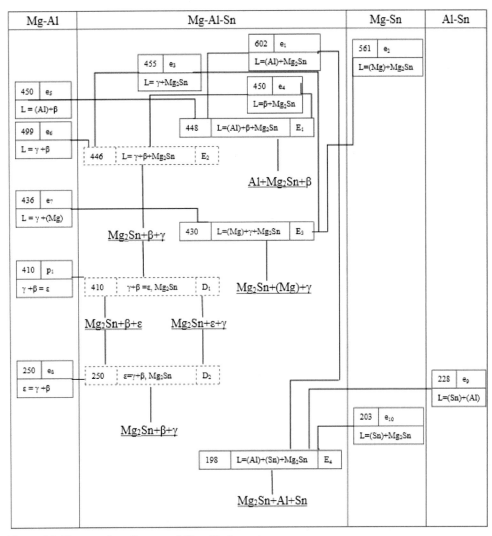

Figure 5.9 The reaction diagram of Mg—Al—Sn system.

composition of the 2# balance sample of $Mg_{40}Al_{40}Sn_{20}$, whose XRD pattern is shown in Fig. 5.15. Since there is no standard PDF card of $Al_{30}Mg_{23}$ phase in the 2004 PDF card library, the XRD pattern (Fig. 5.16) of the $Al_{30}Mg_{23}$ phase is completed by using the software of PCW, according to the atomic site occupation of the $Al_{30}Mg_{23}$ phase (Table 5.4). In comparison with the XRD pattern of 2# sample, this sample contains three phases, including $Mg_{17}Al_{12}$, $Al_{30}Mg_{23}$, and Mg_2Sn. Therefore the phase region of 2# sample is determined in Mg—Al—Sn phase diagram. Fig. 5.17 shows the SEM

Table 5.3 Zero variable reaction.

T(°C)	Reaction	Type	Phase	Al wt.%	Mgwt.%	Snwt.%
602	Liquid $\leftrightarrow$ Mg$_2$Sn+Fcc	e_1	Liquid	87.96	11.12	0.92
			Mg$_2$Sn	0	90.73	9.27
			Fcc	97.77	2.216	0.004
448	Liquid $\leftrightarrow$ Mg$_2$Sn+Fcc+β	E_1	Mg$_2$Sn	0	90.73	9.27
			β	58.47	41.53	0
			Fcc	81.44	18.53	0.03
			Liquid	60.47	39.46	0.07
446	Liquid $\leftrightarrow$ Mg$_2$Sn+β+γ	E_2	Liquid	54.35	45.56	0.09
			Mg$_2$Sn	0	90.73	9.27
			β	58.47	41.53	0
			γ	49.37	50.63	0
430	Liquid $\leftrightarrow$ Mg$_2$Sn+γ+Hcp	E_3	Liquid	29.96	69.52	0.52
			Mg$_2$Sn	0	90.73	9.27
			Hcp	44.98	54.97	0.05
			γ	49.37	50.63	0
198	Liquid $\leftrightarrow$ Mg$_2$Sn+Bct+Fcc	E_4	Liquid	2.89	31.32	65.79
			Bct	3	0	97
			Fcc	100	0	0
			Mg$_2$Sn	0	90.73	9.27
455	Liquid $\leftrightarrow$ Mg$_2$Sn+γ	e_3	Liquid	43.33	56.48	0.19
			Mg$_2$Sn	0	90.73	9.27
			γ	49.37	50.63	0
450	Liquid $\leftrightarrow$ Mg$_2$Sn+β	e_4	Liquid	0.98	8.67	90.35
			Mg$_2$Sn	0	90.73	9.27
			β	58.47	41.53	0
410	γ+β $\leftrightarrow$ Mg$_2$Sn, ε	D1	γ	47.97	52.03	0
~250	ε+β $\leftrightarrow$ Mg$_2$Sn, γ	D2	γ	43.80	56.20	0

and EDS results of the 2# sample, from which this sample is verified to contain three phases.

In summary, the phase distribution of the Mg−Al−Sn ternary system at 300°C is presented in Fig. 5.10. No ternary compound is detected, which is consistent with the phase composition constructed by us and other researchers. The isothermal section at 300°C of this system contains seven single-phase regions: α-Mg, Al, Sn, Mg$_{17}$Al$_{12}$, Al$_{30}$Mg$_{23}$, Al$_3$Mg$_2$, and Mg$_2$Sn, eleven binary phase regions: Al−Al$_3$Mg$_2$, Al3Mg$_2$−Mg$_{17}$Al$_{12}$, Mg−Mg$_{17}$Al$_{12}$, Mg−Mg$_2$Sn, Mg$_2$Sn−Sn, Sn−Al, Mg$_2$Sn−Al$_3$Mg$_2$, Mg$_2$Sn−Mg$_{17}$Al$_{12}$, Mg$_2$Sn−Al, Al$_3$Mg$_2$−Al$_{30}$Mg$_{23}$, and Mg$_2$Sn−Al$_{30}$Mg$_{23}$, and five ternary phase regions: Mg$_2$Sn−Mg$_{17}$Al$_{12}$−Mg, Mg$_2$Sn−Mg$_{17}$Al$_{12}$−Al$_3$Mg$_2$, Mg$_2$Sn−Al−Al$_3$Mg$_2$, Mg$_2$Sn−Al−Sn, and Al$_3$Mg$_2$−Mg$_2$Sn−Al$_{30}$Mg$_{23}$.

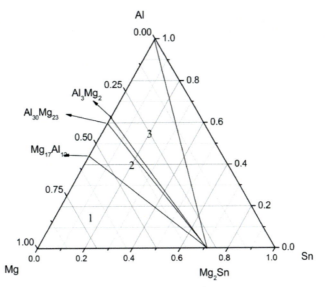

Figure 5.10 The distribution diagram of the balance samples.

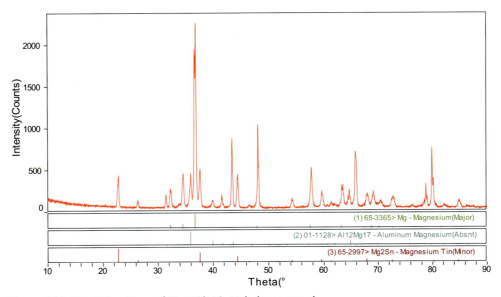

Figure 5.11 The XRD pattern of Mg70Al15Sn15 balance sample.

5.2.2 Alloy design

According to thermodynamic data, the calculated isothermal section at 300°C and 500°C are shown in Fig. 5.18. From the phase diagrams, the solid solubility of Sn and Al in Mg matrix decreases with decreasing temperature. In comparison, the decrease

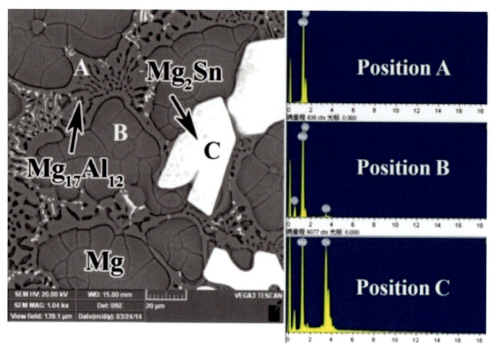

Figure 5.12 SEM image of the Mg70Al15Sn15 sample.

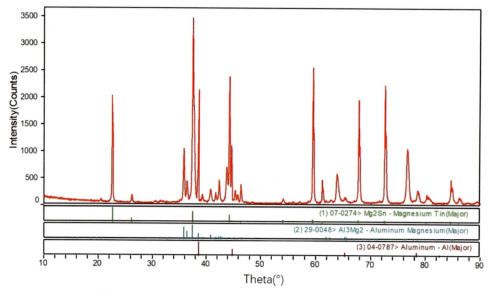

Figure 5.13 X-ray diffraction (XRD) pattern of the $Mg_{25}Al_{55}Sn_{20}$ balance sample.

Figure 5.14 SEM image of the Mg$_{25}$Al$_{55}$Sn$_{20}$ sample.

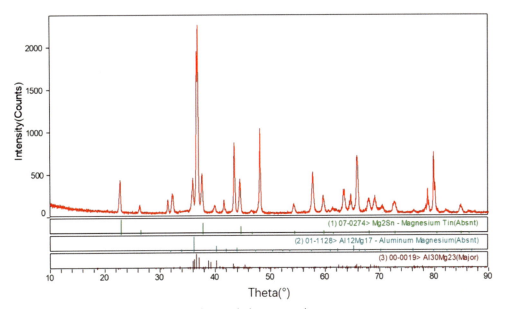

Figure 5.15 XRD pattern of the Mg$_{40}$Al$_{40}$Sn$_{20}$ balance sample.

of the solid solubility of Sn in Mg is larger than that of Al. The solid solubility of Sn in Mg matrix at 500°C is ~9.6%, which reduces to 1.6% at 300°C. The solid solubility of Al in Mg matrix decreases from 7.7% (500°C) to 6.5% (300°C). This system has no ternary compound. The Mg-rich end contains two compounds of Mg$_{17}$Al$_{12}$ and Mg$_2$Sn. The two compounds have no solid solubility with each other. Therefore it is

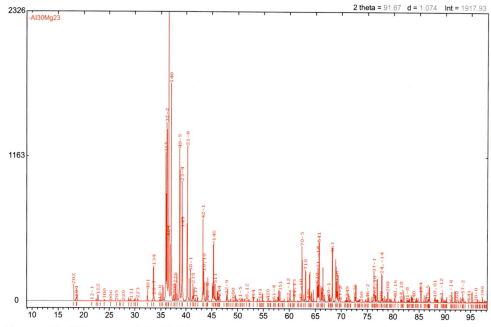

Figure 5.16 XRD pattern of the Al$_{30}$Mg$_{23}$ compound.

Table 5.4 The atomic occupation of the Al$_{30}$Mg$_{23}$ compound.

Al$_{30}$Mg$_{23}$; R$\overline{3}$; Space group: 148; lattice parameter $a = 1.28254$ nm.

Atom	Position	X	Y	z	Occupancy
Mg1	3b	0	0	0.5	1
Mg2	6c	0	0	0.33926	1
Mg3	6c	0	0	0.07712	1
Al1	18f	0.08890	0.23632	0.00171	1
Al2	18f	0.22991	0.26550	0.10453	1
Al3	18f	0.11604	0.13180	0.20143	1
Al4	18f	0.21908	0.03171	0.27175	1
Al5	18f	0.18443	0.44428	0.07118	1
Mg4	18f	0.26287	0.04320	0.12466	1
Mg5	18f	0.43644	0.07282	0.01534	1
Mg6	18f	0.17556	0.39857	0.20706	1

possible to become two strengthening phases of the Mg—Al—Sn alloy. Elsayed studied the age hardening of Mg—10Sn—3Al—1Zn alloy and found that Al can effectively improve the morphology of Mg$_2$Sn in Mg—Sn alloy, and more Mg$_2$Sn strengthening phase with better morphology can be precipitated from Mg—Al—Sn system. At the same time, because the solid solution strengthening of Al in Mg is relatively good,

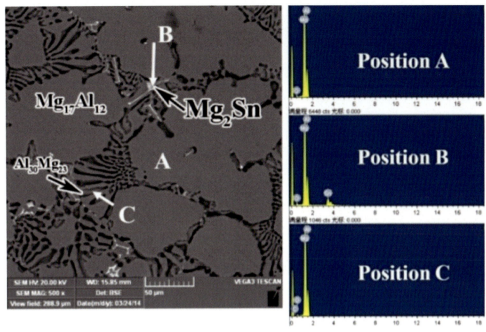

Figure 5.17 SEM image of the Mg$_{45}$Al$_{50}$Sn$_5$.

together with its large solid solubility, Al plays an important role in the solid solution strengthening in Mg—Al—Sn alloy. Therefore the main influence of Sn in Mg—Al—Sn alloy is forming Mg$_2$Sn with Mg to provide the second phase strengthening; when Al is low in the alloy, it mainly provides solid solution strengthening; when Al is high, it mainly plays the role of solid solution strengthening and second phase strengthening (Mg$_{17}$Al$_{12}$). In summary, Mg—Al—Sn alloy contains three phases of Mg, Mg$_{17}$Al$_{12}$, and Mg$_2$Sn, and its strengthening methods are mainly the solid solution strengthening and the second phase strengthening.

Chen et al. [3,4] found that the mechanical properties are the best when Sn is 5 wt.% for as-cast Mg—Sn binary alloy. In addition, the strength change is not obvious when Sn is larger than 5 wt.% in as-extruded Mg—Sn binary alloy. For example, the tensile strength of Mg—7Sn alloy is merely 10 MPa higher than that of Mg—5Sn alloy. The corrosion resistance is significantly improved when 5% Sn is added to AZ alloy. However, the Sn content of the designed Mg—Al—Sn alloy in this chapter is not higher than 5%, because the price of Sn is relatively high.

The concentration cross sections of Mg—1Al—ySn, Mg—3Al—ySn, Mg—6Al—ySn, and Mg—9Al—ySn are shown in Fig. 5.19. Mg—Al—Sn alloys with different phases are designed according to the phase diagrams. The alloy can be designed to be a single-phase alloy containing α-Mg or a binary phase alloy containing α-Mg and

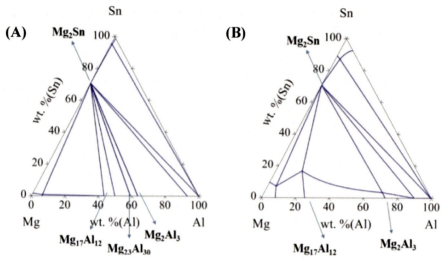

Figure 5.18 Isothermal section of Mg–Al–Sn at (A) 300°C and (B) 500°C.

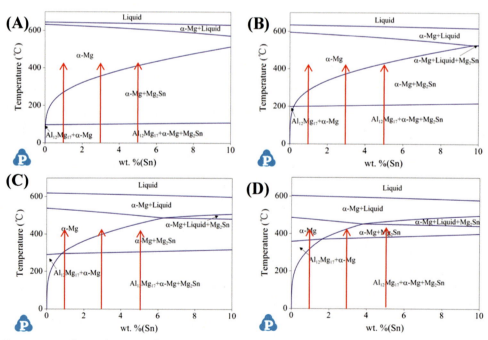

Figure 5.19 Phase diagram of (A) Mg–1Al–ySn, (B) Mg–3Al–ySn, (C) Mg–6Al–ySn, and (D) Mg–9Al–ySn.

Mg_2Sn, based on the concentration cross section of Mg—1Al—ySn alloy. Al and Mg_2Sn play the roles of solid solution strengthening and precipitation strengthening in this alloy, respectively. Based on Mg—3Al—ySn, Mg—6Al—ySn, and Mg—9Al—ySn, the alloy can be designed as a single-phase alloy containing α-Mg or binary phase alloy containing two phases of α-Mg and $Mg_{17}Al_{12}$, or ternary phase alloy containing phases of α-Mg, Mg_2Sn, and $Mg_{17}Al_{12}$.

The alloy composition is determined as Mg—xAl—ySn ($x = 1,3,6,9$; $y = 1,3,5$) after considering the above factors. According to the concentration cross section of the phase diagram, alloys with different phase compositions and phase contents can be obtained by controlling temperature and alloy composition, to analyze and confirm the strengthening mechanism of Mg—Al—Sn alloy and optimize the composition of the Mg—Al—Sn alloy. It was found that the impurity of Fe in Mg alloy can seriously deteriorate its corrosion resistance, and Mn is an effective impurity removal element in Mg alloys. Therefore a small amount of Mn (0.3%) added to Mg—Al—Sn alloy can purify the alloy melt, without affecting the phases of the Mg—Al—Sn alloy. According to the phase diagram, twelve alloys of Mg—xAl—ySn—0.3Mn ($x = 1,3,6,9$; $y = 1,3,5$) were designed. The Mg—Al—Sn—Mn alloy is abbreviated as ATM alloy in this book (for example, Mg—6Al—3Sn—0.3Mn is abbreviated as ATM630).

5.3 Microstructure and property of the Mg—Al—Sn—Mn cast alloy

The yield strength of the alloys generally depends on the solution element, grain size (or dendrite spacing), and the volume fraction, type, and morphology of the contained second phase for the cast alloys. According to the phase diagram, the Mg—Al—Sn—Mn system contains two second phases of $Mg_{17}Al_{12}$ and Mg_2Sn. The content of Sn and Al vary widely in this system. The factors influencing the strength of the alloy mainly come from three aspects: the solid solubility of the element in Mg matrix, the grain size of the alloy, and the content of the second phase.

The following will study the effect of Al and Sn content on the solidification, microstructure, and mechanical property of the cast alloys based on the twelve designed Mg—xAl—ySn—0.3Mn ($x = 1,3,6,9$; $y = 1,3,5$) alloys.

5.3.1 Phase composition and microstructure of Mg—Al—Sn—Mn alloy

(1) Phase composition of the Mg—Al—Sn—Mn alloy

Fig. 5.20 shows the XRDs patterns of as-cast ATM alloys with the same Sn content but different Al contents. When the Sn content is 1%, ATM110 is a single-phase alloy, only containing phase of α-Mg. The XRD pattern of ATM310 contains a relatively weak diffraction peak of $Mg_{17}Al_{12}$, indicating it has a small amount of $Mg_{17}Al_{12}$ phase. When Al contents are 6% and 9%, the XRD patterns appear relatively intensive

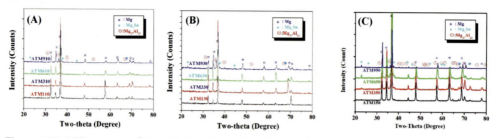

Figure 5.20 XRD patterns of Mg−Al−Sn−Mn casting alloys with (A) 1% Sn, (B) 3% Sn, and (C) 5% Sn.

diffraction peak of $Mg_{17}Al_{12}$. At the same time, when Al content is 9%, there appears the diffraction peak of Mg_2Sn at 2-Theta equal to 23 degrees, indicating the solid solubility of Sn in Mg matrix decreases with increasing Al content and forms Mg_2Sn phase. When the Sn content is 3%, the diffraction peak intensity of $Mg_{17}Al_{12}$ increases with increasing Al content (as shown in Fig. 5.20B). When the Al content is 6%, there appears diffraction peak of Mg_2Sn in the XRD pattern. When the Al content increases to 9%, the diffraction peak of Mg_2Sn is more obvious, indicating the fraction of Mg_2Sn phase increases. From Fig. 5.20C, all the alloys with 5% Sn contain Mg_2Sn phase, and the $Mg_{17}Al_{12}$ and Mg_2Sn contents increase with increasing Al. No Mn-containing phase is detected in all the alloys, probably resulting from the low Mn content.

At the Mg-rich end of the Mg−Al−Sn phase diagram, the solid solubility of Sn in Mg matrix decreases with increasing Al content. Therefore the phase of Mg_2Sn increases with increasing Al in Mg−Al−Sn−Mn alloy (Fig. 5.21).

Fig. 5.22 shows the XRD diffraction patterns of alloys with the same Al content but different Sn contents. When Al content is 1%, the phase of Mg_2Sn increases with increasing Sn, but there is no $Mg_{17}Al_{12}$ peak, indicating all Al is dissolved in the Mg matrix. When Al content is 3%, there is no Mg_2Sn diffraction peak in the alloy with 1% Sn. Nevertheless, when Sn content increases to 3%, relatively weak diffraction peak of Mg_2Sn appears at 2-Theta of 23 degrees. When Sn content further increases to 5%, the diffraction peak intensity of Mg_2Sn increases and relatively weak diffraction peak of $Mg_{17}Al_{12}$ appears. The results indicate that increasing Sn content leads to the decrease in the solid solubility of Al in Mg matrix.

The phase composition of Mg−6Al−ySn−0.3Mn alloy is similar to that of Mg−3Al−ySn−0.3Mn alloy (Fig. 5.22C). No diffraction peak of Mg_2Sn is detected in the alloy with 1% Sn. When the Sn content increases to 3%, relatively weak diffraction peak of Mg_2Sn appears at 2-Theta of 23 degrees. When the Sn content is 5%, the diffraction intensity of Mg_2Sn peak increases. All the alloys contain the phase of $Mg_{17}Al_{12}$, whose diffraction intensity increases with increasing Sn content. Mg−9Al−ySn−0.3Mn alloy contains three phases of $Mg_{17}Al_{12}$, Mg_2Sn, and α-Mg, possessing the same phase composition with Mg−6Al−ySn−0.3Mn alloy (Fig. 5.22C), where the diffraction intensities of both Mg_2Sn and $Mg_{17}Al_{12}$ phases increase with increasing Sn content.

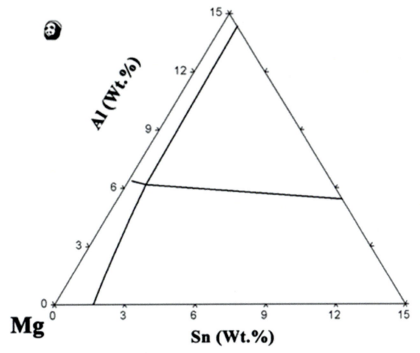

Figure 5.21 The isothermal section at 300°C for the Mg-rich end in the Mg–Al–Sn system.

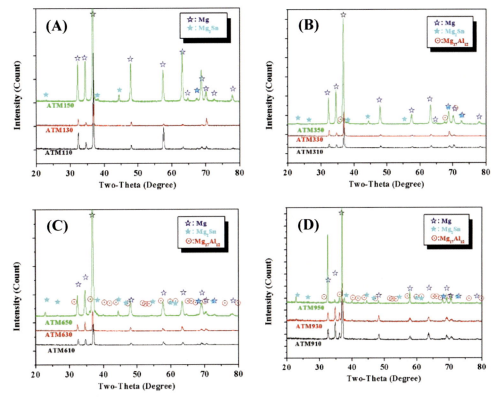

Figure 5.22 XRD patterns of as-cast Mg–Al–Sn–Mn alloys with Al content of (A) 1%, (B) 3%, (C) 6%, and (D) 9%.

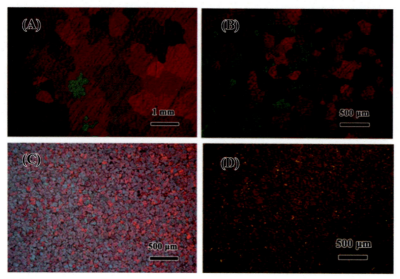

Figure 5.23 Microstructure of as-cast Mg−xAl−1Sn−0.3Mn alloy for (A) ATM110, (B) ATM310, (C) ATM610, and (D) ATM910.

(2) Microstructure of as-cast Mg−Al−Sn−Mn alloy

Fig. 5.23 shows the microstructure of the as-cast Mg−xAl−1Sn−0.3Mn alloy. When the Al content is less than 6%, the grain size decreases with increasing Al content. The average grain size is presented in Table 5.5. Compared with the grain size (>1 mm) of ATM110 alloy, the microstructure of ATM310 alloy is significantly refined (~ 300 μm). Combined with the SEM image shown in Figs. 5.24, AT5.M110 alloy has no obvious contrast; ATM310 alloy has a small amount of Mg_2Sn and $Mg_{17}Al_{12}$ phases at the grain boundaries. No Mg_2Sn is detected in XRD pattern, ascribing to the less content. The grain of ATM610 alloy is significantly refined (~ 68 μm). There are more divorced eutectic $Mg_{17}Al_{12}$ and a small amount of fine Mg_2Sn at the grain boundary. Most of Mg_2Sn are encapsulated by $Mg_{17}Al_{12}$, and they present the companionship. Since the melting point of Mg_2Sn is higher than that of $Mg_{17}Al_{12}$, Mg_2Sn has the possibility of being a heterogeneous nucleation point for the divorced eutectic $Mg_{17}Al_{12}$. Assuming that the high-melting-point Mg_2Sn is the heterogeneous nucleation point for the divorced eutectic $Mg_{17}Al_{12}$, Mg_2Sn will have a certain modification effect on the divorced eutectic $Mg_{17}Al_{12}$.

It is worth noting that a small amount of Al−Mn−Fe phase is found in this system, which is reported as $Al_8(Fe, Mn)_5$, as shown by the EDS results of ATM610 alloy. This result indicates the addition of Mn plays a role in removing the impurity of iron to a certain extent. Because the Mn content is very small, it can be concluded that all alloys contain a small amount of Al−Mn, and $Al_8(Fe, Mn)_5$ phases. It is difficult to

Table 5.5 Grain size of the as-cast alloys.

Samples	Grain size (μm)	Dendrite spacing (μm)
ATM110	1523	\
ATM310	314	\
ATM610	68	\
ATM910	63	\
ATM130	925	25
ATM330	286	13
ATM630	60	\
ATM930	58	\
ATM150	753	27
ATM350	357	11
ATM650	60	\
ATM950	56	\

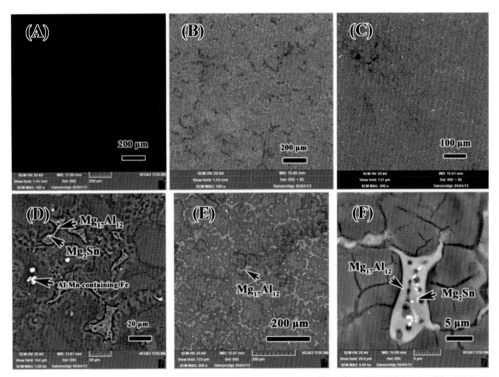

Figure 5.24 SEM images of as-cast Mg−xAl−1Sn−0.3Mn alloy for (A) ATM110, (B) ATM310, (C) ATM610, (D) ATM610 at high magnification, (E) ATM910, and (F) ATM910 at high magnification.

affect the phase relationship and content of the main second phases (Mg_2Sn and $Mg_{17}Al_{12}$) in the alloy. The Mn content is the same in all the alloys. The influence of Mn is ignored for convenience.

For the alloys with Al as the main alloying element, Al atoms are enriched at the front end of the solid/liquid interface during solidification according to the classic principle of atomic diffusion at the solid/liquid interface. With increasing Al concentration, an eutectic reaction occurs when the Al concentration reaches the Mg–Al eutectic point, and $Mg_{17}Al_{12}$ is formed at the solid/liquid interface. According to the principles of physical metallurgy, particles will effectively hinder the movement of the interface at high temperature. Assuming that the particles are spherical, the drag force of the particles on the interface per unit area can be expressed as:

$$B = \frac{nf\gamma}{2r} \qquad (5.1)$$

where, f is the volume fraction of grains, r is the radius of grains, γ is the interface energy, B is the drag force of grains to the interface per unit area, and n is a constant.

Therefore the $Mg_{17}Al_{12}$ phase increases with increasing Al content. That is, the increase in f leads to the increase of B, which ultimately hinders the growth of grains and refines the grains. On the other hand, the grain refinement from solid solution elements can be expressed by the growth limiting factor of $C_0m(k-1)$ (GRF), where m is the slope of the solid/liquid line of the binary phase diagram, C_0 is the concentration of alloying elements, and k is the scale factor. The GRF value of Al is 4.32. The grains are refined as the Al content increases for the Mg–xAl–1Sn–0.3Mn alloys. This refinement mechanism is similar with that of AM and AZ cast alloys.

Fig. 5.25 shows the microstructure of the as-cast Mg–xAl–3Sn–0.3Mn. The same trend with that of Mg–xAl–1Sn–0.3Mn, the grains are continuously refined as increasing Al content. The average grain size is presented in Table 5.5. When Al is 1% and 3%, the alloy has a dendritic microstructure, and the space between dendrites is $\sim$25 and $\sim$13 μm for the two alloys, respectively. When Al is larger than 6%, the alloy has an equiaxed microstructure. According to the SEM-BSE image shown in Fig. 5.26, when Al is 1% and 3%, Sn is mainly concentrated at the grain boundaries or between the dendrites. In addition, ATM330 alloy also contains a small amount of Mg_2Sn and $Mg_{17}Al_{12}$. More Mg_2Sn and divorced eutectic $Mg_{17}Al_{12}$ phases are distributed at the grain boundaries of the ATM630 alloy. According to high-resolution SEM-BSE (Fig. 5.26D), ATM630 alloy contains a coarser Mg_2Sn phase, presenting companionship with divorced eutectic $Mg_{17}Al_{12}$ (Fig. 5.26E shows the element distribution of the associated Mg_2Sn and the divorced eutectic $Mg_{17}Al_{12}$ by line-scan EDS). The ATM930 alloy has more second phases than ATM630 alloy, while the distribution is similar.

As shown in Fig. 5.27, the microstructure of as-cast Mg–xAl–5Sn–0.3Mn alloy is similar with that of alloys with 1% and 3% Sn. With increasing Al content, the grains of

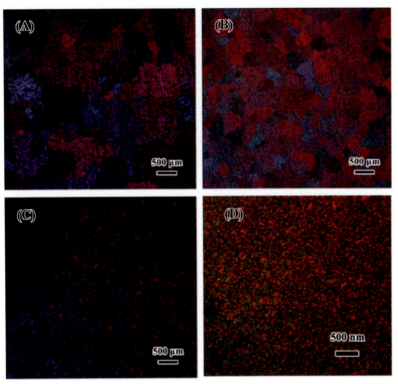

Figure 5.25 Microstructure of as-cast Mg–xAl–3Sn–0.3Mn alloys for (A) ATM130, (B) ATM330, (C) ATM630, and (D) ATM930.

the alloy are refined, and the average grain size is presented in Table 5.5. The microstructure of ATM150 and ATM350 alloy is dendrites. The distance between the dendrites of ATM150 and ATM130 alloy is almost the same (about 2 μm). When Al is larger than 6%, the microstructure of the alloy is equiaxed. Fig. 5.28 shows the SEM-BSE. Sn is mainly concentrated at the grain boundaries and between the dendrites in the ATM150 alloy. ATM350 contains a large amount of Mg_2Sn and $Mg_{17}Al_{12}$ phases, distributed at the grain boundaries, consistent with the XRD result. Coarse Mg_2Sn and divorced eutectic $Mg_{17}Al_{12}$ are associated in ATM650.

In summary, with increasing Al content in as-cast Mg–Al–Sn–Mn alloys with the constant Sn content, the microstructure changes from dendrites to equiaxed grains, the grain is refined, and the second phases (Mg_2Sn and $Mg_{17}Al_{12}$) increase in content. Mg_2Sn and divorced eutectic $Mg_{17}Al_{12}$ phases are distributed at the grain boundaries, with companionship.

By comparing the microstructure of Mg–1Al–ySn–0.3Mn alloys, the microstructure is equiaxed grain when the Al content is 1%. With increasing Sn content, the

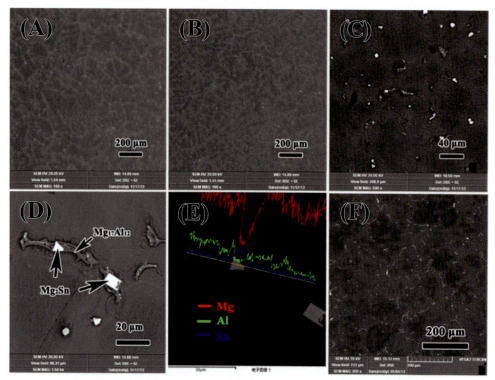

Figure 5.26 SEM images of Mg−xAl−3Sn−0.3Mn alloys for (A) ATM130, (B) ATM330, (C) ATM630, (D) ATM630 with high magnification, (E) EDS line scan of ATM630, and (F) ATM930.

microstructure changes to the dendritic grain. The Sn element is enriched at the grain boundary or between the dendrites (as shown in Figs. 5.26E and 5.28A). The formation of dendrite microstructure and the segregation of Sn element are related to the quickly decreased solid solubility of Sn in the Mg matrix with decreasing temperature. The solute enrichment thus occurs at the crystallization front during the solidification of the alloy, resulting in the constitutional supercooling.

The condition for the growth of dendrites in the supercooled region of the wider composition during the solidification of multi-component alloy is that the actual temperature gradient at the interface is less than or equal to the liquidus temperature gradient in the corresponding alloy phase diagram. Because the Mn content is very small in the alloy, the consumption of a small amount of Al has little effect on the phase relationship and the content of the main second phases (Mg_2Sn, $Mg_{17}Al_{12}$). For simplicity, only the main alloying elements of Mg−Al−Sn is considered during the solidification of the alloys (the influence of Mn is ignored in the discussion of solidification afterwards). Assuming that the solute distribution reaches equilibrium at the solid/liquid interface and

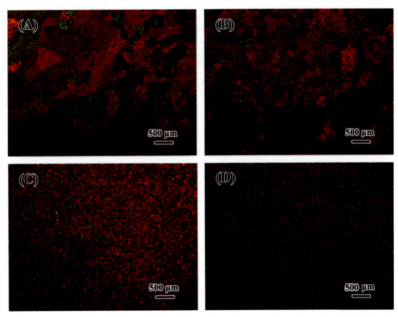

Figure 5.27 Microstructure of as-cast Mg−xAl−5Sn−0.3Mn alloys for (A) ATM150, (B) ATM350, (C) ATM650, and (D) ATM950.

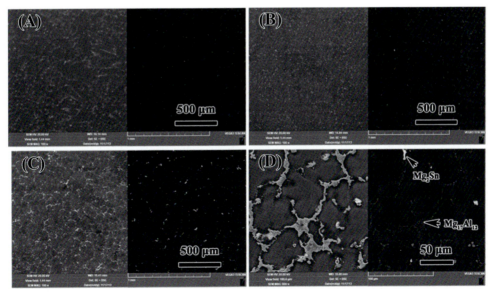

Figure 5.28 SEM images of as-cast Mg−xAl−5Sn−0.3Mn alloys for (A) ATM150, (B) ATM350, (C) ATM650, and (D) ATM950.

the solid phase does not diffuse; the liquid phase is mixed uniformly, no convection, and the criterion for the constitutional supercooling can be derived as follows:

$$\frac{G}{R} \leq -\frac{m_{Al}C_o^{Al}(1 - K_{Al})}{K_{Al}D_{Al}} - \frac{m_{Sn}C_o^{Sn}(1 - K_{Sn})}{K_{Sn}D_{Sn}} \tag{5.2}$$

where G is the temperature gradient of the melt at the interface, R is the grain growth rate, m_{Al} is the liquidus surface slope of the Mg–Al–Sn at the Al content of C_o^{Al}, m_{Sn} is the liquidus surface slope of the Mg–Al–Sn at the Sn content of C_o^{Sn}, C_o^{Al} and C_o^{Sn} are the initial contents of Al and Sn, respectively, in the alloy. K_{Al} and K_{Sn} are the equilibrium partition coefficients and D_{Al} and D_{Sn} are the equilibrium diffusion coefficient of Al and Sn, respectively.

$$\Delta T_o^{Al} = -\frac{m_{Al}C_o^{Al}(1 - K_{Al})}{K_{Al}} \tag{5.3}$$

$$\Delta T_o^{Sn} = -\frac{m_{Sn}C_o^{Sn}(1 - K_{Sn})}{K_{Sn}} \tag{5.4}$$

In the above two formulas, ΔT_o^{Al} and ΔT_o^{Sn} are the crystallization temperature intervals of the Al-rich and Sn-rich solid-liquid surfaces of the Mg–Al–Sn phase diagram, respectively.

Substituting Eqs. (5.3) and (5.4) to Eq. (5.5), then

$$\frac{G}{R} \leq \frac{\Delta T_o^{Al}}{D_{Al}} + \frac{\Delta T_o^{Sn}}{D_{Sn}} \tag{5.5}$$

Therefore when the Al content is 1%, the C_o^{Sn} increases with increasing Sn content with a certain G/R, and the crystallization temperature interval ΔT_o^{Sn} of the alloy also increases. Consequently, the constitutional supercooling increases, which is conductive to the formation of dendrites. If there is any protrusion at the solid/liquid interface during the solidification process, the growth of dendrite is further accelerated. Meanwhile, the excessive solute element of Sn is continuously "discharged" from the surrounding melt. It is more difficult for solute in the concave area diffuses to the liquid phase than that in the convex area. The increase in the solute element of Sn in the concave area is thus very fast. The rapid growth of the convex area results in the continuous enrichment of Sn in the concave area, finally resulting in the enrichment of Sn between the dendrites (as shown in the schematic diagram of Fig. 5.29).

In addition, the formation of new crystal nuclei in the liquid phase close to the front of the interface can be accelerated when the supercooling before the solid/liquid interface exceeds the supercooling degree required for the nucleation of crystal nuclei. At the same time, the dendrite branch is formed into a narrow neck, which is easy to

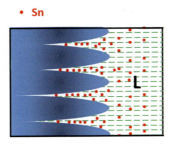

Figure 5.29 Schematic diagram for the growth of dendrite.

be melt and fall off, refining the grain. Therefore in the alloy with 1% Al, its grain size undergoes a certain degree of refinement with increasing Sn content.

In alloys with high Al content (>3%), it is indicated in the phase diagram that the eutectic temperature of Mg—Al—Sn is very low (428°C L→Mg$_2$Sn+Mg$_{17}$Al$_{12}$+α-Mg). With increasing Al content, the C_o^{Al} increases andthe ΔT_o^{Al} increases significantly, resulting in the highly increased supercooling degree. It in turn leads to the formation of narrow necks in dendrite branch, and promotes the microstructure transformation from dendrite to equiaxed crystal.

The grains of all Mg—6Al—ySn—0.3Mn alloys are relatively small, and the grain sizes of ATM610, ATM630, and ATM650 are approximately 125, 110, and 90 μm, respectively. With the increase in Sn content, the grains are refined to a certain extent, but the refinement effect of Sn is weaker than Al. By comparing the SEM-BSE images of alloys with 6% Al and different Sn contents, the Mg$_2$Sn is associated with the divorced eutectic Mg$_{17}$Al$_{12}$, and almost all Mg$_2$Sn is wrapped by Mg$_{17}$Al$_{12}$. Since the GRF value of Sn is very small, the grain refinement effect caused by solid solubility is not noticeable. The alloying element of Sn is mainly precipitated in the form of the second phase at the front of the solid/liquid interface during the solidification. Although the volume fraction of Mg$_2$Sn increases with increasing Sn content, Mg$_2$Sn is wrapped by divorced eutectic crystal of Mg$_{17}$Al$_{12}$ or they are precipitated adjacent to each other. The Mg$_2$Sn is not precipitated individually, so the grain refinement effect of Sn is not obvious.

The microstructure evolution of Mg—9Al—ySn—0.3Mn alloy with Sn content is the same with that of Mg—6Al—ySn—0.3Mn alloy. As the Sn content increases, the grains are refined to a certain extent, the Mg$_2$Sn phase increases in alloy, and almost all Mg$_2$Sn is associated with Mg$_{17}$Al$_{12}$.

In summary, when Al content is smaller than 3%, Sn is mainly enriched at grain boundaries or between dendrites; when Al content is larger than 6%, the microstructure changes from dendrite to equiaxed grains, and the grains are refined with increasing Sn content.

5.3.2 Property of the as-cast Mg−Al−Sn−Mn alloy

(1) Effect of Al content on the property of Mg−Al−Sn−Mn alloy

Fig. 5.30 shows the stress-strain curves of casting alloys with the same Sn content but different Al contents, and the corresponding mechanical properties are presented in Table 5.6. When the Sn content is constant, the yield strength of the alloy increases with increasing Al content. This is resulted from the combined effects of fine grain strengthening, solid solution strengthening, and second phase strengthening.

1) Fine grain strengthening

From the grain size statistics shown in Table 5.5, the grain size of the alloy decreases with increasing Al content. According to the formula of Hall-Petch:

$$\sigma = \sigma_0 + Kd^{-1/2} \tag{5.6}$$

where σ is the yield strength of alloy, σ_0 is the lattice resistance to the dislocation motion (usually related to the number of types of alloying elements dissolved in

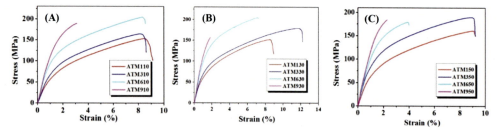

Figure 5.30 Stress-strain curves of casting Mg−Al−Sn−Mn alloys with the constant Sn content and different Al contents.

Table 5.6 The mechanical properties of Mg−Al−Sn−0.3Mn casting alloy at room temperature.

Samples	YS (MPa)	UTS (MPa)	$\varepsilon\%$	f_v %	$C_{Al\ at.\%}$	$C_{Sn\ at.\%}$	Calc. σ(MPa)
ATM110	56	154	8.9	0.5	1.26	0.17	52.57
ATM310	73	165	7.5	1.1	1.51	0.16	70.06
ATM610	108	204	7.4	5.6	2.26	0.16	115.07
ATM910	132	190	1.7	8.7	3.26	0.17	150.53
ATM130	61	151	7.9	1.8	1.26	0.18	54.99
ATM330	78	179	11.1	2.3	1.88	0.17	83.64
ATM630	116	204	6	6.7	2.39	0.16	121.78
ATM930	140	157	0.55	9.1	3.38	0.14	155.98
ATM150	68	161	9.1	2.5	1.26	0.20	56.53
ATM350	85	190	8.1	3.6	2.01	0.19	86.48
ATM650	121	180	2.6	7.2	2.51	0.19	126.37
ATM950	156	184	0.8	10	3.767	0.17	169.75

the matrix and the type, number, and shape of the precipitation-strengthened second phase), equal to the yield strength of Mg solid solution in this system, and d is the average grain diameter, K is the fine grain strengthening coefficient, whose value usually ranges from 170 to 400 MPa $(\mu m)^{1/2}$. The coefficient depends on the density and stability of the dislocation source at grain boundaries and is also affected by the segregation of solute atoms at grain boundaries. Al does not segregate in Mg matrix. K can be considered as a constant. The grain size decreases with increasing Al content. The yield strength of alloy will continuously increase according to Eq. (5.6).

2) Solid solution strengthening

The Al amount of solid solution in Mg matrix increases with increasing Al content. According to the equation:

$$\sigma_0 = kC^{2/3} \tag{5.7}$$

where k is a constant, C is the solid solution amount of the solute atom. The C increases with increasing Al content, so the yield strength of the alloy gradually increases with increasing Al content.

3) Second phase strengthening

From the TEM image of ATM950 alloy shown in Fig. 5.31, besides the microscale primary second phase, the alloy contains nanoscale precipitates, which is $Mg_{17}Al_{12}$ according to the EDS result. Combined with the XRD and SEM results, it is inferred that this precipitated phase increases with increasing Al content. According to the formula:

$$\sigma = \frac{Gb}{2\pi\sqrt{1-v}\left(\frac{0.953}{\sqrt{fv}} - 1\right)dt}\ln\frac{d_t}{b} \tag{5.8}$$

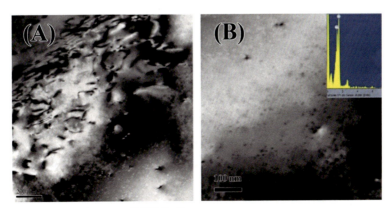

Figure 5.31 TEM images of as-cast ATM950 alloy.

where G is the shear modulus of the matrix, which is generally 16.6 GPa, b is the Burgers vector, about 3.21×10^{-10} m, ν is the Poisson's ratio (0.35), d is the diameter of the precipitated phase, and f_v is the volume fraction of the precipitated phase. The second phase continuously increases with increasing Al content, and the strength of the alloy will increase with the increase of the second phase.

In the Mg−Al−Sn−Mn casting alloy with the same Sn content, the combined effect of the above three strengthening mechanisms significantly increases the yield strength of the alloy as the increase of Al content. The tensile strength and elongation of the alloy depends on whether the alloy contains coarse second phase or not. When the Al content is less than 6%, the alloy does not contain coarse second phases (Mg_2Sn and $Mg_{17}Al_{12}$); the tensile strength of the alloy increases with increasing Al content, but the elongation does not change a lot. When the Al content is more than 6%, the alloy contains coarse second phases (Mg_2Sn and $Mg_{17}Al_{12}$), and the tensile strength and the elongation decrease with increasing Al content.

In summary, as for the casting alloys with the same Sn content and different Al contents, the yield strength of the alloy increases monotonically with increasing the Al content. The tensile strength and elongation of the alloy depend on whether there is coarse second phase in the alloy. The alloy possesses good elongation and its tensile strength increases with increasing Al content when the alloy has no coarse second phase of Mg_2Sn or $Mg_{17}Al_{12}$. When there is coarse second phase in the alloy, the plasticity of the alloy decreases and the tensile strength decreases as well.

(2) Effect of Sn content on the property of Mg−Al−Sn−Mn alloy

Fig. 5.32 shows the line chart of the yield strength of casting alloys with the same Al content and different Sn content. With the increase in the Sn content, the yield strength of the alloy increases to a certain extent, but the magnitude remains small (between 5−10 MPa). In addition, the increment resulted from Sn is smaller than that from Al.

By comparing the metallographic photographs, it can be found that the grain refinement by Sn is weaker than that of Al. Combining the solid solution strengthening of the Mg−Sn binary alloy with the solid solution strengthening effect of the Mg−Al binary alloy, according to formula of 3.9: $\sigma_{Al} = 197.5 \cdot C^{2/3}$, $\sigma_{Sn} = 286 \cdot C^{2/3}$, the solid solution strengthening of Al in Mg matrix is smaller than that of Sn. However, at low temperature, the solid solubility of Al in Mg matrix is large, but the solid solubility of Sn in Mg matrix is very small. Therefore the solid solution strengthening of Sn is weaker than that of Al. Moreover, from the XRD and SEM results, a majority of Sn is segregated between the dendrites in the alloy with low Al content (1% and 3%), and the phase content of Mg_2Sn is low. Although a large amount of Mg_2Sn is formed in alloys with high Al content, a majority of Mg_2Sn is wrapped by the divorced eutectic $Mg_{17}Al_{12}$ phase (the schematic diagram of microstructure of the alloys with high Al content is shown in Fig. 5.33). The divorced eutectic $Mg_{17}Al_{12}$ phase is refined to a certain extent with increasing Sn content.

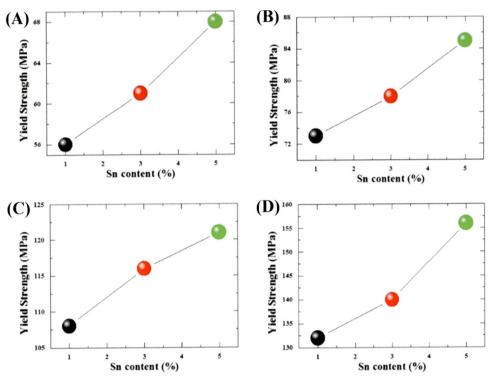

Figure 5.32 Influence of the Sn content on the yield strength of Mg—Al—Sn—Mn system: (A) Mg—1Al—ySn—0.3Mn, (B) Mg—3Al—ySn—0.3Mn, (C) Mg—6Al—ySn—0.3Mn, and (D) Mg—9Al—ySn—0.3Mn.

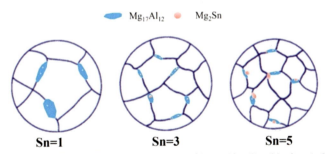

Figure 5.33 Schematic diagram of the microstructure of Mg—Al—xSn—Mn ($x = 1$, 3, and 5) alloy.

The tensile strength of the alloy depends on whether the alloy contains coarse second phases. When the Al content is less than or equal to 3%, the alloy does not contain coarse second phases (Mg_2Sn and $Mg_{17}Al_{12}$). The tensile strength and elongation of the alloy thus do not change obviously with the increase of Sn content. When the

Al content is not lower than 6%, the alloy contains coarse second phase, and the tensile strength and the elongation decrease with increasing Sn content.

(3) Calculation of strength theory

According to the solid solution strengthening theory of the multi-component alloy:

$$\Delta\sigma = \left(\sum_i k_i^{1/n} c_i\right)^n \tag{5.9}$$

where c_i and k_i are the concentration and solid solution strengthening coefficient of solute element i, respectively, and n is 2/3. For the as-cast alloys, the contribution of solid solution strengthening to the yield strength of the alloy can be expressed as:

$$\sigma = \left(k_{Al}^{1.5} c_{Al} + k_{Sn}^{1.5} c_{Sn}\right)^{2/3} \tag{5.10}$$

As for the casting alloys, the fine grain strengthening, solid solution strengthening, and the second phase strengthening have significant effect on the strength of the alloy. The diameter of the precipitated phase is set as 20 nm (ignore the influence of newborn second phase on the strength), so the yield strength of the alloy can be expressed as:

$$\sigma \approx Kd^{-1/2} + \left(k_{Al}^{1.5} c_{Al} + k_{Sn}^{1.5} c_{Sn}\right)^{2/3} + \frac{Gb}{2\pi\sqrt{1-v}\left(\frac{0.953}{\sqrt{f_v}} - 1\right)d_t}\ln\frac{d_t}{b} \tag{5.11}$$

K is 300 MPa $(\mu m)^{1/2}$, $\sigma_{Sn} = 286 \cdot C^{2/3}$, and $\sigma_{Al} = 197 \cdot C^{2/3}$. Taking the grain size, volume fraction of the second phase, and solid solubility presented in Table 5.6 into Eq. (5.11), the strength of the alloy is calculated, which is in good agreement with the experimental data. The calculated results are shown in Table 5.6. The error between the theoretical calculation and the actual one is mainly resulted from the following factors: (1) The used model is simplified, which ignores the influence of newborn second phase on the strength; (2) there are certain errors in the selection of empirical parameters, such as the selection of K value; 3) there are errors in the experimental parameters, such as grain size, volume fraction of the second phase, and solid solubility.

5.3.3 Effect of Sn on the microstructure and property of AM alloy

The microstructure of Mg$_2$Sn assigns to BCC structure, with a lattice constant of $a = 0.6763$ nm, belonging to the FM$\bar{3}$M (225) space group. It has the following close-packed planes and nearly close-packed planes of (111), (220), and (200). The atomic occupancy is shown in Table 5.7.

The microstructure of Mg$_{17}$Al$_{12}$ assigns to FCC structure, with a lattice constant of $a = 1.056$ nm, belonging to the I$\bar{4}$3M (217) space group. It has the following close-

Table 5.7 The atomic occupancy of Mg_2Sn phase.

Mg₂Sn; Fm-3m; Space group: 225; lattice parameter $a = 0.6739$ nm.

Atom	Position	x	y	z	Occupancy
Mg	8c	0.25	0.25	0.25	1
Sn	4a	0	0	0	1

Table 5.8 The atomic occupancy of $Mg_{17}Al_{12}$ phase.

Mg₁₇Al₁₂; I-43m; Space group: 217; lattice parameter $a = 1.05438$ nm.

Atom	Position	X	y	Z	Occupancy
Mg1	2a	0	0	0	1
Mg2	8c	0.3240(15)	0.3240(15)	0.3240(15)	1
Mg3	24g	0.3582(8)	0.3582(8)	0.0393(14)	1
Al	24g	0.0954(14)	0.0954(14)	0.2725(19)	1

packed planes and nearly close-packed planes of (321), (330), and (400). The atomic occupancy is shown in Table 5.8.

The formula for calculating the mismatch is as follows:

$$F_d = |(d_M - d_p)/d_M| \times 100\% \tag{5.12}$$

$$F_r = |(r_M - r_p)/r_M| \times 100\% \tag{5.13}$$

where F_d is plane mismatch, F_r is line mismatch, d_M and d_p are correspondingly the plane spacing between the two close-packed plane or nearly close-packed plane, and r_M and r_p are the atomic distance along the close-packed direction or nearly close-packed direction of the two close-packed planes.

According to the calculated plane mismatch between Mg_2Sn and $Mg_{17}Al_{12}$, the mismatch between the (220) plane of Mg_2Sn and the (400) plane of $Mg_{17}Al_{12}$ is 9.43%; the mismatch between the (220) plane of Mg_2Sn and the (330) plane of $Mg_{17}Al_{12}$ is merely 3.57%. Therefore the habit plane of Mg_2Sn and $Mg_{17}Al_{12}$ may be $(220)_{Mg2Sn}//(400)_{Mg17Al12}$ or $(220)_{Mg2Sn}//(330)_{Mg17Al12}$. The plane mismatch of Mg_2Sn and $Mg_{17}Al_{12}$ is shown in Table 5.9.

By calculating the mismatch along the close-packed direction on the nearly close-packed plane with a mismatch less than 10%, no close-packed direction with a mismatch less than 10% was found on the (220) close-packed plane of Mg_2Sn and the (400) close-packed plane of $Mg_{17}Al_{12}$. Fig. 5.34 shows the atomic arrangement of the (330) plane of the $Mg_{17}Al_{12}$ phase. The three close-packed directions are $<1\bar{1}\bar{1}>^z$

Table 5.9 The mismatch of the possible habit plane for Mg₂Sn and Mg₁₇Al₁₂, ignoring the mismatch larger than 30%.

Plane pairs	$\{200\}_{Mg2Sn}$ $//\{321\}_{Mg17Al12}$	$\{220\}_{Mg2Sn}$ $//\{321\}_{Mg17Al12}$	$\{220\}_{Mg2Sn}$ $//\{400\}_{Mg17Al12}$	$\{220\}_{Mg2Sn}$ $//\{330\}_{Mg17Al12}$
Mismatch	20.78	14.6	9.43	3.59

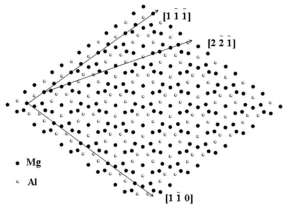

Figure 5.34 The atomic arrangement of (330) plane of Mg₁₇Al₁₂, with a lattice constant of $a = 1.05438$ nm.

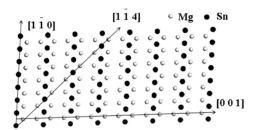

Figure 5.35 The atomic arrangement of (220) plane of Mg₂Sn, with a lattice constant of $a = 0.6739$ nm.

Mg17Al12, $<1\bar{1}0>^z$ Mg17Al12, and $<2\bar{2}\bar{1}>^z$ Mg17Al12. As shown in Fig. 5.35, the three close-packed directions of the (220) plane of the Mg₂Sn are $<1\bar{1}0>^S$ Mg2Sn, $<001>^S$ Mg2Sn, and $<1\bar{1}4>^S$ Mg2Sn; where S is the atomic linear close-packed model and Z is the Z-typed close-packed model.

By calculating the mismatch along the close-packed direction on these two close-packed planes, the close-packed direction with mismatch less than 10% is $<001>^S$ Mg2Sn$//<2\bar{2}\bar{1}>^z$ Mg17Al12 (as shown in Table 5.10). According to the edge-to-edge

Table 5.10 The mismatch of the possible matched crystallographic direction between $Mg_{17}Al_{12}$ and Mg_2Sn.

Direction pairs	$<001>{}^s Mg_2Sn//$ $<111>{}^z Mg17Al12$	$<001>{}^s Mg_2Sn//$ $<121>{}^z Mg17Al12$	$<001>{}^s Mg_2Sn//$ $<110>{}^z Mg17Al12$	$<114>{}^s Mg_2Sn//$ $<221>{}^z Mg17Al12$	$<114>{}^s Mg_2Sn//$ $<111>{}^z Mg17Al12$	$<114>{}^s Mg_2Sn//$ $<110>{}^z Mg17Al12$
Misfit (%)	10.69	8.83	10.69	31.95	34.2	34.2

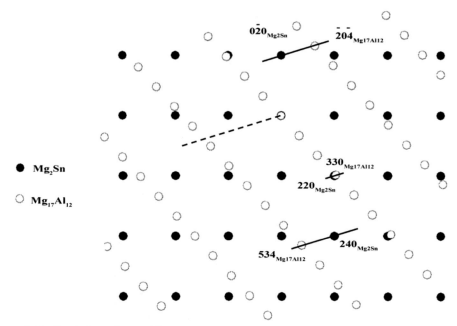

Figure 5.36 Simulation of axis diffraction pattern of the $Mg_{17}Al_{12}$ $<2\bar{2}\bar{1}>$ and Mg_2Sn $<001>$, where the dotted line shows the habit plane.

model, the habit planes of Mg_2Sn and $Mg_{17}Al_{12}$ are $(220)_{Mg2Sn}//(330)_{Mg17Al12}$ and $<001>^S_{Mg2Sn}//<2\bar{2}\bar{1}>^z_{Mg17Al12}$.

By simulating the diffraction pattern (as shown in Fig. 5.36) under the axis of $<001>^S_{Mg2Sn}//<2\bar{2}\bar{1}>^z_{Mg17Al12}$, it is found that the final habit plane is $(220)_{Mg2Sn}//(330)_{Mg17Al12}$ and $<001>^S_{Mg2Sn}//<2\bar{2}\bar{1}>^z_{Mg17Al12}$.

According to the DSC curve shown in Fig. 5.37, Mg_2Sn is formed at 500°C, and $Mg_{17}Al_{12}$ is formed at approximately 450°C. At the same time, there exist habit planes in Mg_2Sn and $Mg_{17}Al_{12}$ alloys. Therefore Mg_2Sn is the heterogeneous nucleation point of $Mg_{17}Al_{12}$, and Mg_2Sn can effectively improve the morphology of the divorced eutectic $Mg_{17}Al_{12}$ phase.

It is not difficult to find that adding a large amount of Sn cannot effectively increase the strength of the Mg−Al−Sn−Mn alloy, according to the results of the properties for the previous casting alloys. In addition, since Sn is expensive, it is of great significance to evaluate the influence of a small amount of Sn on the microstructure and property of the Mg−Al−Mn (AM) alloy.

The fabrication of AM60 and AM90, and ATM610 and ATM910 alloys was applied the same casting procedure. Fig. 5.38 shows the engineering stress-strain curves of these four alloys. The mechanical properties are presented in Table 5.11. The results

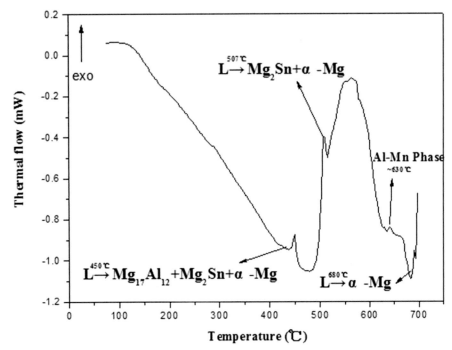

Figure 5.37 The DSC curve of ATM630 alloy.

indicate the Sn-containing alloys has higher strength than the Sn-free alloys. The yield strength and tensile strength of ATM610 are increased by about 13 and 47 MPa, respectively, compared with those of AM60. The yield strength and tensile strength of ATM910 are higher than that of AM90 alloy by about 17 and 25 MPa, respectively. The elongation of Sn-containing alloy is larger than that of Sn-free alloy.

Comparing the grain size (as shown in Fig. 5.39) of ATM and AM alloys, the grains of Sn-containing alloys are significantly smaller than that of the Sn-free alloys. The average grain size of AM60, AM90, ATM610, and ATM910 alloys are 115, 82, 68, and 62 μm, respectively. According to the SEM images of these four alloys in Fig. 5.40, the divorced eutectic $Mg_{17}Al_{12}$ of the Sn-containing alloys is finer and more dispersed than that of Sn-free divorced eutectic $Mg_{17}Al_{12}$. Sn has some modification effect on the divorced eutectic $Mg_{17}Al_{12}$ in AM alloys. Adding a small amount of Sn to the AM alloy can effectively improve the morphology of the divorced eutectic $Mg_{17}Al_{12}$ phase, refine the grains and improve the strength and plasticity of the alloy. Sn is thus a very effective alloying element (Table 5.11).

In the Mg—Al—Sn—Mn casting alloy, since Mg_2Sn is the heterogeneous nucleation site of $Mg_{17}Al_{12}$, Mg_2Sn induces the formation of $Mg_{17}Al_{12}$ at the solid/liquid interface during the solidification. The morphology of $Mg_{17}Al_{12}$ can be effectively

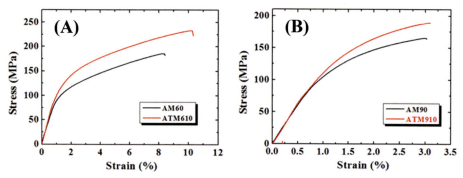

Figure 5.38 Stress-strain curves of (A) AM60 and ATM610, and (B) AM90 and ATM910.

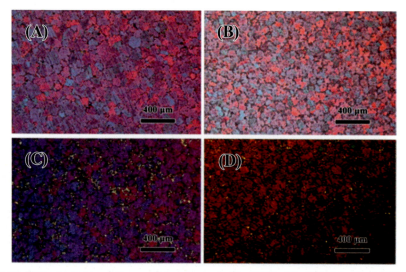

Figure 5.39 Microstructure of the casting alloys of (A) AM60, (B) ATM610, (C) AM90, and (D) ATM910.

improved even with a small amount of Sn. On the other hand, the Mg_2Sn can be wrapped by the divorced eutectic $Mg_{17}Al_{12}$, indirectly refining the grains. Therefore Sn can only play a role in alloys with a large amount of Al; in alloys with less Al content ($\leq$3%), Sn segregates at grain boundaries or dendrites. Although a large amount of Sn is added, the grain refinement and strength improvement of the alloy are not significant. This is also the reason why the Al content should be high in the Mg—Al—Sn—Mn casting alloy. In summary, considering the cost, the composition of Mg—Al—Sn—Mn casting alloy with better comprehensive properties is: Al content ranges from 6% to 9%, Sn content ranges from 1% to 3%, and Mn content is 0.3%.

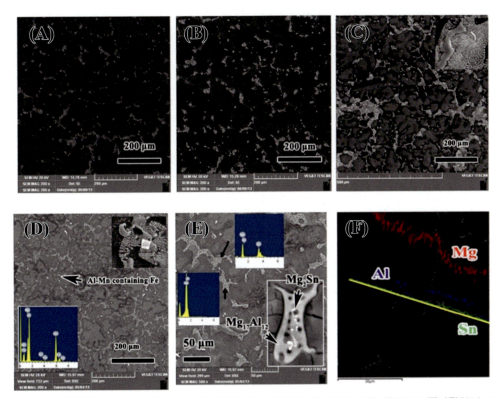

Figure 5.40 SEM images of alloys of (A) AM60, (B) ATM610, (C) AM90, (D) ATM910, (E) ATM910 at high magnification, and (F) line-scan image of ATM910.

Table 5.11 The mechanical properties and grain size of AM60, AM90, ATM610, and ATM910 casting alloys.

Samples	YS (MPa)	UTS (MPa)	$\varepsilon\%$	Grain size (μm)
AM60	95	185	8.5	115
AM90	115	165	3.0	82
ATM610	108	232	7.4	68
ATM910	132	190	1.7	62

5.4 Microstructure and property of the wrought Mg—Al—Sn—Mn alloy

5.4.1 Effect of Al on the microstructure and property of the as-extruded Mg—Al—Sn—Mn alloy

Fig. 5.41 shows the SEM images of alloys under the homogeneous state, in which the second phase was dissolved into Mg matrix after the homogenization treatment. Since

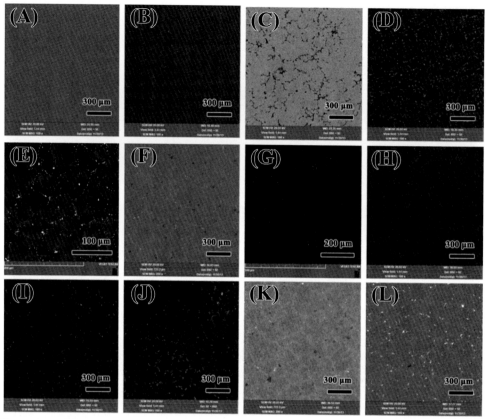

Figure 5.41 SEM images for (A) ATM110, (B) ATM130, (C) ATM150, (D) ATM310, (E) ATM330, (F) ATM350, (G) ATM610, (H) ATM630, (I) ATM650, (J) ATM910, (K) ATM930, and (L) ATM950 in homogeneous state.

the solubility of Al in Mg matrix is relatively high, the diffusion rate of Al in Mg is fast. The melting point of $Mg_{17}Al_{12}$ is relatively low, so it is almost dissolved into Mg matrix. Because Mg_2Sn has a high melting point and thermal stability as well, Mg_2Sn is the heterogeneous nucleation site of $Mg_{17}Al_{12}$ phase in the casting alloy, and distributes in the $Mg_{17}Al_{12}$ phase. After the $Mg_{17}Al_{12}$ phase is dissolved in Mg matrix, a part of fine Mg_2Sn grains are dispersed near the grain boundaries and some coarse Mg_2Sn still remain in the homogenous sample.

Fig. 5.42 shows the XRD pattern of the extruded alloy perpendicular to the extrusion direction. The XRD pattern of Mg–xAl–1Sn–0.3Mn alloy has no Mg_2Sn phase. ATM110 and ATM310 alloys merely contain α-Mg phase. When the Al content in alloy is larger than 6%, the diffraction peak of $Mg_{17}Al_{12}$ appears at 2-Theta of 36 degrees, and the alloy contains α-Mg and $Mg_{17}Al_{12}$ phase. In comparison, the diffraction intensity of $Mg_{17}Al_{12}$ phase in ATM910 alloy is larger than that in

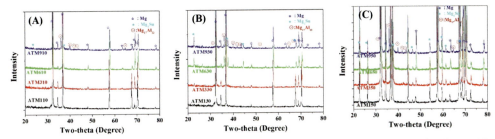

Figure 5.42 XRD pattern of the extruded alloys of (A) Mg—xAl—1Sn—0.3Mn, (B) Mg—xAl—3Sn—0.3Mn, and (C) Mg—xAl—5Sn—0.3Mn.

ATM610 alloy, indicating the $Mg_{17}Al_{12}$ phase increases with increasing Al content. On the other hand, the diffraction peak of α-Mg phase shifts to the right with increasing Al content, indicating the increased solid dissolution of Al in Mg matrix.

According to the XRD pattern of Mg—xAl—3Sn—0.3Mn alloy, the alloy has no $Mg_{17}Al_{12}$ when the Al content is less than 3%; the alloy contains $Mg_{17}Al_{12}$ phase when the Al content is larger than 6%. As the same with the Mg—xAl—1Sn—0.3Mn system, the $Mg_{17}Al_{12}$ phase increases with increasing Al content. It is worth noting that a weak diffraction peak of Mg_2Sn appears at 2-Theta of 23 degrees in ATM330 alloy, and its intensity increases with increasing Al content. The above results indicate the increasing Al content increases the precipitation of $Mg_{17}Al_{12}$ phase, and promotes the precipitation of Mg_2Sn phase as well. The content of Mn is 0.3% in the alloy. The content is relatively small, and the phase containing Mn is not detected in all the alloys.

According to the isothermal section of Mg—Al—Sn alloy at 300°C, the solid solubility of Sn in Mg matrix decreases with increasing Al content in the solid solution area of α-Mg phase. It can be understood that Al occupies the solid solution position of Sn, which promotes the precipitation of Mg_2Sn (consistent with the phenomenon of casting alloys, see Fig. 5.43).

The phase composition and the change of second phase content in the alloys with 5% Sn (Mg—xAl—5Sn—0.3Mn) is similar to the system with Sn content of 3%. All alloys contain Mg_2Sn phase. In addition to the ATM150 alloy that has no $Mg_{17}Al_{12}$ phase, other alloys contain three phases of α-Mg, Mg_2Sn, and $Mg_{17}Al_{12}$.

(1) Microstructure of as-extruded Mg—Al—Sn—Mn alloy

The microstructure of Mg—xAl—1Sn—0.3Mn alloy along the extrusion direction is shown in Fig. 5.44. ATM110 and ATM310 alloys are not completely recrystallized. The alloy contains fine recrystallization microstructure and coarse un-recrystallization microstructure elongated along the extrusion direction. The un-recrystallization area of ATM310 alloy is smaller than that of ATM110 alloy. ATM610 and ATM910 have been completely recrystallized. The average grain size of ATM610 and ATM910 are 10 and 9.6 μm, respectively.

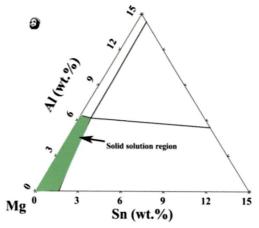

Figure 5.43 Isothermal section at 300°C for Mg−Al−Sn alloy.

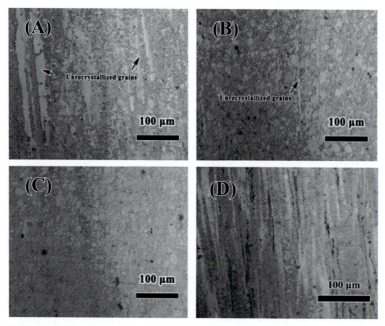

Figure 5.44 Microstructure along the extrusion direction of the Mg−xAl−1Sn−0.3Mn alloys for (A) ATM110, (B) ATM310, (C) ATM610, and (D) ATM910.

Fig. 5.45 shows the SEM image of Mg−xAl−1Sn−0.3Mn alloy along the extrusion direction. The second phase increases with increasing Al content in this system. The alloy contains more $Mg_{17}Al_{12}$ phase when the Al content is 6%, consistent with

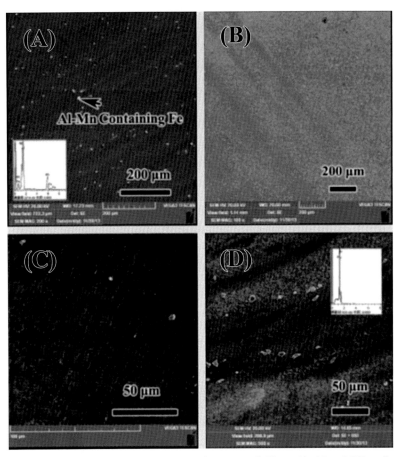

Figure 5.45 SEM images along the extrusion direction of Mg—xAl—1Sn—0.3Mn alloys for (A) ATM110, (B) ATM310, (C) ATM610, and (D) ATM910.

the XRD results. The phase of $Al_8(Fe, Mn)_5$ in the casting alloy remains in the extruded alloy.

In order to further determine the phase of Mn in Mg—Al—Sn—Al alloy, ATM110 was studied using TEM, with the results shown in Fig. 5.46. The alloy contains relatively coarse Al_8Mn_5 phase, non-coherent with the matrix, and contains dot-like Al_8Mn_5 phase (diameter ranges from 20 to 50 nm). The Al_8Mn_5 phase structure is rhombohedral ($a = 1.2645$ nm, $c = 1.5855$ nm). By comparing the standard PDF card (48—1568), the orientation relationship between the precipitated Al_8Mn_5 phase and Mg matrix is $(\bar{2}1\bar{1}0)_{Mg}//(20\bar{2}0)_{Al8Mn5}$, $[2\bar{4}2\bar{3}]_{Mg}//[01\bar{1}1]_{Al8Mn5}$.

In summary, a small amount of Mn in the Mg—Al—Sn—Mn alloy exists as a coarse Al_8Mn_5 phase, a certain amount of Mn exists in the form of the fine precipitated

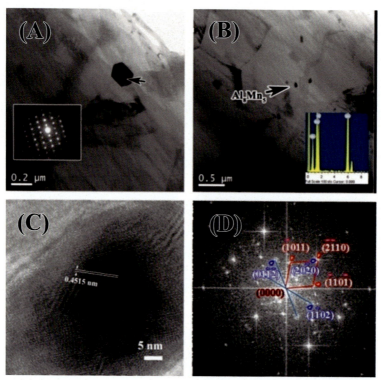

Figure 5.46 Bright-field images of ATM110 alloy for (A) diffraction pattern of the coarse Mn phase, (B) Al_8Mn_5 and its corresponding DES result, (C) Al_8Mn_5 image at high magnification, and (D) Fourier transformation diagram of Al_8Mn_5 and matrix.

Al_8Mn_5 phase (beneficial to improve the strength of the alloy), and a part of Mn and Fe form a compound of $Al_8(Fe, Mn)_5$, removing the impurity element of Fe.

The metallographic image of the $Mg-xAl-3Sn-0.3Mn$ alloy along the extrusion direction is shown in Fig. 5.47. The ATM130 and ATM330 alloys are not completely recrystallized. The alloy contains fine recrystallized grains and coarse un-recrystallized ones. Similar to the $Mg-xAl-1Sn-0.3Mn$ system, the coarse un-recrystallized region is elongated along the extrusion direction. The ATM330 alloy has less un-recrystallized area than ATM130. ATM330 alloy has smaller recrystallized grains than that of ATM130. Grain refinement can effectively improve the mechanical properties of the alloy. It can be speculated that the strength of ATM330 alloy is larger than that of ATM130 alloy. ATM630 and ATM930 alloys are completely recrystallized. The average grain size of ATM630 and ATM930 are 8.5 and 7.8 µm, respectively.

According to Fig. 5.48, the SEM image of $Mg-xAl-3Sn-0.3Mn$ alloy indicates ATM130 alloy contains a small amount of fine Mg_2Sn phase. With increasing Al content, the second phase (Mg_2Sn and $Mg_{17}Al_{12}$) content in $Mg-xAl-3Sn-0.3Mn$ alloy

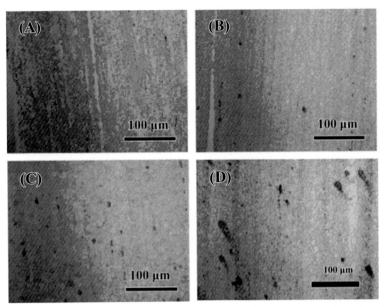

Figure 5.47 Microstructure of Mg—xAl—3Sn—0.3Mn alloy along the extrusion direction for (A) ATM130, (B) ATM330, (C) ATM630, and (D) ATM910.

increases, in consistent with the XRD result. From the high-resolution SEM images of ATM630 and ATM930 alloys, together with the EDS results, it is clear that the morphology of Mg_2Sn is granular (about 1 μm), and the morphology of $Mg_{17}Al_{12}$ is rod-shaped.

The metallographic structure along the extrusion direction of Mg—xAl—5Sn—0.3Mn alloy is shown in Fig. 5.49. The ATM150 and ATM350 alloys and the systems containing 1% and 3% Sn are not completely recrystallized. The alloys contain fine recrystallized grains and coarse un-recrystallized regions. The un-recrystallized region of ATM350 alloy is less than that of ATM150 alloy. ATM650 and ATM950 alloys are completely recrystallized. The average grain size of ATM650 and ATM950 are 8 and 6.8 μm, respectively.

The SEM image of Mg—xAl—5Sn—0.3Mn alloy is shown in Fig. 5.50. ATM150 alloy contains some fine Mg_2Sn phase. The contents of Mg_2Sn and $Mg_{17}Al_{12}$ increase with increasing Al content, which is consistent with the XRD result. ATM950 alloy contains a large amount of elongated strip-shaped $Mg_{17}Al_{12}$ phase and granular (or disk-shaped) Mg_2Sn phase.

In order to get insight into the second phase in the alloys, the ATM950, which contains the maximum amount of the second phase, is carefully studied by using TEM. The results are shown in Fig. 5.51 and Fig. 5.52. The alloy contains relatively coarse $Mg_{17}Al_{12}$ and Mg_2Sn phases. The EDS result indicates the phase with rod-like length of about 1 μm is $Mg_{17}Al_{12}$; the phase with granular diameter of 0.2—1 μm is

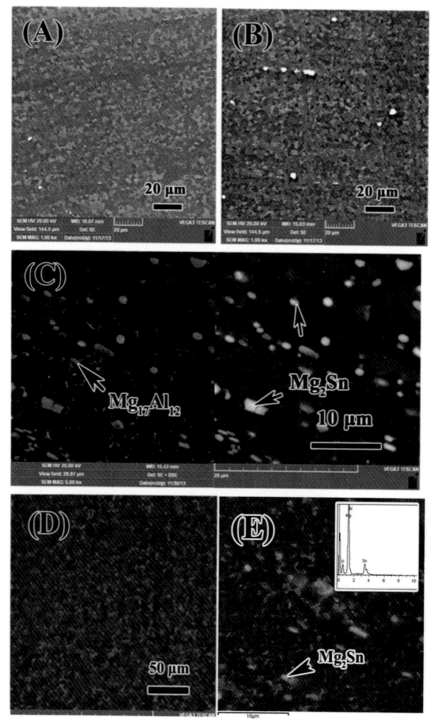

Figure 5.48 SEM images of Mg—xAl—3Sn—0.3Mn alloy for (A) ATM130, (B) ATM330, (C) ATM630, (D) ATM910, and (E) ATM930 and its EDS image.

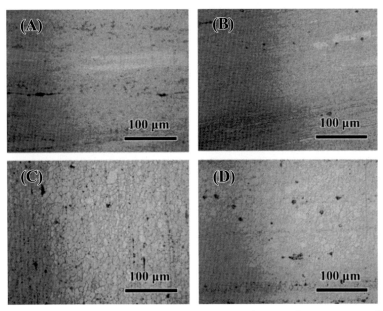

Figure 5.49 Microstructure along the extrusion direction of Mg—xAl—5Sn—0.3Mn alloys for (A) ATM150, (B) ATM350, (C) ATM650, and (D) ATM950.

Mg_2Sn. Some of these second phases are distributed inside the crystal grains, and some are distributed near the recrystallized grain boundaries, in consistent with the SEM results.

The main reason why the morphology of Mg_2Sn is different from the reported morphology of the strip-shaped precipitation along the basal plane is the existence of the relatively coarse and high-melting-point Mg_2Sn phase in the homogenized sample (as shown in Fig. 5.41) before extrusion. These relatively coarse Mg_2Sn are partially broken and welded to form a round or elliptical Mg_2Sn during the extrusion process. Since the solid solubility of Al in Mg matrix is very high, most Al atoms have been dissolved in Mg matrix during the homogenization process, and the diffusion rate of Al in Mg is relatively high. $Mg_{17}Al_{12}$ phase is dynamically precipitated during the extrusion process, forming rod-like $Mg_{17}Al_{12}$ phase (in consistent with the morphology of Mg—Al—Zn extrusion alloy).

The high-resolution images of Mg_2Sn and Mg matrix are shown in Fig. 5.51C. From the Fourier transform spots shown in Fig. 5.51D and E, the black phase in Fig. 5.51C is Mg_2Sn, and the light part is Mg matrix. According to the Fourier transform of Mg_2Sn and Mg matrix in Fig. 5.51E and F, the orientation relationship between Mg_2Sn and Mg matrix can be determined to be $(0001)_{Mg}//(0\bar{3}3)_{Mg2Sn}$ and $[2\bar{1}\bar{1}0]_{Mg}//[\bar{1}22]_{Mg2Sn}$ by comparing the standard diffraction spot manual with PDF card of (7—274). This is consistent with the reported result of high-strength Mg—Sn—Zn—Al by Sasaki [5], which

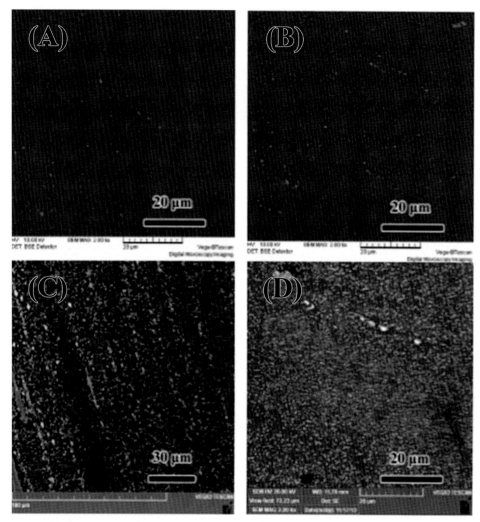

Figure 5.50 SEM images of Mg−xAl−5Sn−0.3Mn alloy for (A) ATM150, (B) ATM350, (C) ATM650, and (D) ATM950.

precipitates along the basal plane. It is worth to note that the Mg_2Sn and Mg matrix have three kinds of interface relationship of coherent, semi-coherent, and non-coherent. As shown in Fig. 5.51G,H,I, the Mg_2Sn morphology gradually increases corresponding to the three kinds of interface relationship. It can be inferred that the Mg_2Sn being coherent and semi-coherent with Mg matrix is dynamically precipitated during the extrusion, while the Mg_2Sn with the non-coherent interface with Mg matrix exists before extrusion.

The further analysis of the bright field image of the alloy reveals that the alloy contains not only the micron second phase, but also the dot-like or rod-like second phase

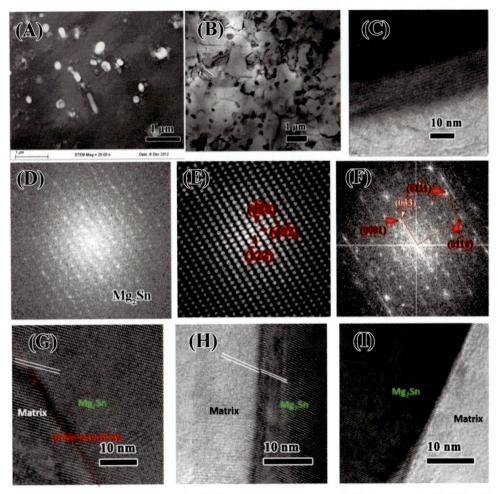

Figure 5.51 (A) S-STM image of ATM950 alloy, (B) bright field image of ATM950, (C) high-resolution image of ATM950, (D) Fourier transform image of Mg_2Sn, (E) Fourier transform image after masked, (F) Fourier transform image of Mg_2Sn and matrix, and (G), (H), and (I) interface between Mg_2Sn and matrix.

with a diameter of about $10-20$ nm, which are evenly distributed inside the grains as shown by the red circle in Fig. 5.52A. The EDS result confirms that this phase is Mg—Al phase. Because Mg—Al has no transition phases such as GP regions, it can be determined as $Mg_{17}Al_{12}$ by combining with the Mg-rich end phase diagram. Fig. 5.52 shows the high-resolution image of the $[11\bar{2}0]$ in Mg matrix. The black zone in the figure is $Mg_{17}Al_{12}$ phase. The high-resolution Fourier transform diagram (Fig. 5.52D) is used to separate $Mg_{17}Al_{12}$ phase and the diffraction spots of the matrix. Compared with the standard diffraction spot manual and the standard PDF card $(1-1128)$, the

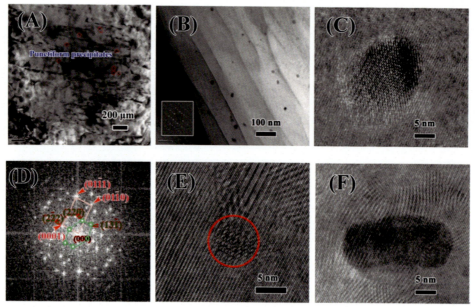

Figure 5.52 (A) S-TEM images of ATM950, (B) the bright field image of ATM950 alloy along [112($-$)0], (C) high-resolution image of $Mg_{17}Al_{12}$, and (D) Fourier transform image of $Mg_{17}Al_{12}$ and Mg matrix.

phase relationship between the $Mg_{17}Al_{12}$ phase and the Mg matrix is $(0001)_{Mg}//$ $(2\overline{2}2)_{Mg17Al12}$ and $[2\overline{1}\,\overline{1}\,0]_{Mg}//[122]_{Mg17Al12}$. This precipitation phase with small morphology can be considered as the $Mg_{17}Al_{12}$ phase that has just precipitated and not grown up during the extrusion process. The high-resolution image of the Mg$-$Al phase at the beginning of nucleation and reuniting is highlighted in the circular area in Fig. 5.52E. It is clear that the lattice has been distorted. The hot extrusion process is a short-term dynamic process. The Mg$-$Al phase has not been precipitated yet in this image. Fig. 5.22F shows the rod-shaped $Mg_{17}Al_{12}$ precipitated phases that have grown to a certain extent. These fine $Mg_{17}Al_{12}$ phases are dynamically precipitated during the extrusion process, which can effectively improve the strength of the alloy.

In summary, the alloys with Al content less than 3% are not completely recrystallized, while the alloys with Al content more than 6% are completely recrystallized among the Mg$-$xAl$-$ySn$-$0.3Mn extruded alloys. The alloying addition of Al promotes the recrystallization. The elements of Al and Mg form the second phase of $Mg_{17}Al_{12}$ with micrometer or nanometer size. The elements of Sn and Mg form the second phase of Mg_2Sn with micrometer size. A part of relatively coarse second phase is distributed around the grain boundary, and some are distributed inside the grains. The nano-sized $Mg_{17}Al_{12}$ second phase is dispersedly distributed inside the grains.

(2) Properties of as-extruded Mg—Al—Sn—Mn alloys

The engineering stress-strain curves of Mg—xAl—1Sn—0.3Mn, Mg—xAl—3Sn—0.3Mn, and Mg—xAl—5Sn—0.3Mn (x = 1, 3, 6, and 9) alloys are shown in Fig. 5.53. The corresponding yield strength, tensile strength, and elongation are presented in Table 5.12. The line chart of change in the yield strength with Al content is shown in Fig. 5.54. It indicates the yield strength of the three alloys decrease first then increase with increasing Al content. The specific change trend is that the yield strength decreases with increasing Al content when the Al content is less than 6%, and the yield strength reaches a maximum when the Al content is 9%.

Among the Mg—xAl—1Sn—0.3Mn alloys, the tensile strength of ATM110 alloy is larger than that of ATM310 alloy. When the Al content is larger than 3%, the tensile

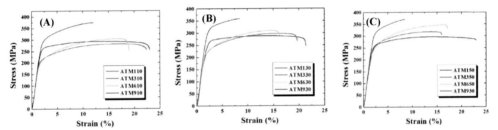

Figure 5.53 The stress-strain curves of (A) Mg—xAl—1Sn—0.3Mn (x = 1, 3, 6, and 9), (B) Mg—xAl—3Sn—0.3Mn (x = 1, 3, 6, and 9), and (C) Mg—xAl—5Sn—0.3Mn (x = 1, 3, 6, and 9).

Table 5.12 Mechanical properties of the Mg—xAl—ySn—0.3Mn (x = 1, 3, 6, and 9) (y = 1, 3, and 5) alloys.

Samples	$\sigma_{0.2}$ (MPa)	UTS (MPa)	ε%
ATM110	260	292	21.0
ATM310	211	281	20.5
ATM610	200	300	19.0
ATM910	281	371	10.2
ATM130	252	288	19.5
ATM330	212	298	18.0
ATM630	209	326	14.3
ATM930	290	358	6.0
ATM150	249	297	21.0
ATM350	242	316	14.0
ATM650	232	346	15.0
ATM950	298	370	6.0
AZ31	180	277	16.0
AZ91	263	357	9.0

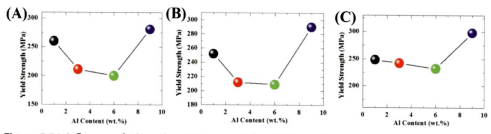

Figure 5.54 Influence of Al on the yield strength for (A) Mg−xAl−1Sn−0.3Mn (x = 1, 3, 6, and 9), (B) Mg−xAl−3Sn−0.3Mn (x = 1, 3, 6, and 9), and (C) Mg−xAl−5Sn−0.3Mn (x = 1, 3, 6, and 9).

strength of the alloy increases with increasing Al content, but the elongation decreases with increasing Al content. For Mg−xAl−3Sn−0.3Mn and Mg−xAl−5Sn−0.3Mn alloys, the tensile strength increases with increasing Al content, and the elongation decreases with increasing Al content.

Among the Mg−xAl−1Sn−0.3Mn alloys, ATM110 alloy is an alloy with both high yield strength and high plasticity, for which the yield strength is 260 MPa, the tensile strength is 292 MPa, and the elongation is 21%; in addition, the system also includes the ATM910 alloy, which has the yield strength of 281 MPa and the tensile strength of 371 MPa. Among the Mg−xAl−3Sn−0.3Mn alloys, there is an ATM130 alloy with a relatively high strength and plasticity, whose yield strength and elongation are 252 MPa and 19.5%, respectively; it has an ATM930 alloy with a yield strength above 290 MPa. For the Mg−xAl−5Sn−0.3Mn alloy, there is an ATM150 alloy with a relatively high strength and plasticity, whose yield strength and elongation are 249 MPa and 21%, respectively; it also has an ATM950 alloy with the yield strength up to 298 MPa and the tensile strength to 370 MPa.

In order to compare the performance with the commercial Mg alloys, AZ31 and AZ91 alloys were prepared by the same melting, heat treatment, and extrusion process. These alloys were compared with ATM110, ATM130, and ATM150 alloys exhibiting relatively good plasticity in Mg−Al−Sn−Mn alloys, as shown in Fig. 5.55. The result shows the yield strength of ATM110 alloy is 80 MPa higher than that of AZ31, and the elongation is 5% higher than that of AZ31; the yield strength of ATM950 is 35 MPa higher than that of commercial high-strength AZ91 Mg alloy. Therefore regardless of the high-plasticity or the high-strength classification, the performance of the current Mg−Al−Sn−Mn system is superior to the existing commercial AZ series alloys.

5.4.2 Effect of Sn on the microstructure and property of as-extruded Mg−Al−Sn−Mn alloy

The investigation of the influence of Sn content on the microstructure and property of the Mg−Al−Sn−Mn alloy is carried out to lay the foundation for the development of low-cost Mg−Al−Sn−Mn alloys. The alloy is divided into four systems of

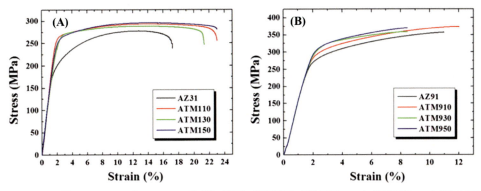

Figure 5.55 The stress-strain curves of (A) AZ31, ATM110, ATM130, and ATM150 and (B) AZ91, ATM910, ATM930, and ATM950.

Mg−1Al−ySn−0.3Mn, Mg−3Al−ySn−0.3Mn, Mg−6Al−ySn−0.3Mn, and Mg−9Al−ySn−0.3Mn in this section. Considering the samilar influence of Sn content on the microstructure and property for these four alloys, only some alloys are analyzed to avoid repetition.

The XRD patterns of the Mg−1Al−ySn−0.3Mn, Mg−3Al−ySn−0.3Mn, Mg−6Al−ySn−0.3Mn, and Mg−9Al−ySn−0.3Mn alloys are shown in Fig. 5.56. The Mg_2Sn phase increases with increasing Sn content. For the Mg−3Al−ySn−0.3Mn system, a relatively weak diffraction peak of $Mg_{17}Al_{12}$ appears when the Sn content is 5%, indicating the addition of Sn decreases the solid solubility of Al in Mg matrix; an obvious diffraction peak of $Mg_{17}Al_{12}$ can be observed in ATM350 alloy. This phenomenon is more pronounced in alloys with 6% and 9% Al content. The $Mg_{17}Al_{12}$ and Mg_2Sn increase with increasing Sn content in Mg−6Al−ySn−0.3Mn and Mg−9Al−ySn−0.3Mn alloys. The phase composition of all alloys has already been analyzed in Section of 5.4.1. It will not be discussed here to avoid repetition.

The incompletely recrystallized Mg−1Al−ySn−0.3Mn system and the completely recrystallized Mg−6Al−ySn−0.3Mn system were chosen to discuss the effect of Sn on the extruded microstructure of the alloy.

From the microstructure of Mg−1Al−ySn−0.3Mn alloy shown in Fig. 5.57, the recrystallized grains become finer and the un-recrystallized region decreases with increasing Sn content, indicating the increase of Sn can promote the recrystallization to a certain extent. From the metallographic images of Mg−6Al−ySn−0.3Mn alloy shown in Fig. 5.58, the alloy of this system is completely recrystallized. The average grain size of ATM610, ATM630, and ATM650 are 10, 8.5, and 8 μm, respectively.

From the SEM images of Mg−6Al−ySn−0.3Mn alloy, the second phase increases with increasing Sn content, which is consistent with the XRD results (Fig. 5.59).

The yield strength of these four alloys are compared in Fig. 5.60, where the yield strength of the Mg−1Al−ySn−0.3Mn alloy decreases with increasing Sn content,

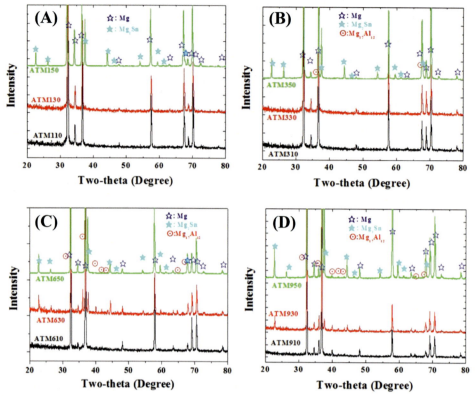

Figure 5.56 XRD patterns of Mg−Al−Sn−Mn alloys for (A) Mg−1Al−ySn−0.3Mn, (B) Mg−3Al−ySn−0.3Mn, (C) Mg−6Al−ySn−0.3Mn, and (D) Mg−9Al−ySn−0.3Mn.

Figure 5.57 Metallographic images of Mg−1Al−xSn alloys for (A) ATM110, (B) ATM130, and (C) ATM150.

whereas the yield strength of the other alloys increase with different degrees as the Sn content increases. The tensile strength of the alloy increases with increasing Sn content, but the increment is limited. The tensile strength is increased by about 10 MPa per 2% Sn. The elongation decreases to some extent with increasing Sn content.

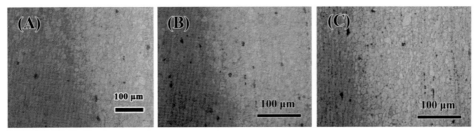

Figure 5.58 Metallographic images of Mg—6Al—xSn alloys for (A) ATM610, (B) ATM630, and (C) ATM650.

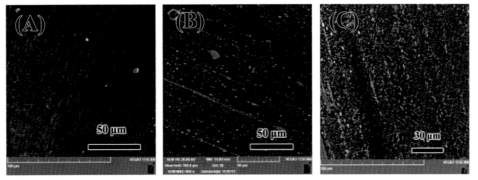

Figure 5.59 SEM images of Mg—6Al—ySn—0.3Mn alloys for (A) ATM610, (B) ATM630, and (C) ATM650.

Comparison of the AM10 and AZ31 with ATM110 and ATM310 alloys at the corresponding Al content is shown in Fig. 5.61. The elongation of the alloy containing Sn element is greatly larger than that of the corresponding AM and AZ alloys, which may be ascribed to the simultaneous addition of Al and Sn decreasing the stacking fault energy of the alloy and increasing the elongation of the alloy. The yield strength of ATM110 is similar to that of AM10. The yield strength of ATM310 alloy is about ∼20 MPa higher than that of AZ31. In summary, the addition of certain amount of Sn into the AM alloys significantly improves the plasticity of the alloy and increases the strength of the alloy to a certain extent.

5.4.3 Effect of extrusion temperature on the microstructure and property of the alloy

Because the strength of as-cast Mg—9Al—ySn—0.3Mn alloy is very high, which represents a high resistance to hot extrusion deformation, it cannot be extruded successfully at 250°C. Therefore the extrusion temperature of the alloy is 300°C and 350°C. The extrusion temperature for other alloys is 250°C, 300°C, and 350°C.

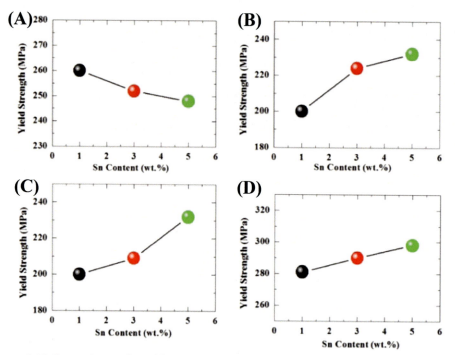

Figure 5.60 Comparison of yield strength of alloys for (A) Mg−1Al−ySn−0.3Mn, (B) Mg−3Al−ySn−0.3Mn, (C) Mg−6Al−ySn−0.3Mn, and (D) Mg−9Al−ySn−0.3Mn.

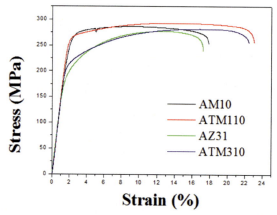

Figure 5.61 The stress-strain curves of the AM10, AZ31, ATM110, and ATM310 alloys.

Fig. 5.62 shows the XRD patterns of the Mg−Al−Sn−Mn extruded alloys at 250°C, 300°C, and 350°C. The test surfaces of the all XRD samples are perpendicular to the extrusion direction. As the extrusion temperature increases, the second phase in the alloy increases to some extent. Here, some alloys are taken as examples to illustrate

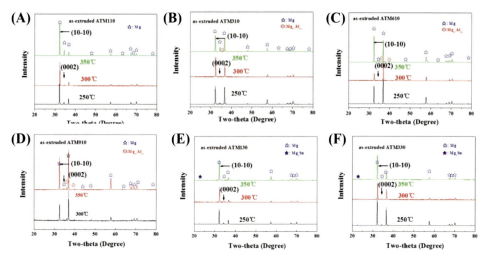

Figure 5.62 XRD patterns of Mg—Al—Sn—Mn alloys at 250°C, 300°C, and 350°C.

this change. The evolution of phase composition of other alloys with the extrusion temperature is similar to the elaborated alloy.

According to the XRD pattern of the alloy prepared under the extrusion temperature of 250°C, 300°C, and 350°C for ATM610 alloy, all alloys contain α-Mg phase without Mg_2Sn phase. No diffraction peak of $Mg_{17}Al_{12}$ is detected in the alloy extruded at 250°C, indicating 6% alloying element of Al is solid dissolved in Mg matrix. When the extrusion temperature increases to 300°C, the weak diffraction peak of $Mg_{17}Al_{12}$ appears at 2-Theta of 36 degrees, indicating the $Mg_{17}Al_{12}$ phase precipitates by extrusion under 300°C for ATM610 alloy. With the extrusion temperature increases to 350°C, the diffraction intensity of the $Mg_{17}Al_{12}$ increases. According to the phase diagram theory, the second phase ($Mg_{17}Al_{12}$ and Mg_2Sn) content should decrease with increasing extrusion temperature (as shown in Fig. 5.63). However, the results of this experiment are opposite to those of the phase diagram. Hot extrusion is a short-term and high temperature dynamic precipitation process with large strains, in comparison with the second phase. The higher the temperature, the more the second phase will precipitate. Therefore the content of the second phase increases with increasing extrusion temperature. If the sample can be close to equilibrium by being kept at 300°C and 350°C for a long time, it can be predicted that the more second phases are precipitated at 300°C than at 350°C.

Figs. 5.64, 5.65, 5.66, and 5.67 show the OM images of Mg—Al—Sn—Mn alloy at 250°C, 300°C, and 350°C. According to the OM images, the influence of extrusion temperature on the microstructure of the alloy can be divided into two categories: (1) alloys that have not been completely recrystallized. The un-recrystallized region decreases with increasing extrusion temperature, and the recrystallized grains grow to a

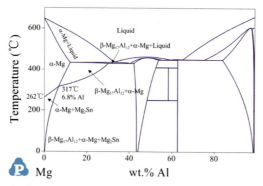

Figure 5.63 The perpendicular cross section of the Mg−1Sn−xAl−0.3Mn alloy.

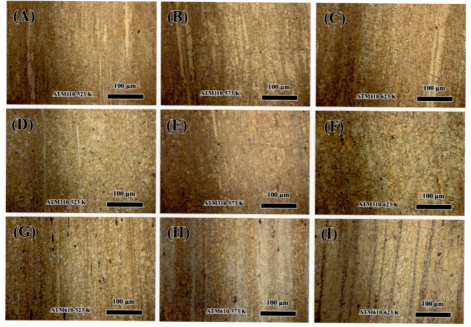

Figure 5.64 Metallographic images parallel to the extrusion direction for the extruded ATM110, ATM310, and ATM610 at the temperature of 250°C, 300°C, and 350°C.

certain extent; (2) alloys that have been completely recrystallized. The grains grow to some extent with increasing extrusion temperature, but the change is small. Here, some alloys are taken as examples to illustrate this effect. The change in the microstructure of the other alloys with the corresponding extrusion temperatures is similar to that observed in this alloy.

Fig. 5.64G,H,I are the metallographic images along the extrusion direction at extrusion temperatures of 250°C, 300°C, and 350°C, respectively, for the ATM610

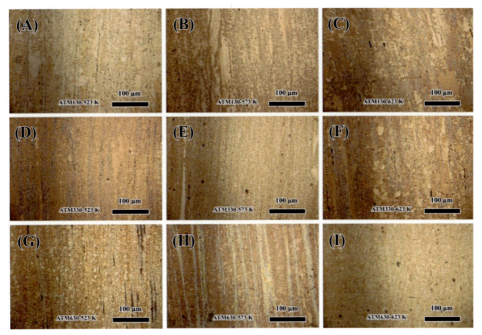

Figure 5.65 Metallographic images of extruded alloys parallel to the extrusion direction for ATM130, ATM330, and ATM630 alloys at 250°C, 300°C, and 350°C.

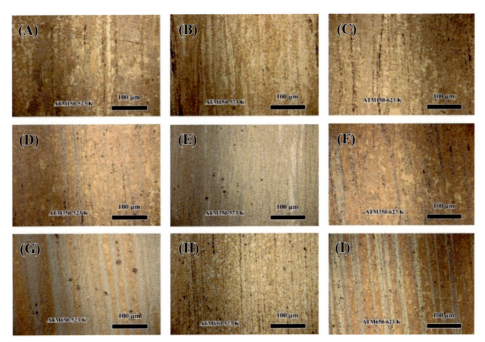

Figure 5.66 Metallographic images of extruded alloys parallel to the extrusion direction for ATM150, ATM350, and ATM650 alloys at 250°C, 300°C, and 350°C.

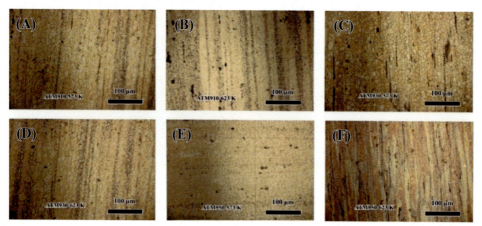

Figure 5.67 Metallographic images of extruded alloys parallel to the extrusion direction for ATM910, ATM930, and ATM950 alloys at 250°C, 300°C, and 350°C.

alloy. It shows that the ATM610 alloy can be completely recrystallized after extrusion under the three temperatures. The average grain size increases with increasing extrusion temperature. The average grain size of the alloy extruded under 250°C, 300°C, and 350°C are 7.6, 8.4, and 9.9 μm, respectively.

Fig. 5.65A, B, and C are metallographic images along the extrusion direction of the alloy prepared by ATM130 alloy at the temperature of 250°C, 300°C, and 350°C. The ATM130 alloy is not completely recrystallized during the extrusion at the three temperatures. The recrystallized structure increases with increasing extrusion temperature, and the non-recrystallized region decreases with increasing extrusion temperature.

The grain size of the alloy prepared by extrusion under different temperatures is similar for all the alloys. This is because the content of second phases (Mg_2Sn and $Mg_{17}Al_{12}$) increases with increasing extrusion temperature (XRD results indicate the second phase content of the alloy prepared at 350°C is larger than that at 300°C). According to the results for the Mg-7.6Al-0.4Zn by Jung [6], the increase of the second phase before extrusion can restrict the growth of the recrystallized grains during the extrusion process. In this experiment, the alloy composition and heat treatment are the same before extrusion, so the state of the alloy before extrusion should be the same. It is generally believed that the nucleation position of the recrystallization is the grain boundary of the parental crystal, because there are many defects at the grain boundary, making it easier for dislocations to accumulate. When the dislocation density reaches a certain value, the nucleation position is provided (critical nucleation rate of the recrystallization). The recrystallization nucleation position in this experiment is schematically shown in Fig. 5.68A. Since the alloy is accompanied by the dynamic precipitation of the second phase (Mg_2Sn and $Mg_{17}Al_{12}$) during the hot extrusion

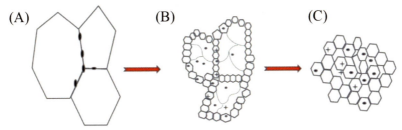

Figure 5.68 Schematic diagram illustrating the grain nucleation and growth and precipitation of the second phase during the extrusion process.

process, there will also be a second phase precipitated during the nucleation of the recrystallization (as shown in Fig. 5.68B). The precipitation of the second phase is accompanied with the nucleation and growth of the recrystallization, so the precipitation of the second phase will restrict the growth of the recrystallized grains (Fig. 5.68C). Roberts and Ahlbolum [7] proposed that the recrystallization nucleation rate and the strain rate of the alloy are related to the deformation temperature. The recrystallization nucleation rate can be expressed:

$$\dot{n} = C\dot{\varepsilon}\exp\left(\frac{-Q_{\text{act}}}{\text{RT}}\right) \tag{5.14}$$

$$\rho_C = \left(\frac{20r_i\dot{\varepsilon}}{3bMl\iota}\right)^{1/3} \tag{5.15}$$

where C is a constant, γ_i refers to the grain boundary energy of the large angle grain boundary, M is the grain boundary mobility, Q_{act} is strain activation energy, l refers to the average free path of dislocation, and ι is the dislocation energy per unit length.

The dislocation density will increase when the temperature T increases, so the nucleation rate of recrystallization is higher when extruded at 350°C than that at 300°C. At the same time, the growth process of the recrystallized grains is the process of grain boundary migration caused by the interface energy difference between the new grain and parental grain. The precipitation of the second phase will effectively restrict the migration of grain boundaries. Although relatively high extrusion temperature can cause the growth of the recrystallized grains, the precipitated second phase content extruded at high temperature is larger than that at low temperature, restricting the growth of the recrystallized grains. The grain size extruded at different temperatures is not significantly different from each other for the alloys with high alloying content (Al ≥ 6% and Sn = 5%).

From Table 5.13, the change in yield strength of the alloy with increasing extrusion temperature can be divided into two groups: for alloys with Al content less than 3%, the yield strength decreases with increasing extrusion temperature; for alloys with Al content larger than 6%, the yield strength increases with increasing extrusion temperature.

Table 5.13 Mechanical properties, grain size, and un-recrystallized region fraction for Mg−Al−Sn−0.3Mn alloy extruded at 250°C, 300°C, and 350°C.

As-extruded samples	Ext. Temp. (K)	YS (MPa)	UTS (MPa)	ε (%)	Y/T	Estimated fraction of un-recrystallized grain (%)	Average size of recrystallized grains (μm)
ATM110	523	263	290	18.0	0.90	70	2.1
ATM110	573	260	292	21.0	0.89	63	2.3
ATM110	623	241	273	15.6	0.88	60	2.7
ATM310	523	229	293	16.0	0.78	38	3.0
ATM310	573	213	281	20.5	0.76	35	4.5
ATM310	623	201	271	20.0	0.74	34	4.2
ATM610	523	241	319	12.0	0.76	15	5.0
ATM610	573	203	300	19.0	0.68	/	7.1
ATM610	623	199	313	13.0	0.64	/	7.6
ATM910	573	285	373	12.0	0.76	/	9.6
ATM910	623	290	370	11.0	0.78	/	9.8
ATM130	523	249	288	18.0	0.86	67	1.6
ATM130	573	253	288	19.5	0.85	60	2.0
ATM330	523	225	306	15.4	0.74	28	3.0
ATM330	573	217	298	18.0	0.73	32	3.5
ATM330	623	203	295	16.4	0.69	26	4.0
ATM630	523	220	318	8.6	0.69	/	4.8
ATM630	573	226	326	14.3	0.69	/	6.8
ATM630	623	234	322	12.4	0.73	/	7.3
ATM930	573	290	355	8.5	0.82	/	7.8
ATM930	623	292	360	8.8	0.81	/	7.7
ATM150	523	235	276	21.0	0.85	30	1.8
ATM150	573	248	297	23.0	0.84	27	2.0
ATM150	623	232	283	17.5	0.82	23	1.9
ATM350	523	221	300	13.0	0.74	15	2.0
ATM350	573	242	316	16.0	0.76	16	2.3
ATM350	623	230	315	13.0	0.73	13	2.0
ATM650	523	221	320	10.0	0.69	/	8.5
ATM650	573	232	346	17.0	0.67	/	8.0
ATM650	623	233	310	9.0	0.75	/	9.0
ATM950	573	300	370	8.2	0.81	/	6.2
ATM950	623	306	371	7.8	0.82	/	7.3
AZ31	573	187	279	16.0	0.67	/	/

5.5 Hot deformation parameter and constitutive equation of Mg−Al−Sn−Mn alloy

The high-temperature flow stress model is used to study the change of metal flow stress at different processing temperatures and strain rates during the high-temperature plastic processing. The model is widely applied in steel and Al alloys. Because the plasticity of Mg alloy at room temperature is poor, most Mg alloy sheets need to be produced by hot rolling (e.g., AZ31, AM30, and LA141 sheets, etc.). During high-temperature deformation processes, the flow stress of the Mg alloy changes with the change of strain rate and temperature, and exhibits a serious processing softening phenomenon. Therefore it is very important to study the high-temperature properties of Mg alloys.

The studies in Section 5.4 indicate that the four deformed Mg alloys of ATM110, ATM130, ATM310, and ATM330 have relatively high plasticity, and may be developed as sheets with excellent comprehensive properties. This section will establish constitutive equations for the high-temperature flow stress models of these four alloys to provide theoretical basis for the application of these new deformed Mg alloys showing good plasticity.

5.5.1 Hot deformation parameter and constitutive equation of ATM110 alloy

(1) True stress–strain curve

Fig. 5.69 shows the true stress–strain curve of ATM110 sample at different temperatures and different strain rates. At the same deformed temperature, the flow stress continuously increases with increasing strain rate. This is mainly because the deformation time required for unit strain during deformation process becomes shorter with increasing stain rate, the dislocation is increased, the softening time emerging from

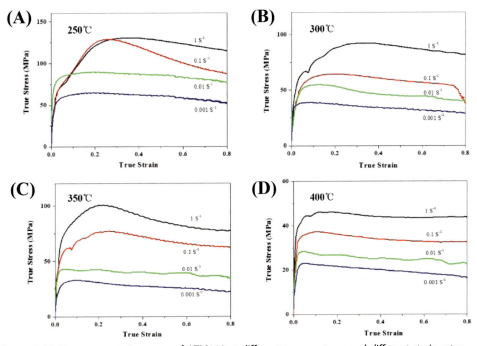

Figure 5.69 True stress-strain curves of ATM110 at different temperatures and different strain rates.

dynamic recrystallization shortens, and the critical shear stress of the alloy increases, finally resulting in the increase of the flow stress of the alloy. On the other hand, the flow stress continues to decrease with increasing temperature under the same strain rate. This is mainly because the thermal vibration of metal atoms increases with increasing temperature, and the interaction between metal atoms weakens; the slip system of the material continues to increase, the slip resistance decreases, the deformability increases, and the flow stress of the alloy is reduced. In addition, the softening caused by dynamic recrystallization also increases with increasing temperature, further decreasing the flow stress of the alloy.

When the strain rate is at the lowest of 0.001 s^{-1}, the flow stress of the alloy increases to the maximum and then gradually decreases to a steady state at temperature of 300°C−400°C. It can be concluded that the dynamic recrystallization and work hardening of the alloy reached a dynamic balance under this condition. While at 250°C, the material may not reach the steady state due to factors such as incomplete dynamic recrystallization of the alloy. It is worth noting that the peak flow stress is almost the same when the strain rate is 1 and 0.1 s^{-1} at the temperature of 250°C. It may be because the deformation temperature of 250°C is relatively low, the dynamic recrystallization is not complete. The softening time caused by dynamic recrystallization is short, and the flow strain of the alloy increases. At the same time, the stain rate is relatively high, and the dislocation greatly increases, which ultimately results in the similar peak stress under the two stain rates. As the strain increases, because of the strain accumulation, the softening time caused by dynamic recrystallization for the alloy with a strain rate of 0.1 s^{-1} is larger than that of 1 s^{-1}, resulting in the reduced flow stress for alloy with strain rate of 0.1 s^{-1}. Because the strain rate is relatively high for alloys with a strain rate of 1 s^{-1}, the dislocation significantly increases, and the softening time caused by dynamic recrystallization is very short. Therefore the flow stress decreases insignificantly with increasing strain.

(2) Hot deformation parameter and constitutive equation

According to Fig. 5.69, there exists a certain relationship between flow stress and deformation temperature during hot compression process. Sellars and Tegart [8,9] proposed that under the condition of hot deformation, the hyperbolic sine model of the dynamic recrystallization activation energy and deformation temperature can be used to express the above parameter, as shown below:

$$\dot{\varepsilon} = A\left[\sinh(\alpha\sigma)^{n}.\exp\left(-\frac{Q}{RT}\right)\right] \tag{5.16}$$

where A, n, and α are material constants, R is the gas constant ($8.314 \text{ J/mol} \cdot \text{K}$), T is absolute temperature, Q is the dynamic recrystallization activation energy, commonly known as softening activation energy, reflecting the balance between the work hardening and dynamic softening during high-temperature deformation process. The following two equations can be obtained by calculating the logarithm of Eq. (5.16):

$$\ln\dot{\varepsilon} = \ln A + n_1 \ln|\sigma| - \frac{Q}{RT} \tag{5.17}$$

$$ln\dot{\varepsilon} = lnA + \beta|\sigma| - \frac{Q}{RT} \tag{5.18}$$

According to Eq. (5.17) and Eq. (5.18), $n_1 = d\ln\dot{\varepsilon}/d\ln|\sigma|$, $\beta = d\ln\dot{\varepsilon}/d|\sigma|$, $\alpha = \beta/n_1$. The peak flow stress and corresponding strain during the hot compression process are selected from Fig. 5.69. The linear relationship between $\ln\dot{\varepsilon}$ and $\ln\sigma$ a different temperatures is shown in Fig. 5.70A. The least squares regression is used to get the slope of each fitted straight line, which is n. The linear relationship between $\ln\dot{\varepsilon}$ and $\ln\sigma$ at different deformation temperatures is shown in Fig. 5.70B. The least squares regression is used to obtain the slope of each fitted line and the slope is β. According to the results, $n_1 = 8.2$, $\beta = 0.1535$, and $\alpha = 0.0652$.

Eq. (5.16) can be expressed as:

$$\ln\dot{\varepsilon} = \ln A + n[\ln \sinh(\alpha\sigma)] - \frac{Q}{RT} \tag{5.19}$$

When $\dot{\varepsilon}$ is known, the above equation can be expressed as:

$$Q = R.\frac{\partial \ln\dot{\varepsilon}}{\partial \ln[\sinh(\alpha\sigma)]}\Big|_T . \frac{\partial \ln[\sinh(\alpha\sigma)]}{\partial(1/T)}\Big|_{\dot{\varepsilon}} \tag{5.20}$$

Eq. (5.20) can be simplified as:

$$Q = R\cdot n'\cdot D \tag{5.21}$$

where R is the gas constant, $n' = \frac{\partial \ln\dot{\varepsilon}}{\partial \ln[\sinh(\alpha\sigma)]}\Big|_T$, and $D = \frac{\partial \ln[\sinh(\alpha\sigma)]}{\partial(1/T)}\Big|_{\dot{\varepsilon}}$.

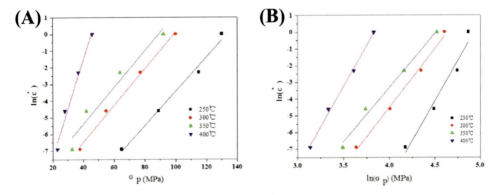

Figure 5.70 Relationship between $\ln\dot{\varepsilon}$ and (A) $\ln\sigma$ and (B) σ.

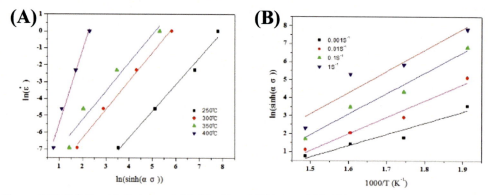

Figure 5.71 Relationship between ln[sinh($\alpha\sigma$)] and (A) ln$\dot{\varepsilon}$ and (B) *1/T*.

The linear relationship of ln$\dot{\varepsilon}$ − ln[sinh($\alpha\sigma$)] under different strain rates is shown in Fig. 5.71A, in which the slope n' of the fitted line is 2.32. The linear relationship of ln[sinh($\alpha\sigma$)] − 1/T under different strain rates is shown in Fig. 5.71B, in which the slope D of the fitted line is 8.314. Therefore $Q = R \cdot D \cdot n' = 186.031$ kJmol^{-1}.

The effects of deformation temperature and strain rate on flow stress can be expressed by introducing the Zener-Hollomon (Z) parameter:

$$Z = \dot{\varepsilon}.\exp\left(\frac{Q}{RT}\right) \tag{5.22}$$

The physical meaning of Z is the temperature-compensated strain rate factor. By comparing the above equation and Eq. (5.14):

$$Z = \dot{\varepsilon}\exp\left(\frac{Q}{RT}\right) = A\sinh(\alpha\sigma)^n \tag{5.23}$$

Taking the logarithm at the two sides of Eq. (5.23),

$$\ln Z = \ln A + n\ln[\sinh(\alpha\sigma)] \tag{5.24}$$

where n is the stress index, and A is a constant related to materials.

Taking the obtained Q to Eq. (5.23), Z can be expressed as:

$$Z = \dot{\varepsilon}.\exp\left(\frac{186.031}{RT}\right) \tag{5.25}$$

The relationship between ln$Z$ and lnsinh($\alpha\sigma$) is shown in Fig. 5.72. The slope of the line obtained by the linear fitting is n. The line intercept is ln$A$. From the figure, the n and ln$A$ of ATM110 are 1.528 and 20.07, respectively.

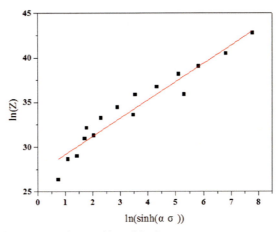

Figure 5.72 Relationship between ln$Z$ and ln$sinh(\alpha\sigma)$.

The flow stress can be expressed as:

$$\sigma = \frac{1}{\alpha}\ln\left\{\left(\frac{Z}{A}\right)^{\frac{1}{n}}+\left[\left(\frac{Z}{A}\right)^{\frac{2}{n}}+1\right]^{\frac{1}{2}}\right\} \tag{5.26}$$

Taking the above constants to Eq. (5.26), the flow stress can be expressed as:

$$\sigma = \frac{1}{0.0652}\ln\left\{\left(\frac{Z}{5.2\times10^8}\right)^{\frac{1}{1.528}}+\left[\left(\frac{Z}{5.2\times10^8}\right)^{\frac{2}{1.528}}+1\right]^{\frac{1}{2}}\right\} \tag{5.27}$$

According to Eq. (5.27), strain temperature, strain rate, and the peak flow stress of ATM110 alloy during the hot deformation process can be calculated.

The calculation of the hot deformation parameters and constitutive equations for ATM130, ATM310, and ATM330 alloys will also use the same method with the ATM110, which will not be discussed in detail.

5.5.2 Hot deformation parameter and constitutive equation of ATM130 alloy

Fig. 5.73 shows the true stress-strain curve of ATM130 sample under different temperatures and strain rates. With the constant strain rate, the flow stress decreases with increasing temperature. The reason for this phenomenon is the same with that for ATM110 alloy: the thermal vibration of metal atoms increases with increasing temperature, and the interaction among metal atoms weakens; the slip system of the material increases, the deformability increases, and the slip resistance decreases, reducing the flow stress of the alloy. In addition, the softening caused by dynamic recrystallization

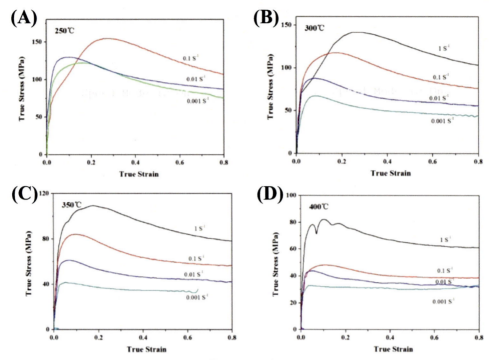

Figure 5.73 True stress-strain curves at different temperatures and strain rates.

also increases with increasing temperature, further decreasing the flow stress of the alloy. On the other hand, at the fixed deformed temperature, the flow stress increases with increasing strain rate (the same reason as the ATM110 alloy). This is mainly because with increasing strain rate, the deformation time required by unit strain during the deformation process becomes short, dislocation increases, the softening time for dynamic recrystallization shortens, and the critical shear stress of the alloy increases, ultimately resulting in the increase in the flow stress of the alloy.

When the stain rate is at the lowest of $0.001 \ s^{-1}$, the flow stress of the alloy increases to the maximum and then gradually decreases to a steady state at the temperature of $300°C - 400°C$. It can be concluded that the dynamic recrystallization and work hardening of the alloy reach a dynamic balance under this condition. However, at the temperature of $250°C$, the material does not reach the steady state, which may be ascribed to the factors such as the incomplete dynamic recrystallization of the alloy.

The relation between $ln\dot{\varepsilon}$ and $ln\sigma$, and σ is shown in Fig. 5.74. From the similar calculations, $n_1 = 7.56$, $\beta = 0.118$, $\alpha = 0.0156$ are determined for ATM130 alloy.

The relationship between $ln sinh(\alpha\sigma)$ and $ln\dot{\varepsilon}$ and $1/T$ is shown in Fig. 5.74. The slope of the fitted line in the figure is $n' = 5.41$, $D = 2.86$. Therefore $Q = R \cdot D \cdot n' = 128.639 \ kJ/mol$ for ATM130 alloy.

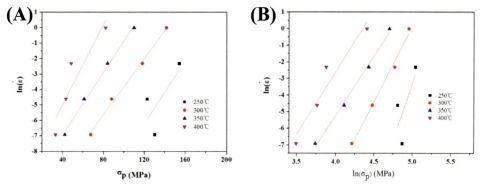

Figure 5.74 Relationship between lnsinh($\alpha\sigma$) and (A) ln$\dot{\varepsilon}$ and (B) $1/T$.

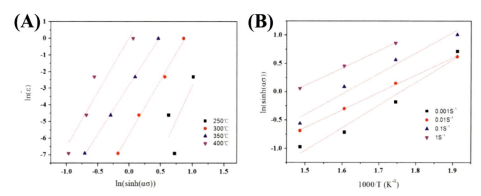

Figure 5.75 Relationship between ln$Z$ and lnsinh($\alpha\sigma$).

The relationship between ln$Z$ and lnsinh($\alpha\sigma$) is shown in Fig. 5.75. The slope of the fitted line is n, the line intercept is ln$A$. From the figure, $n = 5.2$ and $A = 2.67 \times 10^6$ for ATM130 alloy.

The flow stress can be expressed as:

$$\sigma = \frac{1}{0.0156} \ln \left\{ \left(\frac{Z}{2.67 \times 10^6} \right)^{\frac{1}{5.2}} + \left[\left(\frac{Z}{2.67 \times 10^6} \right)^{\frac{2}{5.2}} + 1 \right]^{\frac{1}{2}} \right\} \tag{5.28}$$

According to Eq. (5.28), strain temperature, strain rate, and the peak flow stress of ATM130 alloy during the hot deformation process can be calculated. These results offer valuable opportunities in engineering applications.

5.5.3 Hot deformation parameter and constitutive equation of ATM310 alloy

Fig. 5.76 shows the true stress–strain curve for ATM310 sample under different temperatures and strain rates. The flow stress increases with increasing stain at the

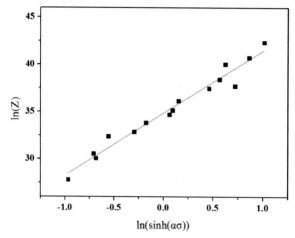

Figure 5.76 True stress-strain curves at different temperatures and different stain rates.

constant temperature. On the other hand, the flow stress decreases with increasing temperature under the fixed strain rate. This is consistent with ATM110 and ATM130 alloys. The main reason is that with increasing temperature, the thermal vibration of metal atoms increases, and the interaction among metal atoms weakens; the slip system of the material increases, the deformability increases, and the slip resistance decreases, reducing the flow stress of the alloy. In addition, the softening caused by dynamic recrystallization increases with increasing temperature, further decreasing the flow stress of the alloy. On the other hand, the flow stress increases with increasing strain rate. This is mainly because with increasing strain rate, the deformation time required by unit strain during the deformation process decreases, the dislocation increases, the softening time provided by dynamic recrystallization decreases, and the critical shear stress of the alloy increases, finally resulting in the increase of the flow stress of the alloy.

When the stain rate is relatively low (0.01 and 0.001 s^{-1}), the flow strain of the alloy increases to the maximum and then gradually decreases to a steady state at all temperatures. It can be concluded that a dynamic balance is reached between the dynamic recrystallization and work hardening under this condition. At a relatively high strain rate, the softening time corresponding to the dynamic recrystallization decreases, and the flow stress of the alloy increases; at the same time, the strain rate is relatively high, the dislocation rapidly increases, and ultimately the flow stress of the alloy does not reach the steady state.

The relationship between $\ln\dot{\varepsilon}$ and $\ln\sigma$, and σ are shown in Fig. 5.77. According to the estimations, $n_1 = 9.378$, $\beta = 0.10585$, and $\alpha = 0.01128$ are determined for ATM310 alloy.

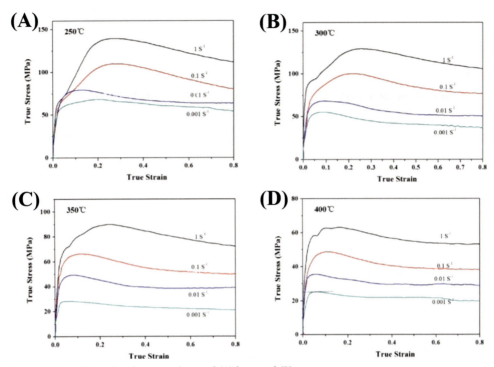

Figure 5.77 Relationship between ln$\dot{\varepsilon}$ and (A) lnσ, and (B) σ.

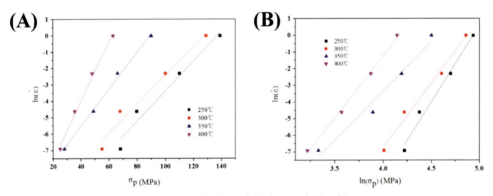

Figure 5.78 Relationship between lnsinh$(\alpha\sigma)$ and (A) ln$\dot{\varepsilon}$, and (B) $1/T$.

The relationship between lnsinh$(\alpha\sigma)$ and ln$\dot{\varepsilon}$, and $1/T$ are shown in Fig. 5.78. The slope of $n' = 6.79$ and $D = 3.439$ for the fitted line in the figure. Therefore $Q = R \cdot D \cdot n' = 194.138$ kJmol^{-1}.

The relationship between ln$Z$ and lnsinh$(\alpha\sigma)$ is shown in Fig. 5.79. The slope of the fitted line is n, the line intercept is ln$A$. From the figure, the $n = 6.58$ and $A = 3.86 \times 10^{11}$ for ATM310.

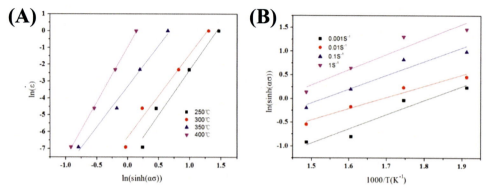

Figure 5.79 Relationship between lnZ and lnsinh($\alpha\sigma$).

The flow stress can be expressed by:

$$\sigma = \frac{1}{0.01128}\ln\left\{\left(\frac{Z}{3.86\times}10^{11}\right)^{\frac{1}{6.58}}+\left[\left(\frac{Z}{3.86\times10^{11}}\right)^{\frac{2}{6.58}}+1\right]^{\frac{1}{2}}\right\} \qquad (5.29)$$

According to Eq. 5.29, strain temperature, strain rate, and the peak flow stress of ATM310 alloy during the hot deformation process can be calculated.

5.5.4 Hot deformation parameter and constitutive equation of ATM330 alloy

Fig. 5.13 shows the true stress–strain curve of ATM330 sample at different temperatures and strain rates. The change trend of the curve is similar to the three alloys discussed previously. At the constant strain temperature, the flow stress increases with increasing strain rate. This is mainly because that with increasing strain rate, the deformation time required by unit strain during the deformation process shortens, the dislocation increases, the softening time resulted from dynamic recrystallization also decreases, and the critical shear stress of the alloy increases, ultimately resulting in the increase in the flow stress of the alloy. On the other hand, at the fixed strain rate, the flow stress decreases with increasing temperature. With increasing temperature, the thermal vibration of metal atoms increases, and the interaction among metal atoms decreases; the slip system of the material continues to increase, the deformability increases, the slip resistance decreases, and the flow stress of the alloy is decreased. In addition, the softening caused by dynamic recrystallization increases with increasing temperature, further decreasing the flow stress of the alloy.

When the strain rate is at the lowest value of 0.001 s^{-1}, the flow stress of the alloy increases to the maximum and then gradually decreases to a steady state at the temperature of 300°C−400°C. It can be concluded that the dynamic recrystallization and work hardening of the alloy reach a dynamic balance under this condition (Fig. 5.80).

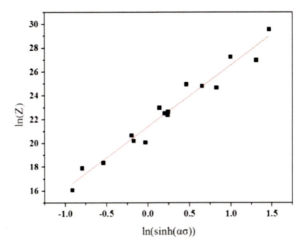

Figure 5.80 True stress-strain curves at different temperatures and different strain rates.

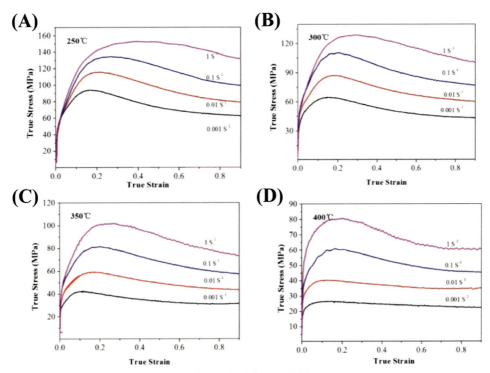

Figure 5.81 Relationship between the $\ln\dot{\varepsilon}$ and (A) $\ln\sigma$, and (B) σ.

The relationship between $\ln\dot{\varepsilon}$ and $\ln\sigma$, and σ are shown in Fig. 5.81. According to the results, $n_1 = 9.3559$, $\beta = 0.1155$, and $\alpha = 0.12345$ are obtained for ATM330 alloy.

The relationship between lnsinh($\alpha\sigma$) and ln$\dot{\varepsilon}$, and $1/T$ are shown in Fig. 5.82 at different strain rates. The slope of $n' = 6.3329$ and $D = 3.10426$ for the fitted line in the figure. Therefore $Q = R \cdot D \cdot n' = 163.44$ kJmol^{-1}.

The relationship between ln$Z$ and lnsinh($\alpha\sigma$) is shown in Fig. 5.83. The slope of the fitted line is n, and the line intercept is ln$A$. From the figure, $n = 6.202$ and ln$A = 27.5$ for ATM330 (Fig. 5.84).

The flow stress can be expressed by:

$$\sigma = \frac{1}{0.12345} \ln \left\{ \left(\frac{Z}{8.72 \times} 10^{11} \right)^{\frac{1}{6.202}} + \left[\left(\frac{Z}{8.72 \times 10^{11}} \right)^{\frac{2}{6.202}} + 1 \right]^{\frac{1}{2}} \right\} \tag{5.30}$$

According to Eq. (5.30), strain temperature, strain rate, and the peak flow stress of the ATM330 alloy can be calculated during the hot deformation process, providing valuable opportunities in engineering applications.

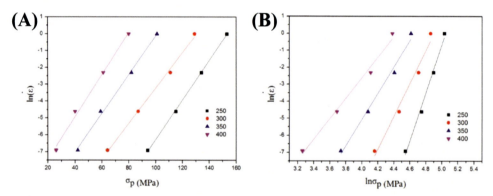

Figure 5.82 Relationship between lnsinh($\alpha\sigma$) and (A) ln$\dot{\varepsilon}$, and (B) $1/T$.

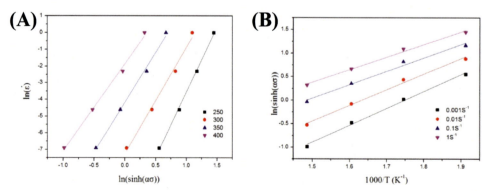

Figure 5.83 Relationship between ln$Z$ and lnsinh($\alpha\sigma$).

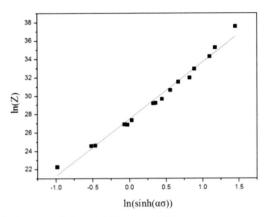

Figure 5.84 Relationship between $\ln\dot{\varepsilon}$ and (A) $\ln\sigma$, and (B) σ.

5.6 Summary

(1) For the extruded Mg—Sn alloy, the tensile strength is greatly increased in all three directions with increasing Sn content. The tensile strength along TD sheet increases from 266 to 331 MPa when the Sn content increases from 0.5 wt.% to 2.5 wt.%. With increasing Sn content, both the elongation and n of extruded Mg—Sn alloy are increased.

(2) The Mg—Al—Sn ternary phase diagram is constructed using the existing thermodynamic data, which has been verified by using the balance samples. No ternary compounds are detected in the Mg—Al—Sn system. The isothermal section of the system at 300°C contains 7 single-phase regions: α-Mg, Al, Sn, $Mg_{17}Al_{12}$, $Al_{30}Mg_{23}$, Al_3Mg_2, and Mg_2Sn; 11 binary-phase regions: Al—Al_3Mg_2, Al_3Mg_2—$Mg_{17}Al_{12}$, Mg—$Mg_{17}Al_{12}$, Mg—Mg_2Sn, Mg_2Sn—Sn, Sn—Al, Mg_2Sn—Al_3Mg_2, Mg_2Sn—$Mg_{17}Al_{12}$, Mg_2Sn—Al, Al_3Mg_2—$Al_{30}Mg_{23}$, and Mg_2Sn—$Al_{30}Mg_{23}$; and 5 ternary-phase regions: Mg_2Sn—$Mg_{17}Al_{12}$—Mg, Mg_2Sn—$Mg_{17}Al_{12}$—Al_3Mg_2, Mg_2Sn—Al—Al_3Mg_2, Mg_2Sn—Al—Sn, and Al_3Mg_2—Mg_2Sn—$Al_{30}Mg_{23}$.

(3) The Mg—Al—Sn—Mn alloy mainly contains three phases of α-Mg, $Mg_{17}Al_{12}$, and Mg_2Sn and a small amount of Mn-containing compounds of Al_8Mn_5 and $Al_8(Mn, Fe)_5$. In the Mg—Al—Sn—Mn casting alloy, the addition of alloying elements can effectively refine the grains and improve the strength of the alloy. The effect of Sn element is weaker than that of Al. The addition of 1% Sn to AM alloys with Al content larger than 6% can improve the morphology of the coarse divorced eutectic $Mg_{17}Al_{12}$ phase, refining the grains to some extent and effectively improving the strength and plasticity of the alloy. The composition of the casting Mg—Al—Sn—Mn alloy with low cost and good comprehensive properties is: Al content larger than 6%, Sn content ranging from 1%—3%, and the Mn content about 0.3%. The yield strength of such alloy is 110—140 MPa, the tensile strength is up to 200 MPa, and the elongation changes within 1%–8%.

(4) In the extruded Mg–xAl–ySn–0.3Mn ($x = 1$, 3 and $y = 1$, 3, and 5) alloys that have not been completely recrystallized, the coarse un-recrystallized region contains relatively strong texture of $(10\bar{1}0)$ and (0001), the normal direction of (0001) base plane and the $<11\bar{2}0>$ slip direction of the un-recrystallized grain tend to be perpendicular to the extrusion direction. The weak recrystallization texture of $(10\bar{1}0)$ causes that the (0001) base plane and the $<11\bar{2}0>$ slip direction of the recrystallized grain tend to be parallel to the extrusion direction. Increasing the content of alloying elements or increasing extrusion temperature can promote recrystallization, while weaken the texture, and ultimately decrease the tensile yield strength of the alloy. On the other hand, it can effectively increase the tensile yield strength of the alloy. The Mg–xAl–ySn–0.3Mn ($x = y = 1$ or 3) alloys exhibit excellent comprehensive properties with yield strength >200 MPa and elongation $\sim 20\%$.

(5) In the as-extruded Mg–xAl–ySn–0.3Mn ($x = 6$ and 9, $y = 1$, 3, and 5) alloy showing complete recrystallization, the effects of composition and extrusion temperature on the recrystallized grain size become small; solid solution strengthening and second phase strengthening are the main strengthening mechanisms for the alloys. Increasing the content of alloying elements or increasing extrusion temperature can effectively improve the tensile yield strength of the alloys; the Mg–9Al–ySn–0.3Mn ($y = 1$, 3, and 5) alloy shows relatively large yield strength (>280 MPa).

(6) In the as-extruded Mg–Al–Sn–Mn alloy, the existing Mg$_2$Sn before extrusion has a non-coherent relationship with the Mg matrix; the Mg$_2$Sn phase precipitated dynamically during the extrusion process is coherent or semi-coherent with the Mg matrix. The orientation of the coherent relationship is: $(0001)_{Mg}//(0\bar{3}3)_{Mg2Sn}$, $[2\bar{1}\bar{1}0]Mg//[\bar{1}22]_{Mg2Sn}$. The relationship between the precipitation phase of Mg$_{17}$Al$_{12}$ and the Mg matrix is: $(0001)_{Mg}//(2\bar{2}2)_{Mg17Al12}$, $[2\bar{1}\bar{1}0]_{Mg}//[122]_{Mg17Al12}$. The relationship between the precipitation phase of Al$_8$Mn$_5$ and the Mg matrix is: $(\bar{2}\bar{1}10)_{Mg}//(20\bar{2}0)_{Al8Mn5}$, $[\bar{2}4\bar{2}3]_{Mg}//[01\bar{1}1]_{Al8Mn5}$.

(7) The average dynamic recrystallization activation energy of ATM110 alloy is evaluated to be 186.031 kJ mol^{-1}. The peak flow stress can be expressed via $\sigma = \frac{1}{0.0652}\ln\left\{\left(\frac{Z}{5.2\times10^8}\right)^{\frac{1}{1.528}} + \left[\left(\frac{Z}{5.2\times10^8}\right)^{\frac{2}{1.528}} + 1\right]^{\frac{1}{2}}\right\}$. The average dynamic recrystallization activation energy of ATM130 alloy is 128.639 kJ mol^{-1}. The peak flow stress can be calculated via $\sigma = \frac{1}{0.0156}\ln\left\{\left(\frac{Z}{2.67\times10^6}\right)^{\frac{1}{5.2}} + \left[\left(\frac{Z}{2.67\times10^6}\right)^{\frac{2}{5.2}} + 1\right]^{\frac{1}{2}}\right\}$. The average dynamic recrystallization activation energy of ATM310 alloy is 194.138 kJ mol^{-1}. The peak flow stress can be expressed as $\sigma = \frac{1}{0.01128}\ln\left\{\left(\frac{Z}{3.86\times10^{11}}\right)^{\frac{1}{6.58}} + \left[\left(\frac{Z}{3.86\times10^{11}}\right)^{\frac{2}{6.58}} + 1\right]^{\frac{1}{2}}\right\}$.

The average dynamic recrystallization activation energy of ATM330 alloy is 163.44 kJ mol^{-1}. The peak flow stress can be quantified as

$$\sigma = \frac{1}{0.12345}\ln\left\{\left(\frac{Z}{8.72\times10^{11}}\right)^{\frac{1}{6.202}} + \left[\left(\frac{Z}{8.72\times10^{11}}\right)^{\frac{2}{6.202}} + 1\right]^{\frac{1}{2}}\right\}.$$

(8) Sn-containing alloys generally have good comprehensive properties, but the improvement of plasticity is relatively small. This is consistent with the previous calculations for alloy design, further confirming the importance of solution elements to the CRSS.

References

[1] Doernberg E, Kozlov A, Schmid-Fetzer R. Experimental investigation and thermodynamic calculation of Mg–Al–Sn phase equilibria and solidification microstructures [J]. Journal of Phase Equilibria and Diffusion 2007;28(6):523−35.

[2] Y.-B. Kang, AD Pelton. Modeling short-range ordering in liquids: The Mg–Al–Sn system [J], 2010, (34):180−188.

[3] Chen D, Ren YP, Guo Y, et al. Microstructures and tensile properties of as-extruded Mg-Sn binary alloys [J]. Transactions of Nonferrous Metals Society of China 2010;020(007):1321−5.

[4] Shi BQ, Chen RS, Ke W. Solid solution strengthening in polycrystals of Mg–Sn binary alloys [J]. Journal of Alloys & Compounds 2011;509(7):0−3362.

[5] Sasaki TT, Yamamoto K, Honma T, et al. A high-strength Mg–Sn–Zn–Al alloy extruded at low temperature [J]. Scripta Materialia 2008;59(10):1111−14.

[6] Jung JG, Park SH, Yu H, et al. Improved mechanical properties of Mg–7.6Al–0.4Zn alloy through aging prior to extrusion [J]. Scripta Materialia 2014;93:8−11.

[7] Roberts W, Ahlblom B. A nucleation criterion for dynamic recrystallization during hot working [J]. Acta Metallurgica 1978;26(5):801−13.

[8] Jonas JJ, Sellars CM, Tegart WJM. Strength and structure under hot-working conditions [J]. Metallurgical Reviews 1969;14(1):1−24.

[9] Luton MJ, Sellars CM. Dynamic recrystallization in nickel and nickel-iron alloys during high temperature deformation [J]. Acta Metallurgica 1969;17(8):1033−43.

CHAPTER 6

Microstructure and mechanical properties of the high-strength Mg−Gd−Y−Zn−Mn alloy

In recent years, attributed to the comprehensive strengthening effects of the long period stacking order (LPSO) phase and aging precipitation, the strength of the developed Mg−RE−Zn−Zr wrought Mg alloy can reach 500 MPa, but the plasticity is very low, and the elongation is generally about 3% to 5%. In addition, there is still a lack of research on the plastic deformation of Mg−RE−Zn alloys. The previous studies have shown that the addition of Mn can improve the plasticity, and the extrusion and rolling deformation can effectively control the microstructure and properties of Mg alloys. This can be applied for mass production of bars, pipes, profiles, and plates, which is of great significance for promoting the production and application of Mg alloys. Therefore this chapter replaces Zr with Mn and controls LPSO phase to develop high-strength and high-toughness wrought Mg alloys. The effects of extrusion and rolling deformation on the microstructure and properties of the high-strength Mg−Gd−Y−Zn−Mn alloy are introduced.

6.1 Effect of extrusion on the microstructure and mechanical properties of Mg−Gd−Y−Zn−Mn alloy

6.1.1 Microstructure of the as-cast Mg−Gd−Y−Zn−Mn alloy before and after homogenization annealing

As shown in Fig. 6.1, the main phases of the as-cast Mg−Gd−Y−Zn−Mn alloy consist of α-Mg, LPSO [$Mg_{12}Zn(Y,Gd)$], and eutectic phase [$(Mg,Zn)_3(Gd,Y)$]. After heat treatment of high-temperature homogenization annealing, the diffraction peak of the eutectic phase disappears, while the diffraction peak of the LPSO phase increases. It shows that the eutectic phase dissolved or transformed into LPSO phase during the homogenization annealing process. After the homogenization annealing, the eutectic phase disappears and the LPSO phase content increases in the alloy.

Fig. 6.2 shows the metallographic images of the microstructure of the alloy before and after the homogenization annealing at 540°C for 4 h. There exist some dendritic grains in the as-cast alloy, and the overall grain microstructure is uneven. After the homogenization annealing, the grains grow significantly, and all the grains evolve into the equiaxed grains.

High Plasticity Magnesium Alloys
DOI: https://doi.org/10.1016/B978-0-12-820110-7.00006-9

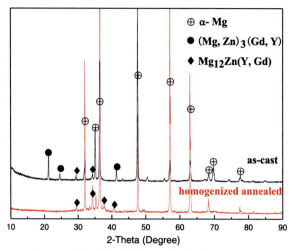

Figure 6.1 XRD patterns of as-cast Mg alloy before and after homogenization annealing.

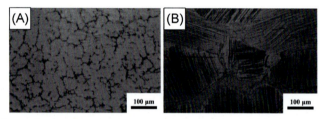

Figure 6.2 Metallographic images of microstructure of the (A) as-cast and (B) as-annealed alloys.

From the backscatter scan images of the as-cast alloy shown in Fig. 6.3A, there are the second phases with two contrasting in the as-cast alloy, namely, white and gray appearance. The white phase with larger size presents skeletal morphology, the remaining white and gray phases show block-like morphology. The two second phases are distributed in a network at grain boundaries and between dendrites. As seen in Fig. 6.3B, the second phase in the homogenized annealed alloy is mainly the gray phase, a part of which is distributed at grain boundary in a block shape, and the other part is distributed in grain interiors in a layered form. The orientation of the layered phase in each grain is the same. The energy-dispersive spectrometer (EDS) composition test was performed on the white and gray phase in the as-cast and homogenized annealed alloys, with the results shown in Table 6.1.

According to the content ratio of each element in Table 6.1 and the X-ray diffraction (XRD) analysis results, it can be inferred that the white phase and the gray phase in the as-cast alloys are the eutectic phase of $(Mg,Zn)_3(Gd,Y)$ and the LPSO phase of $Mg_{12}Zn(Y,Gd)$, respectively, while the gray phase in a block shape in the as-annealed

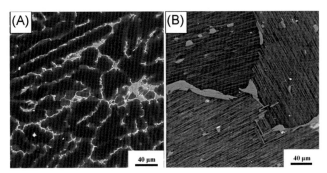

Figure 6.3 Scanning electron microscopy (SEM) images of microstructures of the (A) as-cast and (B) as-annealed alloys.

Table 6.1 EDS results of the second phase in the as-cast and as-annealed alloys.

Alloy state	Second phase	Mg (at%)	Zn (at%)	Y (at%)	Gd (at%)	Phase
As–cast	White bright	79.88	6.09	6.62	7.41	(Mg,Zn)₃RE eutectic phase
	Gray bright	90.42	3.31	3.61	2.66	LPSO phase
Homogenized annealed	Gray bright	88.68	4.19	4.25	2.88	LPSO phase

alloys is the LPSO phase of $Mg_{12}Zn(Y,Gd)$. For the layered phase in the as-annealed alloys, the layered phase is 14H-typed-LPSO phase based on the TEM analysis results of the lamellar phase in the alloys of the similar system.

6.1.2 Effect of extrusion ratio on the microstructure and mechanical properties of the extruded Mg−Gd−Y−Zn−Mn alloy bar

Fig. 6.4 shows the metallographic images of the microstructures of extruded alloy bar. As seen in Fig. 6.4A and B, after the extrusion with an extrusion ratio of 8, there are several lamellar structures in the alloy perpendicular and parallel to the extrusion direction. Only a small amount of new recrystallized grains appear at the boundary of the lamellar structure, which consists of lamellar LPSO phase and α-Mg matrix. The above results indicate that no obvious recrystallization occurs during the extrusion process with the extrusion ratio of 8. During the hot extrusion, the large amount of fine layered LPSO phases hinders the recrystallization and subsequently the growth of the recrystallized grains by separating the Mg matrix and restricting the diffusion of atoms among the Mg matrix. Besides, the extrusion ratio is relatively low, resulting in insufficient distortion energy to promote the nucleation and growth of recrystallization. As shown in Fig. 6.4C and D, the extruded alloy with an extrusion ratio of 11 is still dominant by the layered

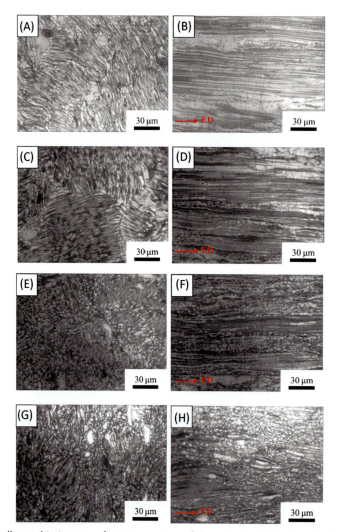

Figure 6.4 Metallographic images of microstructure of the extruded alloy bar (A—B: extrusion ratio is 8, C—D: extrusion ratio is 11, E—F: extrusion ratio is 27, and G—H: extrusion ratio is 42).

structure. As the extrusion ratio increases from 8 to 11, the number of the recrystallized grains increase, which mainly distributed at the grain boundaries of the original grains. When the extrusion ratio is 27, the extrusion ratio is relatively large, which leads to the severe deformation of the alloy during the extrusion process. The increase in the deformation energy storage leads to intense recrystallization, and finally a large amount of equiaxed recrystallized grains appear in the alloy, as observed in Fig. 6.4E and F. As the extrusion ratio is further increased to 42, the alloy is dominant by the recrystallized grains, and only a few lamellar phases can be observed, as seen in Fig. 6.4G and H. In general, as the extrusion ratio increases, the recrystallized microstructure gradually increases, and the layered deformed microstructure gradually decreases, which is attributed to the large amount of recrystallization. The recrystallization grain has no obvious refinement with increasing extrusion ratio. Even at the maximum extrusion ratio of 42, the increase in the size of recrystallized gains is not obvious, which is attributed to the easy growth of partially recrystallized grains at high extrusion temperature.

Fig. 6.5 shows the backscattered scan images of microstructures of the extruded bar. The second phase of the alloy under various processes is mainly the gray phase, namely, the LPSO phase. After extrusion, the LPSO phase is mainly distributed along the extrusion direction. Compared with the original homogenized annealed alloy, the size of the bulk-shaped LPSO phase in the alloy after extrusion is smaller and the distribution is more uniform. Under the action of extrusion, the bulk-shaped LPSO phase is elongated and thinned or cut apart by dislocations, and then redistributed under the action of extrusion deformation and recrystallization. However, with increasing the extrusion ratio, the bulk LPSO phase has no obvious refinement. The elongated LPSO phase in the SEM image indicates the bulk LPSO phase has undergone plastic deformation. In addition, due to the interaction between the LPSO phase and dislocations, the LPSO phase is cut by the dislocations, resulting in the segmented bulk-shaped LPSO phase. It can be induced that the LPSO phase is a second phase with certain toughness. It is indicated that when the alloy with LPSO phase is deformed by an external force, the bulk LPSO phase can be deformed by a corresponding dislocation slip mechanism. In the process of extrusion deformation, the bulk LPSO phase will undergo corresponding deformation, and then the alloy will undergo coordinated deformation to release the stress, while the brittle phase will break directly under the strong stress. Since the content of the LPSO phase in the alloy is not high, the plasticity is mainly contributed from the Mg matrix. Therefore with increasing extrusion ratio, the deformation of the LPSO phase is not large, resulting in the similar size for the bulk LPSO phase in the alloys with different extrusion ratios.

In the alloy with an extrusion ratio of 8, many lamellar LPSO phases are twisted, as shown in Fig. 6.5B. The torsion deformation is an important deformation mechanism for the LPSO phase, which is beneficial to coordinate the deformation of the matrix alloy. This can be observed in a large amount of the related researches on the plastic deformation of Mg alloy containing LPSO phase. The layered structure can be damaged after twisting. At this time, the restriction on the recrystallization is weakened, and the deformation energy storage in the twisted area is relatively large, which makes the twisted deformation area easy to recrystallization. When the extrusion ratio is increased to 11 and 27, the twisting deformation is rarely observed in the alloy, because the degree of deformation increases and the recrystallization is promoted, and the twisting deformation area preferentially recrystallizes and then disappears. In addition, with increasing the extrusion ratio, the layered LPSO phase is gradually decreased. On the one hand, the twisting deformation area increases with increasing deformation, and the layered LPSO phase decreases after the

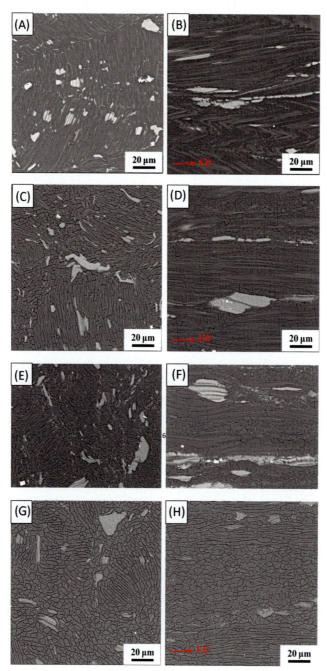

Figure 6.5 SEM images of microstructures of the extruded alloy bar (A–B: extrusion ratio is 8, C–D: extrusion ratio is 11, E–F: extrusion ratio is 27, and G–H: extrusion ratio is 42).

recrystallization. On the other hand, the Mg matrix in the layered LPSO phase is easy to recrystallize with increasing deformation. The appearance of the new grains may disrupt the layered LPSO phase, resulting in the decrease of the layered LPSO phase. However, with the increase of recrystallization grains, the layered LPSO phase does not directly dissolve and disappear. By carefully observing the equiaxed grains area, it can be found that a large number of interrupted LPSO phases still exist at the interface of new grains.

The mechanical properties of the extruded bar alloy are shown in Fig. 6.6 and Table 6.2. Compared with the homogenized annealed alloy before extrusion, the yield strength and elongation of the alloy after extrusion deformation are greatly improved. On the one hand, the casting defects are eliminated after the extrusion deformation. On the other hand, the alloy microstructure is refined after the extrusion deformation, and the layered LPSO phase is distributed along the extrusion direction. It is indicated that the LPSO phase distributed along the extrusion direction plays a role of fiber reinforcement to the alloy. In contrast, with increasing extrusion ratio, the yield strength of the alloy gradually

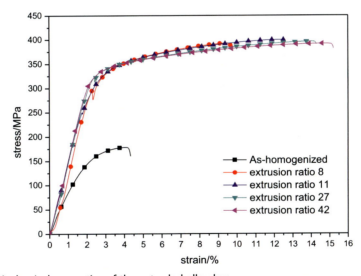

Figure 6.6 Mechanical properties of the extruded alloy bar.

Table 6.2 Mechanical properties of the extruded alloy bar.

Alloy	Tensile strength/MPa	Yield strength/MPa	Elongation/%
As-annealed	178	126	2.4
Extrusion ratio of 8	392	292	7.2
Extrusion ratio of 11	400	298	9.8
Extrusion ratio of 27	396	310	11.8
Extrusion ratio of 42	392	312	12.4

increases. As for the alloys with extrusion ratio of 8 and 11, the grain microstructure is elongated along the extrusion direction and undergoes different deformation without significant recrystallization refinement. The increase in the strength of the alloy after extrusion could be attributed to (1) the fiber reinforcement of a large amount of fine layered LPSO phases distributed along the extrusion direction; (2) the layered phase separates the Mg matrix and restricts the coordination of deformation among the Mg matrix, thereby increasing the strength of the alloy; (3) a certain degree of work hardening. When the extrusion ratio increases from 8 to 11, the degree of deformation of the alloy increases, the parallelism between the lamellar phase and the extrusion direction increases, the strengthening effect of the lamellar phase increases, and the corresponding work hardening increases, thus the yield strength is increased. For the alloys with extrusion ratios of 27 and 42, a large amount of new recrystallization appears after extrusion, and a certain degree of recrystallization refinement occurs. The increase in the strength of the alloy after extrusion is mainly due to the grain refinement strengthening of the fine recrystallized grains, a certain degree of work hardening, and the strengthening of the lamellar phase in the deformed grain. When the extrusion ratio increases from 27 to 42, the deformation increases, the recrystallization grains increase, and the yield strength is slightly increased. In comparison with the alloys with extrusion ratios of 27 and 42, the alloys with extrusion ratio of 8 and 11 have relatively low degree of deformation, no recrystallization refinement occurs, so their yield strength is relatively low. In addition, as the extrusion ratio increases, the plasticity of the alloy increases. When the alloy is deformed, the stress concentration usually occurs near the grain boundary due to the dislocation plugging. In comparison with the alloy with the extrusion ratio of 8, the alloy with an extrusion ratio of 11 recrystallizes at the grain boundaries of the original grains, which releases the stress in this area, so that the alloy with an extrusion ratio of 11 has relatively high plasticity. When the extrusion ratio is increased to 27, a large number of fine recrystallization grains appear in the alloy, which is conductive to the more uniform deformation of the alloy and further improves the plasticity. When the extrusion ratio is further increased to 42, the recrystallization increases. Although the recrystallization grains grow slightly, the deformation among the equiaxed grains is much more coordinated and the increase of the recrystallization gains further increases the plasticity.

It is noteworthy that the tensile strength of the alloy with different extrusion ratios is almost the same. Comparing the stress-strain curves carefully, it can be found that the stress increases faster with the increase of strain for the alloys with extrusion ratios of 8 and 11 in the process of plastic deformation. This indicates that the alloys with extrusion ratios of 8 and 11 have a stronger effect on restricting the dislocation movement during plastic deformation. Comparing the microstructure of alloys with different extrusion ratios, it can be seen that the alloys with extrusion ratio of 8 and 11 have more deformation microstructure consisting of Mg matrix and layered LPSO phase. Studies have shown that the LPSO phase can significantly improve the mechanical properties of the alloy. When tensile testing at room temperature, the plastic deformation mainly occurs in the Mg matrix, and the Mg matrix is separated by a large number of layered LPSO phases. When the plastic deformation reaches a certain level, it is difficult for the dislocation to continue to move after encountering the LPSO phase, resulting in a significant increase in stress with the increase of plastic deformation. In alloys with extrusion ratio of 27 and 42, the layered LPSO phase is destroyed, and a large number of recrystallized grains are dominant. During the deformation, the grains can coordinate with each other and release the stress well, resulting in a small increase of stress with the increase of strain.

6.1.3 Microstructure and mechanical properties of the extruded Mg−Gd−Y−Zn−Mn alloy sheet

Fig. 6.7 shows the microstructure of the extruded alloy sheet. The alloy has mainly a layered deformation microstructure. It can be seen in the cross section that there is a relatively large amount of equiaxed recrystallized grains appearing at the junction of different layered structures. In the longitudinal section, the layered microstructure tends to be parallel to the extrusion direction. The extrusion ratio of the sheet is 11. Compared with the microstructure of the bar alloy with the same extrusion ratio of 11, the sheet has more recrystallization grains and larger grain size. It can be inferred that under the same extrusion temperature and extrusion ratio, the recrystallization degree of the sheet is higher than that of the solid round bar. In the extrusion process of the sheet, the heat generated by friction and deformation is large, which causes a large increase in the extrusion temperature and ultimately leads to the increase and growth of recrystallized grains.

Furthermore, the microstructure of the extruded alloy sheet was characterized by SEM backscattering electron scanning to observe the second phase, as shown in Fig. 6.8. The extruded alloy is mainly composed of a second phase with one contrast, namely, the LPSO phase as described above. Similar to the extrusion alloy bar, the bulk LPSO phase in the alloy sheet is elongated along the extrusion direction during the extrusion, and the bulk LPSO phase is segmented. In addition, there are also many layered LPSO phases that tend to be parallel to the extrusion direction. Generally speaking, the size and distribution of LPSO phase in the alloy sheet is almost the same as that in the extruded alloy bar.

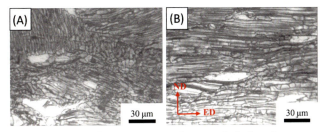

Figure 6.7 Metallographic images of microstructure for extruded alloy sheet.

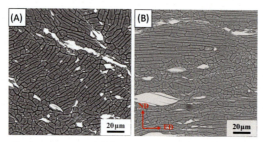

Figure 6.8 SEM images of microstructure of the extruded sheet alloy.

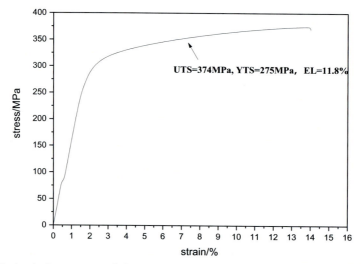

Figure 6.9 Mechanical properties of the extruded alloy sheet. The UTS, YTS, and EL represent the ultimate tensile strength, yield strength and elongation, respectively.

Fig. 6.9 shows the tensile properties of the alloy sheet at room temperature. Compared with the extruded alloy bar with the same processing parameters, the alloy sheet has lower strength and higher plasticity. Because of the larger grain size of the extruded alloy sheet, the grain refinement strengthening is weak, resulting in the relatively low strength. On the other hand, the actual temperature of the sheet in extrusion is relatively high, and the recrystallization degree is large, which slightly increase the plasticity of the alloy.

6.1.4 Aging treatment of the extruded Mg−Gd−Y−Zn−Mn alloy

The high-strength Mg−Gd−Y−Zn alloy has a significant aging strengthening effect. That is, a large amount of phases can be precipitated through appropriate aging treatment to improve the strength of the alloy. To improve the mechanical properties of the extruded alloy, the aging process at 200°C is explored for the extruded alloy to

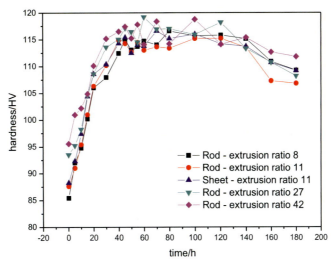

Figure 6.10 Aging hardening curves of the as-extruded alloy at 200°C.

make an appropriate aging process. As shown in Fig. 6.10, with increasing the aging time at the initial stage, the microhardness of the alloy increased rapidly, indicating that the alloy has an obvious aging hardening effect and the aging hardening rate is similar. After aging for about 45 h, the hardness of the alloy with different extrusion processes almost reaches the maximum values. With continuous increase in the aging time, the hardness of the alloy fluctuates up and down. Until after 120 h, the hardness of the alloy shows a downward trend. On the whole, there is no obvious difference among the aging hardening for bar alloys with different extrusion ratios, only a slight difference in the microhardness values. The hardness of the extruded bar alloy increases slightly with increasing extrusion ratio, from 85, 87, and 93 to 95 HV. The corresponding peak aging hardness value (take the average hardness of 45 h to 120 h) is about 115, 114, 117, and 116 HV, respectively. It can be seen that the difference in the microhardness values is small, and the extrusion ratio has little effect on the aging hardening effect of the alloy. In addition, the microhardness of extruded alloy sheet increases from 88 to 115 HV, and the peak aging hardness increases by 27 HV, which is the same as the alloy bar with the extrusion ratio of 11.

According to the aging hardening curve in Fig. 6.10, the aging process for all extruded alloys is set to be 200°C × 48 h. Fig. 6.11 and Table 6.3 show the results of mechanical properties of the alloy after aging treatment. In comparison with the mechanical properties of the as-extruded alloy, the tensile strength of the alloy after aging is increased by 80−100 MPa, showing a significant aging strengthening. Nevertheless, the plasticity is greatly decreased. For the extruded alloy bars with an extrusion ratio of 11, the alloy after aging treatment has the highest tensile strength of

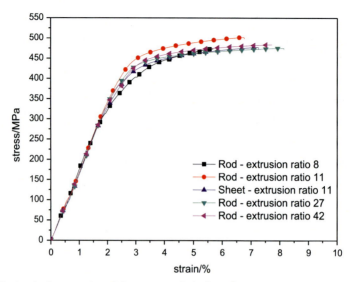

Figure 6.11 Mechanical properties of the as-extruded alloy after aging treatment at 200°C for 48 h.

Table 6.3 Mechanical properties of the as-extruded alloy after aging treatment.

Alloy	Tensile strength/MPa	Yield strength/MPa	Elongation/%
Rod(8)	473	342	2.8
Rod(11)	502	410	3.8
Sheet(11)	477	390	4.4
Rod(27)	475	392	5.4
Rod(42)	484	390	5.0

502 MPa, but its plasticity is poor and the elongation is only 3.8%. In comparison, the alloys with extrusion ratio of 27 and 42 have relatively high plasticity, especially the alloy with extrusion ratio of 42 shows good comprehensive mechanical properties. For the extruded alloy sheet, the tensile strength is increased by 103 MPa after aging and the increment is similar to that of the bar alloy with the same extrusion ratio.

After aging, a large number of precipitation phases can be formed in Mg—RE alloys containing Gd and Y elements, which significantly improves the strength of the alloys. As shown in Fig. 6.12, the bar alloy with an extrusion ratio of 11 after aging at 200°C for 48 h was observed and analyzed by TEM. In addition to the layered LPSO phase mentioned above, a large amount of black lens-like second phase are precipitated in the alloy. The transmitted electron beam is along the belt axis direction of $[11\bar{2}0]$, and the corresponding selected area electron diffraction pattern is obtained. Combined relevant references, the electron diffraction pattern is calibrated to determine that the layered phase is the 14H-type LPSO phase and the lens-like

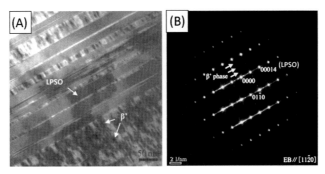

Figure 6.12 (A) Bright-filed TEM image and (B) selected area electron diffraction pattern of [11$\bar{2}$0] direction of the precipitated phase for bar alloy with an extrusion ratio of 11 after aging at 200°C for 48 h.

precipitation phase is β′ phase. Due to the large amount of dispersed β′ phase in the Mg matrix after aging, the movement of dislocations is greatly hindered, so that the strength of the alloy is significantly improved.

6.2 Effect of rolling on the microstructure and mechanical properties of the Mg—Gd—Y—Zn—Mn alloy

The metal sheet can be directly used as a component, and it is also a necessary material for stamping structural parts. It plays an important role in the practical applications. Rolling is the main method for producing sheet. It is easy to crack during rolling deformation due to the poor plasticity of Mg alloys. The rolling deformation process of Mg alloys is not mature enough. For the high-strength Mg—RE—Zn alloy, it is necessary to develop the sheet rolling processes to promote its production and application. Although there have been reports on the preparation of Mg—RE—Zn alloys with relatively high strength by appropriate rolling methods, the research on the rolling process of high-strength Mg—RE—Zn alloys is still insufficient. Therefore this chapter studies the rolling deformation of Mg—Gd—Y—Zn—Mn alloy to provide guidance for the rolling process of Mg—RE—Zn alloy.

As reported, the poor plasticity of Mg—Gd—Y—Zn—Mn alloy is due to the large number of second phases in the alloy. In extrusion deformation with low extrusion ratio at 450°C, the recrystallization is not easy to occur, and the recrystallization is the main softening mechanism for Mg alloys during rolling deformation. The plasticity of the alloy significantly decreases during rolling deformation. Since the plasticity of the alloy can be improved by recrystallization, carking is easy to occur, and rolling process is more difficult. This chapter first explores the effect of the rolling temperature and reduction on the microstructure of the as-annealed alloy, determines the appropriate process parameters, and then proceeds with appropriate rolling deformation to study the microstructure and properties of the as-annealed alloy after rolling. C. Xu et al. [8] compared the rolling deformation of the as-cast and as-annealed Mg−Gd−Y−Zn−Zr alloy. It is found that the as-cast alloys are easy to get uniform recrystallization grains and have higher plasticity. However, the second phase in the as-cast alloy after rolling is still mainly eutectic phase, and the strength of the alloy is relatively low. It is indicated that when the Mg−Gd−Zn alloy is solid solution treated at 500°C, the eutectic phase can be dissolved in the alloy, and the LPSO phase can be precipitated. The type and morphology of the LPSO phases in the alloy have different effects on the recrystallization of the hot deformed alloy. To this end, this chapter carried out appropriate "rolling and heat treatment" on the as-cast alloy. It combines the effects of deformation and heat treatment to adjust the grain structure, LPSO phase in the alloy, rolling deformation, and ultimately to improve the comprehensive mechanical properties of the alloy through process control.

6.2.1 Exploration of rolling process parameters of the as-annealed Mg−Gd−Y−Zn−Mn alloy

During the hot deformation process, the metal will recrystallize and produce a new grain structure, thereby improving the microstructure of the alloy. It is generally believed that the driving force for recrystallization is the lattice distortion energy produced by alloy deformation. The greater the deformation degree, the larger lattice distortion energy, and the easier the recrystallization. At the same time, the surface energy that can be converted to a new grain boundary is large, which means the grain will be efficiently refined. In addition, the nucleation and growth of recrystallization are related to the diffusion of atoms. The higher the temperature, the easier the atom diffusion in the alloy, which promotes the recrystallization nucleation and growth. However, when the temperature is too high, it will cause the second growth of grains, making the gains coarser. In other words, the temperature and reduction have important effects on the microstructure of the alloy after rolling. Therefore this section explores the rolling temperature and cumulative reduction of Mg−Gd−Y−Zn−Mn alloy.

Fig. 6.13 shows the samples obtained by rolling at different temperatures and cumulative reductions, rolled at 450°C, 500°C, and 520°C, respectively. No obvious cracks are observed in the alloys with reduction of 20% and 35%. For the alloy with a reduction of 50%, severe edge cracking occurs when rolled at 450°C, and there are many edge cracks, with the crack depth of 3−4 mm. Slight edge cracking occurs when rolling at 500°C, and there are few edge cracks, with the crack depth of only 1−2 mm. No obvious edge cracking is observed in the alloy rolled at 520°C. Obviously, with increasing rolling temperature, the edge cracking of rolled alloy is improved. When the deformation temperature increases, the thermal motion of atoms increases, and the binding force between the cylindrical and conical surfaces in Mg alloy decreases. That is, the critical shear stress is reduced, and the sliding of prismatic and pyramidal surfaces will be

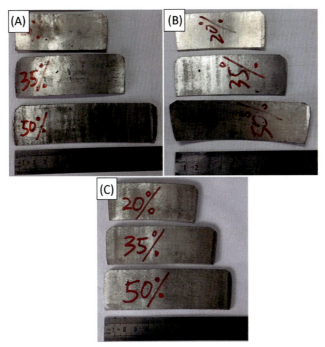

Figure 6.13 Sample images after rolling under different temperatures: (A) 450°C, (B) 500°C, and (C) 520°C.

easily activated. The increase in the slip mechanism makes the deformation of the alloy easier to coordinate and alleviates the premature cracking of the alloy due to stress concentration. On the other hand, with increasing deformation temperature, the alloy is prone to dynamic recovery, recrystallization, and softening, and the plasticity is improved. Therefore the alloy is not easy to crack when the rolling deformation is carried out at a high temperature with a relatively large reduction.

Fig. 6.14 shows the metallographic images of the longitudinal section of the as-rolled alloy at 450°C. Compared with the initial state, it can be observed that the layered phase in the alloy with different reductions has been twisted and deformed to different degrees. With the reduction increases, the degree of torsion deformation increases. When the reduction reaches 50%, the layered phase in the alloy begins to tend to be parallel to the rolling direction. It is worth to note that no sign of recrystallization occurs in the alloys with different reductions, and no grain of recrystallization is observed. As an important softening mechanism in the thermal deformation of Mg alloys, the recrystallization is beneficial to restore the plasticity of the alloy to continue to increase the deformation. During the rolling process at 450°C, the alloy fails to improve its plasticity through recrystallization effectively. With continuous deformation, the alloy suffers relatively severe damage.

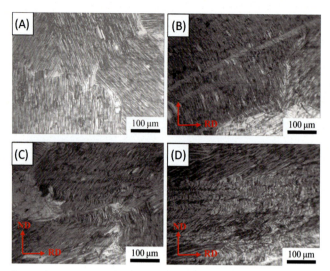

Figure 6.14 Metallographic images of microstructure of alloy rolled at 450°C: (A) initial state, (B) with a reduction of 20%, (C) with a reduction of 35%, and (D) with a reduction of 50%.

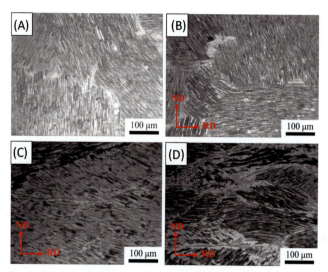

Figure 6.15 Metallographic images of microstructure of alloy rolled at 500°C: (A) initial state, (B) with a reduction of 20%, (C) with a reduction of 35%, and (D) with a reduction of 50%.

Fig. 6.15 shows the metallographic microstructure of the longitudinal section of the alloy rolled at 500°C. When the reduction is 20%, the layered phase in the individual gains of the alloy undergoes a slight twisting deformation. When the reduction is 35%, a very small amount of recrystallized grains appear in the alloy. When the reduction is 50%, a small amount of recrystallized grains appear in the severely

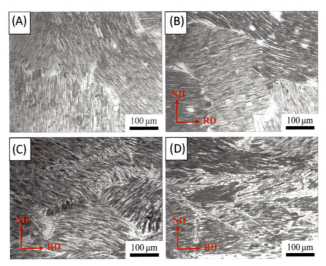

Figure 6.16 Metallographic images of microstructure of alloy rolled at 520°C: (A) initial state, (B) with a reduction of 20%, (C) with a reduction of 35%, and (D) with a reduction of 50%.

deformed area. At the same time, the layered phase tends to be parallel to the rolling direction. When rolling at 500°C, as the rolling reduction increases, the alloy begins to recrystallize and the plasticity of the alloy is improved during the deformation. Therefore only slight edge cracking occurs when the reduction reaches 50%.

Fig. 6.16 shows the metallographic images of the longitudinal section of the alloy rolled at 520°C. When the reduction is 20%, the layered phase in the alloy is slightly twisted and deformed as in the case of rolling at 450°C and 500°C. When the reduction is 35%, it is similar to the alloy rolled at 500°C, appearing a small amount of recrystallized grains. When the reduction is 50%, the recrystallized grains occurs in the severely deformed area of the alloy. In comparison with that rolled at 500°C, there are more recrystallized grains, indicating the increase in the degree of recrystallization.

Based on the microstructure of the rolled alloy at different temperatures, it can be seen that the microstructure is mainly different in the recrystallization degree. It is not easy to recrystallize when rolling at 450°C. When the cumulative reduction exceeds 35% at the temperatures of 500°C and 520°C, the alloys begin to recrystallize. The recrystallization degree is relatively high at 520°C. However, the recrystallization degree of the rolled alloy is not high under the above conditions, mainly because the layered LPSO phase can restrict the recrystallization of the alloy.

By comparing the microstructure of rolled alloy with different reductions, it is inferred the microstructure evolution of the as-annealed alloy during the rolling process as shown in Fig. 6.17. When compressive stress is applied along the direction of the layered LPSO phase, the LPSO phase can be twisted and deformed. Before rolling, the layered phase is randomly distributed in the homogeneous annealed alloy. During

Figure 6.17 Schematic of the microstructure evolution of the homogenized annealed alloy during rolling process.

Table 6.4 Mechanical properties of the as-rolled alloy under different processes.

Rolling temperature	Reduction/%	As-rolled			Rolling state-400°C × 0.5 h		
		UTS/MPa	YTS/MPa	EL/%	UTS/MPa	YTS/MPa	EL/%
450°C	20	188		Fracture	243	202	1.3
	35	179		Fracture	184		fracture
	50	233		Fracture	252	242	0.6
500°C	20	230	202	0.6	220	185	1.2
	35	255		Fracture	256	218	0.8
	50	278		Fracture	262	242	0.5
520°C	20	245	200	0.9	226	183	1.6
	35	268	226	0.7	266	214	1.4
	50	258		Fracture	275	224	1.4

the rolling deformation, the grains are elongated along the rolling direction. To cooperate the deformation of the grains, the layered phase gradually tends to be parallel to the rolling direction. While, the layered LPSO phase with a larger angle to the rolling direction is subjected to compressive stress along the direction of the layered phase, and has a twisting deformation. The further increase in the deformation can increase the twisting deformation, and the layered phase structure will be destroyed. This region begins to recrystallize owing to the relatively large distortion energy. In addition, in the vicinity of the grain boundaries and the bulk phase of the homogenized annealed alloy, the recrystallization is also likely to occur due to the large distortion energy and stress concentration.

The mechanical properties of each rolled alloy were tested to discuss the effects of rolling temperature and reduction on the properties. The results are shown in Table 6.4. All the as-rolled alloys are obviously brittle, and have fractured without obvious plastic deformation. The poor plasticity can be attributed to that the alloy does not undergo significant recrystallization during rolling deformation, and the alloy has a high degree of work hardening, making it more brittle. The plasticity of the alloy is slightly improved after the stress-relief annealing at 400°C for 0.5 h. In terms of mechanical properties, the strength of the alloy after rolling is improved compared to the initial as-annealed alloy. Among them, the tensile strength of the alloy rolled to 50% at 520°C increases from 178 to 275 MPa. However, compared with the extrusion processing results mentioned above, the performance improvement is still not satisfactory. This is mainly because the deformation degree of the rolling is still very low. A large number of layered phases in the as-extruded alloy with an extrusion ratio of 8 have tended to be parallel to the extrusion direction. That is, there are already obvious processing flow lines. The layered LPSO phase has a great effect on the strength of the alloy. However, in the alloy with a reduction of 50%, the orientation of the layered phase is still random, and there is no obvious processing flow line. The strengthening effect of the layered phase is relatively poor.

Comparison of the mechanical properties of the alloy under various process conditions shows that the strength of the alloy increases with the increase of deformation at the same rolling temperature. The results indicate the severe rolling deformation can optimize the microstructure and then improve the mechanical properties. Therefore the reduction should be further increased to improve the mechanical properties of the alloy. The plasticity of the alloy is also required. It is worth to note that under the same deformation, although the yield strength increases with decreasing rolling temperature, the alloy with a high rolling temperature has a relatively high tensile strength due to its good plasticity. That is, the alloy rolled at 520°C has relatively good comprehensive mechanical properties. In addition, compared with the previous microstructure, the alloy rolled at 520°C will be significantly recrystallized. The recrystallization can soften the alloy and enable the alloy to be further rolled. In summary, for the rolling process of the homogenized annealed alloy, the appropriate rolling temperature is 520°C.

6.2.2 Rolling deformation of the homogenized annealed Mg−Gd−Y−Zn−Mn alloy and its microstructure and mechanical properties

The results show that the homogenized annealed alloy is not easy to crack during rolling at 520°C. However, the recrystallized grains of the alloy grow abnormally at high temperature, which fails to achieve the purpose of refining the microstructure. Therefore this section compares the rolling processes at 520°C−450°C to know whether it is possible to get finer and more uniform microstructure by decreasing the temperature of the second half of the rolling deformation. It is expected to further improve the properties of the alloy. To facilitate discussion, each annealed rolled sample is numbered, as presented in Table 6.5.

Table 6.5 Number of the homogenized annealed rolled alloy.

Sample number	Process
H60	The homogenized annealed alloy is rolled to reduction of 60% at 520°C
H76I	H60 → 400°C × 0.5 h rolling at 450°C accumulated reduction of 76%
H76II	H60 → 400°C × 0.5 h rolling at 520°C accumulated reduction of 76%

Figure 6.18 Optical photographs of the H76I and H76II samples.

After the edge cracks of H60 alloy are removed during the rolling process, the H76I and H76II alloys are obtained by continuous rolling at 450°C and 520°C, respectively. Fig. 6.18 shows the optical photographs of H76I and H76II samples. Overall, the alloys after continuous rolling of H60 have different degrees of edge cracking and two-end fracture. The depth of the edge crack of H76I is about 4−6 mm, and the one of the H76II is about 3−4 mm. In comparison, the edge cracking and two-end fracture of the rolled sample at 450°C are more serious.

The metallographic images of the cross-section and longitudinal cross-section microstructures of the rolled alloy after stress-relief annealing at 400°C for 0.5 h are shown in Fig. 6.19. H60 alloy still has a lot of un-recrystallized layered microstructure, and the layered microstructure tends to be parallel to the rolling direction in the longitudinal section. Compared with the alloy with 50% rolling reduction at 520°C in the previous section, it is found that more lamellar phase twisting regions appear in H60 alloy, and more fine equiaxed grains appear in these regions. The H60 alloy was continuously rolled at 450°C. In comparison, the twisting deformation of the H76I alloy increases, but the recrystallized grain does not increase significantly. From the longitudinal cross-sectional structure, it is found that there are more grain boundaries with an angle of around 45 degrees with the rolling direction. It is inferred that during the rolling process at 450°C, the degree of recrystallization is weak, the work hardening is gradually increased, and the plastic deformation coordination is gradually deteriorated. Under the action of the rolling shear force, a similar shear band area is formed. By careful observation, part of boundary has a certain width, that is, the formation of microcracks. In H76II alloy, which is continuously rolled at 520°C, the equiaxed grain

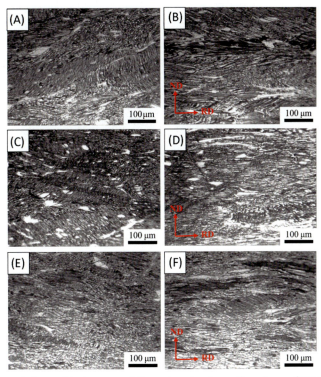

Figure 6.19 The metallographic images of the microstructures for alloys of H60, H76I, and H76II after stress-relief annealing (A and B: H60, C and D: H76I, and E and F: H76II).

microstructure dominates. In comparison with H60 alloy, the lamellar structure in H76 II alloy decreases, while the recrystallization is increased with a slight increase in the size of partial recrystallization grains. This indicates that during the continuous rolling at 520°C, with increasing deformation, a high degree of recrystallization occurs and partial recrystallization grains are easily to grow at 520°C.

The microstructure of the rolled alloy is further observed by SEM, as shown in Fig. 6.20. On the whole, each rolled alloy mainly has a second phase with one contrast, namely, the LPSO phase. As for the bulk LPSO phase, part of them are elongated and separated along the rolling direction, which is similar with the result of extrusion deformation. Overall, there is no significant difference in the size and distribution for the bulk phase in the alloys of H60, H76I, and H76II. This can be attributed to the fact that the alloy is mainly deformed by the Mg matrix when the force is applied, and the deformation of the bulk phase is relatively small, while the degree of the rolling deformation under 60% and 76% reduction is not much different. For the layered LPSO phase, the twisted deformation of the layered phase in some recrystallized regions could be observed in H60 alloy. The layered phase after the twisting deformation still exists at

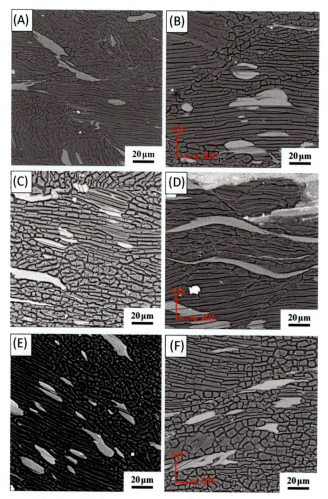

Figure 6.20 The SEM images of the microstructures for alloys of (A) and (B) the H60, (C) and (D) the H76I, and (E) and (F) the H76II after stress-relief annealing.

the boundary of the recrystallized grains. In the H76I alloy continuously rolled at 450°C, the lamellar phase on the longitudinal section is cut by the shear band under the action of rolling shear force. In the H76II alloy, which is continuously rolled at 520°C, the layered phase is relatively small, due to the relatively large degree of recrystallization, but the fine LPSO phase can also be observed at the junction of equiaxed grains. In addition, the sizes of the recrystallization grains of the alloy of H60, H76I, and H76II are about 4–6, 4, and 4–10 μm, respectively. There is no refinement in the recrystallized grains in the alloy that is continuously rolled at 450°C. The size of partially recrystallized grains in the alloys that are continuously rolled at 520°C is slightly larger. This indicates that it is easy for alloy to crack when it is continuously rolled at 450°C due to

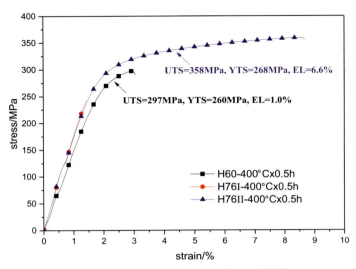

Figure 6.21 Mechanical properties of the alloys of H60, H76I, and H76II after stress-relief annealing.

the lack of recrystallization, which fails to achieve the purpose of refining the microstructure and improve the properties of the alloy. When it is continuously rolled at 520°C, with increasing the deformation degree, the recrystallization is easy to occur, but the high temperature leads to coarser recrystallized grains.

The tensile mechanical properties of H60, H76I, and H76II alloys after stress-relief annealing at 400°C for 0.5 h are evaluated, as shown in Fig. 6.21. The tensile strength and plasticity of the H60 alloy are relatively poor with UTS of 297 MPa and EL of only 1.0%. However, its yield strength is up to 260 MPa, which is not much different from the extruded sheet alloy with 275 MPa in the previous chapter. Comparing the microstructure of these two alloys, their recrystallized grain sizes are almost the same, but the streamline of the layered structure on the longitudinal section for the H60 alloy is weak, indicating the deformation process of the H60 alloy is still weak. In addition, a large number of layered structure deformation leads to the poor plasticity, which indirectly determines its low tensile strength. Due to the presence of microcracks in H76I alloy, the brittle fracture occurs directly during the stretching. In contrast, the H76II alloy presents better comprehensive mechanical properties, with UTS of 358 MPa and EL of 6.6%. This is mainly because compared with the H60 alloy, the recrystallized grains in H76II alloy are increased, and its layered microstructure is decreased, significantly improving its plasticity and finally resulting in the good comprehensive mechanical properties. However, the yield strength of the H76II alloy is merely increased by 8 MPa. After further rolling deformation, although a large amount of fine recrystallization is produced and the fine grain strengthening is enhanced, the deformation microstructure of the alloy containing the fine layered LPSO phase is reduced. Due to the weakening of the work

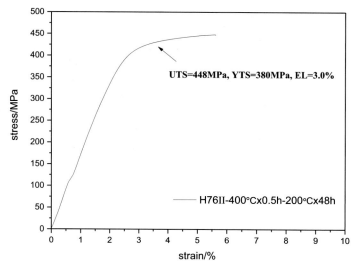

Figure 6.22 Mechanical property of aged H76II alloy.

hardening effect and strengthening effect of the layered phase, the yield strength of the alloy has little increase. In comparison with the extruded sheet alloy, the mechanical properties of the H76II alloy are still poor, mainly due to the relatively large recrystallized grain size in the H76II alloy and the weaked strengthening effect of layered phase.

Further, the H76II alloy with good comprehensive property is subjected to aging treatment at 200°C for 48 h, and the mechanical properties are shown in Fig. 6.22. After aging treatment, the strength of the alloy is significantly improved. The tensile strength is increased from 358 to 448 MPa, increased by 90 MPa. The yield strength is increased from 268 to 380 MPa, with an increment of 112 MPa. The aging treatment effect is similar to that of the extruded sheet alloy, indicating the rolled alloy also has significant aging strengthening effect.

6.2.3 "Rolling+solid solution+rolling" process of as-cast Mg−Gd−Y−Zn−Mn alloy and its microstructure and mechanical properties

The rolling experiments on the homogenized annealed alloys indicate that the presence of the layered LPSO phase restricts the recrystallization of the alloy. It is easy to cause alloy to crack when the degree of the rolling deformation is large. The alloy can only be rolled at a relatively high temperature. This section applies the process of "rolling+solid solution+rolling" on the as-cast alloy to control the LPSO phase in the alloy to avoid the influence of layered LPSO phase on the rolling deformation. For the convenience of discussion, the rolled alloys are numbered, as shown in Table 6.6.

Table 6.6 Number of the as-cast rolled alloy.

Sample number	Process
C60	The as-cast alloy is rolled to with a reduction of 60% at 520°C
C76I	C60 → 500°C × 4 h rolling at 450°C accumulated reduction of 76%
C76II	C60 → 500°C × 4 h → rolling at 520°C accumulated reduction of 76%

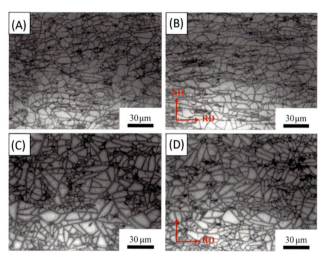

Figure 6.23 Metallographic images of the microstructure for as-cast rolled alloys before and after solid solution treatment. (A) and (B) the C60 and (C) and (D) the C60—500°C × 4 h.

Fig. 6.23 shows the metallographic microstructure of C60 alloy before and after solid solution treatment at 500°C for 4 h. The grain microstructure of the C60 alloy is irregular polygonal, and no elongated grain microstructure is found on the longitudinal section. This indicates that the as-cast alloy undergoes relatively complete recrystallization during the rolling process. In addition, it is noted that the grain size of the alloy is not uniform, with grain size ranging from 4 to 16 μm, which is due to the ununiform composition of the alloy without homogenization treatment. After the solution treatment at 500°C for 4 h, some grains in the alloy grow obviously.

XRD analysis is performed on the C60 alloy before and after solid solution treatment, and the results are shown in Fig. 6.24. In comparison, the diffraction peak of the eutectic phase $(Mg,Zn)_3(Gd,Y)$ is weakened after rolling 60% at 520°C, but it does not disappear completely. Therefore the eutectic phase in the as-cast alloy can be dissolved when rolling at a high temperature of 520°C. After the solid solution treatment of C60 alloy at 500°C for 4 h, the diffraction peak of the eutectic phase completely disappears. From the XRD pattern, the phase constitute of C60—500°C × 4 h alloy is α-Mg matrix and LPSO phase.

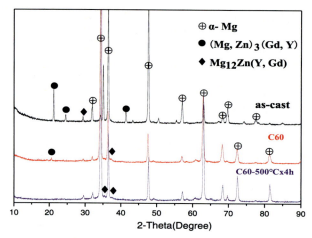

Figure 6.24 XRD patterns of the as-cast alloys before and after rolling process.

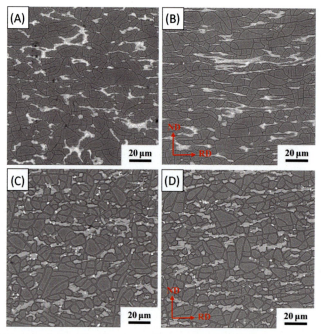

Figure 6.25 SEM images of the microstructures for as-cast rolled alloys before and after solid solution treatment. (A) and (B) C60 and (C) and (D) C60−500°C × 4 h.

The backscattered electron imaging is performed on the microstructure of the C60 alloy before and after solid solution treatment, and the morphology and distribution of the second phase are analyzed. As shown in Fig. 6.25, from the cross section, the second phase in the C60 alloy is distributed on grain boundaries in the form of a semicontinuous network. The size of the second phase with irregular block shapes is uneven. From the longitudinal section, the second phase is distributed along the rolling direction. However, in the as-cast alloy before rolling, the second phase is distributed in the form of network, and the distribution has no obvious orientation. Therefore the network-like second phase is interrupted and distributed along the rolling direction under the rolling force. After the solid solution treatment at 500°C for 4 h, the second phase in the alloy mainly presents the morphology of block. In comparison, the second phase is still distributed at the grain boundaries after solid solution treatment, but the distribution becomes more dispersed, and the size of the second phase is relatively uniform. In addition, in the longitudinal sectional microstructure, some bulk LPSO phases are discontinuously distributed in the form of band along the rolling direction in the alloy after solid solution treatment, but this trend is not obvious.

The second phase in the as-cast rolled alloy is further analyzed by EDS, and the results are shown in Fig. 6.26 and Table 6.7. There are two contrasting second phases in the C60 alloy, namely, white phase and gray phase. Combined with the EDS composition test and previous XRD results, it can be inferred that the white phase is eutectic phase and the gray phase is LPSO phase. It is noted that the white eutectic phase is mainly distributed in the relatively coarse bulk phase, which is mainly the gray LPSO phase. The coarse bulk phase in the as-cast alloy is mainly the white eutectic

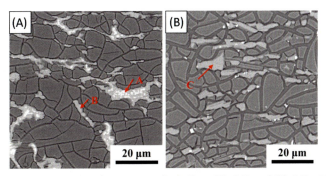

Figure 6.26 EDS detection point in the as-cast rolled alloys (A) C60 and (B) C60−500°C × 4 h.

Table 6.7 EDS results of the as-cast rolled alloy.

Alloy state	Detection point	Mg (at%)	Zn (at%)	Y (at%)	Gd (at%)	Phase composition
C60	A	81.24	1.77	5.92	11.08	(Mg,Zn)₃RE eutectic phase
	B	86.03	5.43	4.62	3.93	LPSO phase
C60−500°C × 4 h	C	87.14	5.81	4.27	2.78	LPSO phase

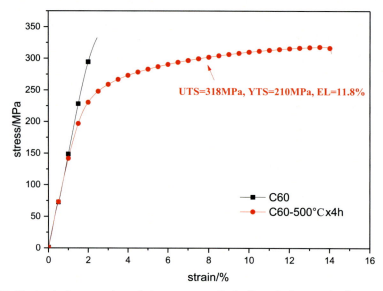

Figure 6.27 Mechanical properties of the as-cast rolled alloys before and after solid solution treatment.

phase. It can be inferred that the atoms diffuse in the eutectic phase of the as-cast alloy during the rolling process at 520°C, and it gradually transforms to the LPSO phase when the micro-area composition reaches a certain condition. After solid solution treatment of C60 alloy at 500°C for 4 h, the remaining white phase disappears in the alloy, and the main second phase is gray phase. Similarly, the gray phase can be inferred as LPSO phase according to the EDS and XRD results.

Fig. 6.27 presents the mechanical properties of the C60 alloy before and after solid solution treatment. The C60 alloy suffers direct brittle fracture during the tensile process. The plasticity of the alloy is significantly improved after solid solution treatment at 500°C for 4 h. According to the analysis, after rolling process of C60 alloy, because of the faster cooling rate, the internal stress in the alloy is relatively large, and the work hardening effect is great. Besides, the second phase is distributed in the form of semicontinuous network and its distribution is not uniform, resulting in the easy fracture of the alloy during tensile test. After the solid solution treatment, because of the weakening of the work hardening effect and the slight growth of grains, the strength of the alloy is slightly decreased, and the plasticity is significantly improved. In comparison with the H60−400°C × 0.5 h alloy described in the previous section, the yield strength of the C60−500°C × 4 h alloy is lower, but its plasticity is higher, resulting in a higher tensile strength. Overall, the C60−500°C × 4 h alloy has good comprehensive mechanical properties, and the relatively high plasticity is conductive to further rolling.

Figure 6.28 Optical photographs of the samples of C76I and C76II.

Similar to the process in the previous section, after cutting off the edge crack parts of the C60 alloy, continuous rolling is performed at 450°C and 520°C. Fig. 6.28 shows the optical photographs of the C76I and C76II samples obtained by continuous rolling. Overall, after further rolling, both ends of the C76I and C76II alloys are fractured to different degrees. In comparison, the C76I alloy shows more edge cracks with size of 2—3 μm, while the C76II alloy has few edge cracks. Therefore after rolling at the high temperature of 520°C, the deformation ability is improved, and it is not easy to occur brittle fracture. Compared with the homogenized annealed rolled alloy in the previous section, the C60—500°C × 4 h alloy has less cracking after further rolling, indicating its rolling deformation ability is improved. It is believed that the C60—500°C × 4 h alloy has the relatively high plasticity and low yield strength, namely, the low resistance to deformation, which is beneficial to rolling deformation. On the other hand, it is relatively easy to occur recrystallization softening, which is beneficial to improving the plasticity of the alloy. It is conductive to rolling with high degree of deformation.

Fig. 6.29 shows the metallographic images of the microstructures of C76I and C76II alloys after stress-relief annealing at 400°C for 0.5 h. In comparison with the C60—500°C × 4 h before continuous rolling, the microstructures of the C76I and C76II alloys are significantly refined. From the longitudinal sectional grain microstructure, no obvious processing streamline is observed. Among them, in addition to the equiaxed recrystallization grains, the C76I alloy continuously rolled at 450°C has more strip-shaped grains. Based on the research, the Mg—RE—Zn alloy after solid solution treatment can precipitate the layered LPSO phase during the solution cooling

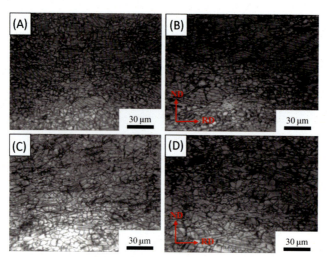

Figure 6.29 Metallographic images of microstructures of C76I and C76II alloys after stress-relief annealing. (A) and (B) C76Iand (C) and (D) C76II.

process. On the one hand, a small amount of layered LPSO phase has been precipitated during the solid solution treatment. On the other hand, the layered LPSO phase has also been precipitated in the alloy during rolling at 450°C and annealing process. The layered LPSO phase hinders the deformation of the grains and restricts the recrystallization of the alloy, finally forming the strip-like grain microstructure. The C76II alloy continuously rolled at 520°C is mainly composed of equiaxed grains, and only a small amount of strip-like grains is observed. The RE and Zn elements in the alloy have high solid solubility at 520°C. It is not easy to precipitate layered LPSO phase. By rolling at 520°C, it is easy for alloy to recrystallize and form equiaxed grains, so only a few strip-like grains appear. In comparison, the recrystallized grain size of the C76II alloy is relatively large, and there is phenomenon of abnormal growth of individual grains. Therefore too high rolling temperature is not conductive to the refinement and homogenization of the grain microstructure.

The microstructure of the C76I and C76II alloys after stress-relief annealing at 400°C for 0.5 h is observed by SEM, as shown in Fig. 6.30. There is mainly a second phase with gray contrast in the two alloys, namely, the LPSO phase. The size and distribution of the LPSO phase in the two alloys have no significant difference, indicating that the continuous rolling temperature has little effect on the second phase. In comparison with the alloy before continuous rolling, the size of the LPSO phase is not significantly changed. However, from the longitudinal section, the distribution of the LPSO phase along the rolling direction is enhanced. Compared with the previous homogenized annealed rolled alloys, the size of the LPSO phase is relatively small and the distribution is more uniform.

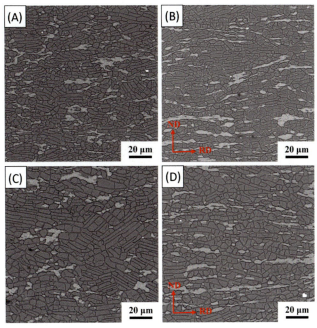

Figure 6.30 SEM images of the microstructures of (A) and (B) the C76I and (C) and (D) the C76II alloy after stress-relief annealing.

The mechanical properties of C76I and C76II alloys after stress–relief annealing at 400°C for 0.5 h are tested, with the results shown in Fig. 6.31. In comparison with C60 alloy, the strengths of C76I and C76II alloys have been improved, which are due to the obvious refinement of the alloy after continuous rolling. In comparison, although the C76I alloy has higher strength, it has lower plasticity. According to the analysis, the C76I alloy continuously rolled at 450°C has finer grains, and the layered LPSO phase is precipitated, beneficial to improving the strength of the alloy. However, it contains layered LPSO phase and more strip-like grains, which is not conductive to the deformation of the alloy, resulting in the poor plasticity. The C76II alloy continuously rolled at 520°C has higher comprehensive mechanical properties, and the elongation reaches 11.6%. Compared with the H76II alloy in the previous section, the C76II alloy has lower strength but higher plasticity. Comparing the structures of the two alloys, it is found that the grains of C76II alloy is slightly coarser, while the H76II alloy has higher layered LPSO phase strengthening effects, so the H76II alloy possesses higher strength. However, because the C76II alloy has more equiaxed grains, smaller sized and more uniformly distributed second phase, the deformation is easier to coordinate, and the plasticity is relatively high.

The C76I and C76II alloys were subjected to the aging treatment at 200°C for 48 h, and the mechanical properties of the alloy in aging state were tested and shown in Fig. 6.32. After aging, the strength of the two alloys has been greatly improved, the

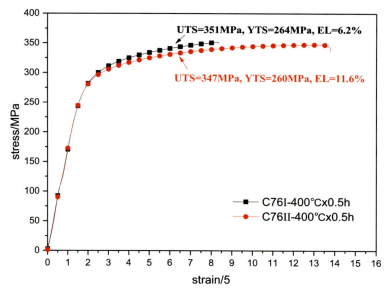

Figure 6.31 Mechanical properties of C76I and C76II alloys after stress-relief annealing.

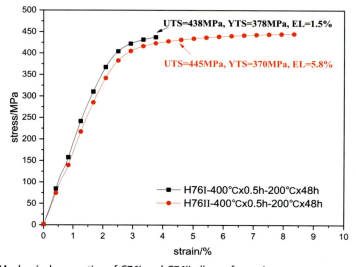

Figure 6.32 Mechanical properties of C76I and C76II alloys after aging treatment.

yield strength of the C76I and C76II alloy increased by 114 and 110 MPa respectively, while the tensile strength increase by 87 and 98 MPa, respectively. In comparison with the H76II alloy in the previous section, the increase in the strength of the C76II alloy after aging treatment is similar, while the tensile strength of the C76I alloy is not ideal due to the large plastic loss after aging treatment.

6.3 High-plasticity mechanism of Mg−Gd−Y−Zn−Mn alloy

6.3.1 Effect of long-period stacking order structure phase

Mg−Gd−Y−Zn−Mn alloy has two types of long-period stacking ordered structure phases of 18R and 14H [1,2,3], and the atomic stacking sequence of 18R LPSO phase is ACACBABABACBCBCBACA, and the atomic stacking sequence of 14H LPSO phase is ABABCACACACBABA. The Zn and Y atoms are distributed on a specific close packed surface in the ordered manner. The overall structure of the LPSO phase is essentially a multilayer fault structure consisting of basic faults [4].

At present, it is generally believed that the deformation of LPSO phase is mainly carried out in the form of kink deformation. Hagihara et al. [5] observed the existence of kink bands in Mg−Zn−Y alloy, and believed that the basal plane slip of $(0001) < 11\bar{2}0 >$ is the most important way of dislocation slip in the LPSO phase (Fig. 6.33).

Fig. 6.34 shows the metallographic images of as-cast Mg−Gd−Y−Zn−Mn alloys with different Gd and Y contents. The four alloys exhibit the dendritic structure uniformly distributed in the matrix. The GWZM1 alloy with the maximum Y content and minimum Gd content possesses the highest volume fraction of the second phase. However, in the GWZM2 alloy, as the increase of the Gd content, the size of the second phase is slightly increased, presenting a bulk shape. With further increasing the Gd content while decreasing Y content, the second phase becomes fine and dispersed again. The GWZM4 alloy possesses the smallest second phase.

Fig. 6.35 presents the XRD patterns of as-cast Mg−Gd−Y−Zn−Mn alloys with different Gd and Y contents. When the Gd content is low and the Y content is high (GWZM1 and GWZM2), the main phase of the alloy is α-Mg phase, $Mg_{12}YZn$ phase (LPSO phase), and $Mg_{24}(Gd,Y,Zn)_5$ phase (eutectic phase). With the increase of Gd content and decrease of Y content, the $(Mg,Zn)_3(Gd,Y)$ phase (W phase) appears in GWZM3 and GWZM4 alloys, but the peak of the eutectic phase begins to

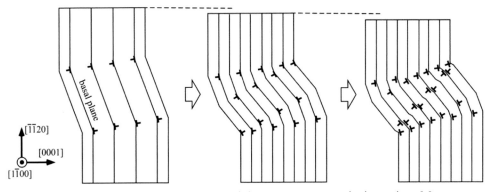

Figure 6.33 Kink deformation zone formed by dislocation motion on the base plane [5].

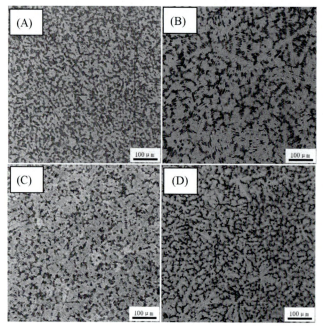

Figure 6.34 Metallographic images of the as-cast Mg−Gd−Y−Zn−Mn alloys with different Gd and Y contents, (A) GWZM1, (B) GWZM2, (C) GWZM3, and (D) GWZM4.

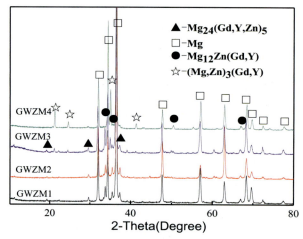

Figure 6.35 XRD patterns of as-cast Mg−Gd−Y−Zn−Mn alloys with different Gd and Y contents.

decrease. In the GWZM4 alloy, no peak of eutectic phase is observed, indicating there is no eutectic phase.

Fig. 6.36 shows the SEM images of as-cast Mg−Gd−Y−Zn−Mn alloys with different Gd and Y contents. In the four alloys, the second phase is dispersedly distributed in

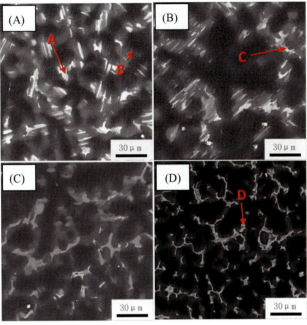

Figure 6.36 SEM images of as-cast Mg–Gd–Y–Zn–Mn alloys with different Gd and Y contents, (A) GWZM1, (B) GWZM2, (C) GWZM3, and (D) GWZM4.

the matrix, where the morphology of second phase in the two alloys with high Y and low Gd (GWZM1 and GWZM2 alloys) is different from that in the alloys with high Gd and low Y (GWZM3 and GWZM4 alloys). In GWZM1 and GWZM2 alloys, the second phase mainly has two morphologies, one is the white bulk phase and the other is the gray bulk phase. Some of these white phases are distributed in blocks around the gray bulk phase. Some are directly attached to the gray phase and distributed in the form of stripes. According to the EDS results, combined with the XRD patterns, there is no Zn element in the white bulk phase (point A), which is eutectic phase of $Mg_5(Gd,Y)$, while the ratio of Mg to RE in the gray phase (point C) is close to 11, which is LPSO phase. In the GWZM3 alloy, in addition to the white bulk and gray bulk phases, a lot of small white bright skeleton-like phases also appear, and most of these white bright phases are attached around the original white bulk phase. The number of these white bright skeleton-like phases is further increased in GWZM4 alloy, and the white and gray bulk phases cannot be observed in the alloy. The EDS spectrum results show that there are a high content of RE and Zn elements in the phase (point D). Combined with the XRD pattern, this white bright skeleton-like phase is $(Mg,Zn)_3(Gd,Y)$ phase (W phase). According to the SEM and XRD results, with decreasing Y content and increasing Gd content, the content of LPSO phase in the as-cast Mg–Gd–Y–Zn–Mn alloy decreases while the number of skeleton-like W phase increases.

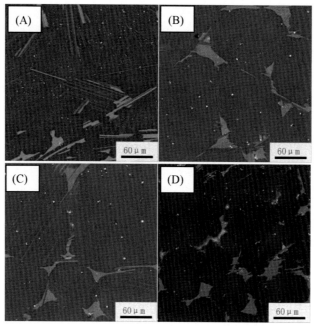

Figure 6.37 SEM images of Mg–Gd–Y–Zn–Mn alloys with different Gd and Y contents after furnace cooling annealing at 540°C for 4 h, (A) GWZM1, (B) GWZM2, (C) GWZM3, and (D) GWZM4.

To eliminate the eutectic phase and the hard and brittle W phase in the as-cast alloy, and make them completely transform into the LPSO phase, the four alloys were treated by furnace cooling after annealing at 540°C for 4 h. Fig. 6.37 shows the SEM images of the four alloys after heat treatment. Only the gray bulk phase at the grain boundaries and the layered phase in the grains exist in the alloys after heat treatment, and the layered phases are parallel to each other. Both the eutectic phase and the W phase disappear in the as-cast alloy.

For the Mg–Gd–Y–Zn–Mn alloy, it was first insulated at 540°C, so that the atoms of Gd, Y, and Zn in the $Mg_{24}(Gd,Y,Zn)_5$ and $(Mg,Zn)_3(Gd,Y)$ phases can diffuse into the matrix. However, because the LPSO phase is a high-temperature stable phase, the atom diffusion does not occur at this temperature. Therefore only LPSO phase exists at the grain boundary after insulation at 540°C. Afterwards, during the high-temperature annealing process, the Gd, Y, and Zn atoms that have originally solid dissolved in the matrix react with the Mg matrix to regenerate the LPSO phase of $Mg_{12}Zn(Y, Gd)$. This regenerated LPSO phase usually grows into the grain interiors from the front of grain boundaries and penetrates the entire grain, appearing with a morphology of the layered phase.

Fig. 6.38 presents images of the microstructure of four heat-treated alloys extruded at 450°C with an extrusion ratio of 11. In the alloys with high Y and low Gd

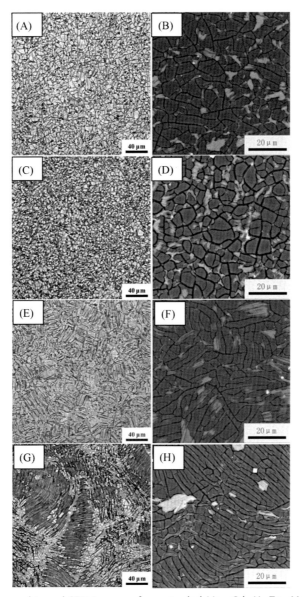

Figure 6.38 Metallographic and SEM images of as-extruded Mg—Gd—Y—Zn—Mn alloys with different Gd and Y contents, (A) and (B) GWZM1, (C) and (D) GWZM2, (E) and (F) GWZM3, and (G) and (H) GWZM4.

(GWZM1 and GWZM2), the microstructure is mainly composed of equiaxed grains. The grain size of GWZM1 alloy is about 12 μm, while the grain size of the GWZM2 is about 8 μm. On the other hand, in the alloys with high Gd and low Y (GWZM3 and GWZM4), after extrusion, obvious dynamic recrystallization occurs in the alloys, but it presents two kinds of microstructures with different morphologies, one is equiaxed grains with uniform size, and the grain size is about 10 μm, and the other is the twisted and elongated microstructure arranging in parallel with the width of about 2−4 μm, belonging to an un-recrystallized zone. In the GWZM4 alloy, the main microstructure is twisted morphology, and there are few recrystallized structures. From the SEM images of the four as-extruded alloys, the gray bulk LPSO phase in heat-treat state has been dispersedly distributed in the alloys after extrusion treatment. With increasing the Gd content and decreasing the Y content, the size of the bulk LPSO phase in the alloy after extrusion is slightly increased.

Fig. 6.39 shows the mechanical properties of four as-extruded Mg−Gd−Y−Zn−Mn alloys with different Gd and Y contents, and the corresponding results are presented in Table 6.8. With increasing the Gd content and decreasing the Y content, the mechanical properties of four alloys present a monotonous increasing trend, but the elongation is slightly decreased. The strength of GWZM1 and GWZM2 alloy are significantly lower than that of GWZM3 and GWZM4 alloys. The tensile strength of the GWZM4 alloy is 405 MPa, and its elongation is 10%. While the comprehensive mechanical properties of the GWZM3 alloy is the best, with a tensile strength of 400 MPa and elongation of 12.6%.

As discussed above, after extrusion, only bulk LPSO phase exists in GWZM1 and GWZM2 alloy, while the twisted microstructure can be observed in GWZM3 and GWZM4 alloy. On the one hand, the existence of the twisted microstructure can

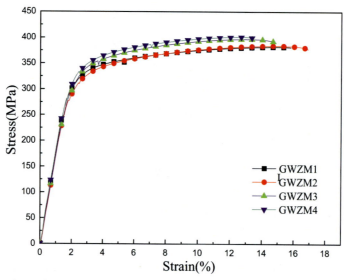

Figure 6.39 Mechanical property curve of as-extruded Mg−Gd−Y−Zn−Mn alloys with different Gd and Y contents.

Table 6.8 Mechanical property of as-extruded Mg−Gd−Y−Zn−Ni−Mn alloys after furnace cooling at 540°C for 4 h.

	Tensile properties		
Sample	σ_{UTS} (MPa)	σ_{TYS} (MPa)	EL (%)
GWZM	381	288	13.4
GWZN1	382	290	14.5
GWZN2	398	295	12.6
GWZN3	402	298	10.8

restrict the dislocation motions, increasing the strength of the alloy; on the other hand, it reduces the volume fraction of the recrystallized grains in the alloy and the elongation of the alloy. For the dispersedly distributed fine bulk LPSO phase after extrusion, it can become the nucleation site of recrystallized grains. After the formation of the recrystallized grains, these LPSO phases are dispersedly distributed at the grain boundary, finally increasing the plasticity of the alloy.

In the Mg−Gd−Y−Zn−Mn alloy, the equiaxed grains can improve the plasticity of the alloy, and the twisted and elongated microstructure can increase the strength of the alloy. However, both the equiaxed grains and the deformed structure are determined by the microstructure before extrusion and the followed heat treatment. When the content of the layered phase is high, the twisted microstructure in the alloy after extrusion increases, and it is not easy to recrystallize.

6.3.2 Combined effects of LPSO phase and precipitation phase

The LPSO phase and precipitation hardening phase influence and restrict each other in the alloy. After a certain deformation and heat treatment process, the types and morphologies of the phases are much more complicated. The actual toughening effect is ascribed to the combined effects of the morphology, size, quantity, phase, and distribution of these phases. It is also affected by the factors of alloying, dislocation configuration, twining, etc., so the research is more complicated.

Fig. 6.40 shows the mechanical properties of Mg−8.3Gd−4.2Y−1.4Zn−1.1Mn alloys under different aging time at 200°C, and the corresponding results are shown in Table 6.9. The E+200x100h alloy shows the highest mechanical properties with UTS of 538 MPa, TYS of 390 MPa, and EL of 13.1% [6]. After aging, the strength of the alloy is significantly increased. Although the plasticity of the aging alloy is greatly decreased compared to that of the extruded alloy, it is found that during the aging process, with increasing aging time, both the strength and plasticity of the alloy increase. The traditional strengthening and toughening mechanism reveals that while the precipitation strengthening improves the strength of the alloy, its plasticity would be decreased, and only fine grain strengthening can increase both the strength and the

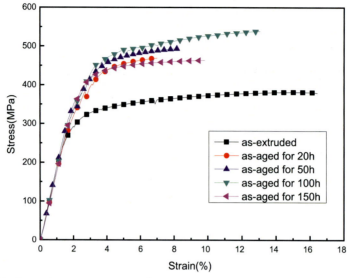

Figure 6.40 Mechanical property curves of Mg−8.3Gd−4.2Y−1.4Zn−1.1Mn alloys with different aging time at 200°C.

Table 6.9 Mechanical properties of Mg−8.3Gd−4.2Y−1.4Zn−1.1Mn alloys with different aging time at 200°C.

	Tensile properties		
Sample	σ_{UTS} (MPa)	σ_{TYS} (MPa)	EL (%)
As-extruded	388	282	14.3
E+200x20h	470	375	4.6
E+200x50h	495	385	5.9
E+200x100h	538	390	10.0
E+200x150h	465	360	8.1

plasticity. However, from the aging heat treatment experiment, the gain size of the alloy does not change obviously. The grain size in both the extrusion state and aging state is about 5 μm, which seems to contradict the traditional strengthening and toughening mechanism.

The Mg−Gd−Y−Zn−Mn alloy has the most excellent mechanical properties after aging at 200°C for 100 h, which is mainly attributed to the presence of LPSO phase, precipitation hardening phase, and a large amount of stacking fault in the alloy. To clarify the specific role of LPSO phase, precipitation hardening, and stacking fault in the alloy during the deformation process, this section selects the sample after aging for 100 h, and observes the morphology near the fracture using TEM, with the result

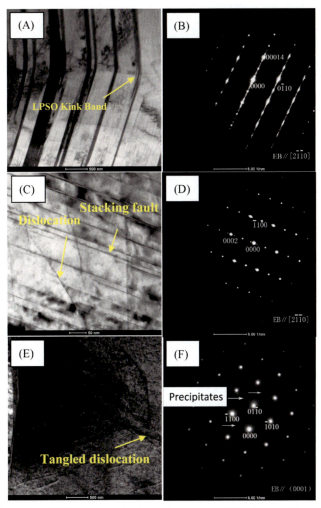

Figure 6.41 TEM bright-field image and selected electron diffraction of the second phase and stacking fault in the alloy during the deformation process. (A) and (B) LPSO phase, (C) and(D) stacking fault, and (E) and (F) precipitation hardening phase.

shown in Fig. 6.41. Overall, these three micro–phases have their own deformation methods, which are coordinated and promoted each other.

6.3.3 Effect of Mn element

As mentioned above, Mn element has the effect of "solid solution strengthening plasticization." The role of Mn is generally reflected in two aspects of solid solution and precipitation. Its grain refinement effect is mainly reflected in the process of hot deformation, but the effect of refining the as-cast alloy is poor. Normally, the formation of

X phase requires RE/Zn ≥ 1.33 in the local area, while the formation of W phase requires 0.32 ≤ RE/Zn ≤ 1.33 [7]. This also indicates in the alloy containing Zr, the Zn content increases and RE content is lower in the local area. The appearance of this phenomenon may be related to the activity of Zr. Zr can react with element of RE (this phenomenon does not appear in Mn-containing alloy), forming the compound of Zr and precipitating, which decreases the RE yield and makes the content of RE in the local area of the alloy is low, so that the W phase appears. The effect of the solid solution of Mn has to be effectively combined with controlling the second phases such as LPSO phase.

From Fig. 6.42, after 4 h of solution treatment at 540°C and furnace cooling, it can be seen that the structure and morphology are not significantly different from those of the Mg−Gd−Y−Zn alloy. Both the fine form-like and network-like second phases in the as-cast alloy are eliminated. The bulk phase at the grain boundary grows up, while the fine layered second phase appears inside grains. After heat treatment of Mg−Gd−Y−Zn−Zr, the gray LPSO phase is mainly rod-shaped. In the Mg−Gd−Y−Zn−Zr alloy, in addition to the bulk LPSO phase, there appears a large amount of white particles, where the contents of Zr and RE are high. It indicates that the addition of Zr can promote the separation of RE element, which is not conductive to the complete formation of LPSO phase in the alloy.

Fig. 6.43 shows the SEM images of as-extruded Mg−Gd−Y−Zn−Zr/Mn alloy. From Fig. 6.43, the distribution and morphology of the second phase in the alloy and its grain size are changed after the hot extrusion. The complete recrystallization has occurred. The grain size of the alloy is about 5 μm. There is no obvious difference in the grain size between the alloys. However, in comparison with the original Mg−Gd−Y−Zn alloy, it has been refined to a certain degree. By further observing the distribution of the second phase in the alloy, the remaining gray bulk phases after heat treatment have been crushed by extrusion and dispersed at the grain boundaries. The second phase is the most broken after adding Mn element, and its distribution is

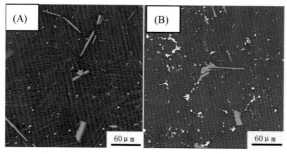

Figure 6.42 SEM images of heat-treated (A) Mg−Gd−Y−Zn−Mn and (B) Mg−Gd−Y−Zn−Zr alloys.

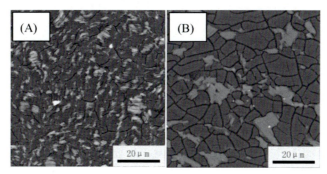

Figure 6.43 SEM images of as-extruded (A) Mg−Gd−Y−Zn−Mn and (B) Mg−Gd−Y−Zn−Zr alloys.

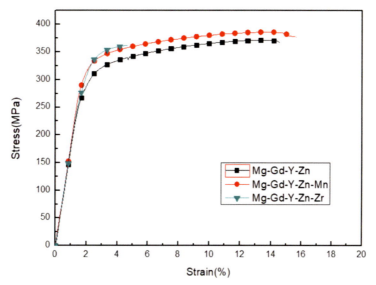

Figure 6.44 Mechanical property curves of as-extruded Mg−Gd−Y−Zn−Zr/Mn alloys at room temperature.

the most dispersed. After the addition of Zr, the size of the second phase is larger than that in the alloy without the addition and with the addition of Mn, and it is still distributed in the form of bulk.

According to Fig. 6.44, the tensile strength of the as-extruded alloy is not very different from that of the original Mg−Gd−Y−Zn alloy. However, after the addition of Zr, the plasticity of the alloy is significantly decreased which is attributed to the formation of intermetallic compound formed with Zr during smelting with Zn, reducing the metallurgical quality and subsequently the plasticity of the alloy. At the same time, when Zr element is added to the extruded alloy, the size of the bulk second phase at the grain boundary is also increased, leading to the low strength and plasticity of the alloy after extrusion. The tensile strength of the Mg−Gd−Y−Zn−Mn alloy reaches 378 MPa, and the elongation reaches 13%. In addition, compared with the Mg−Gd−Y−Zn alloy, the addition of Mn not only improves the strength, but also increases the plasticity, with obvious strengthening and toughening effects.

6.4 Summary

About the production and application of high-strength Mg−RE−Zn alloy, this chapter mainly introduces the extrusion and rolling deformation of Mg−Gd−Y−Zn−Mn alloy to provide a reference for plastic deformation of the high-strength Mg−RE−Zn alloy containing Mn. The main conclusions are shown as follows:

1. As the homogenized annealed alloy possesses a large number of layered LPSO phases, recrystallization hardly occurs in the alloy when the extrusion ratio is not higher than 11. It is mainly composed of the layered deformed microstructure. With the increase in extrusion ratio, the recrystallized microstructure increases. When the extrusion ratio is increased to 42, the alloy is mainly composed of equiaxed microstructure. With the increase in the extrusion ratio, both the yield strength and plasticity of the as-extruded alloy increase, but there is no significant difference in tensile strength. After aging treatment, the tensile strength of each alloy is different. The aged bar alloy with an extrusion ratio of 11 possesses the highest mechanical properties with UTS of 502 MPa, YTS of 410 MPa, and EL of 3.8%. The aged bar alloy with an extrusion ratio of 42 possesses higher comprehensive mechanical properties with UTS of 484 MPa, YTS of 390 MPa, and EL of 5.0%. The sheet alloy with an extrusion ratio of 11 is further produced, which is also mainly composed of layered deformed microstructure. After aging treatment, its mechanical properties are good with UTS of 477 MPa, YTS of 390 MPa, and EL of 4.4%.

2. As for the homogenized annealed alloy with a large number of layered LPSO phases, recrystallization hardly occurs during rolling at 450°C, and it is easy to occur fracture. When the rolling temperature is increased to 520°C, it is relatively easy to occur recrystallization. After multiple rolling at 520°C, until the reduction is 76%, a large number of equiaxed grains appear in the alloy. Its mechanical properties are improved greatly. The UTS, TYS and EL of the alloy are 448, 380 MPa and 3.0%, respectively.

3. After "rolling (reduction of 60%) and solid solution treatment" for the as-cast alloy, the alloy is recrystallized and refined. Only the bulk LPSO phase is precipitated. The phase size is relatively small and the distribution is more uniform. Similarly, continuously rolling at 520°C to the reduction of 76%, the cracking of the alloy is weak. The alloy is mainly composed of equiaxed microstructure with UTS of 445 MPa, YTS of 370 MPa, and EL of 5.8%. In comparison with the process of

"homogeneous annealing+rolling," the process of "rolling+solid solution+rolling" is much easier to perform rolling with large deformation, so that it is expected to obtain uniform and fine microstructure and produce high-performance rolled sheet.

4. When the Mg—Gd—Y—Zn—Mn alloy is aged at 200°C for 20 h—100 h, with increasing the aging time, both the strength and plasticity of the alloy are improved. After aging for 100 h, the strength of the alloy reaches 538 MPa, the elongation is about 10%, and the comprehensive mechanical properties are excellent.

References

[1] Wang JF, et al. High-strength and good-ductility Mg-RE-Zn-Mn magnesium alloy with long-period stacking ordered phase. Materials Letters 2013;93:415—18.

[2] Wang K, et al. Enhanced mechanical properties of Mg—Gd—Y—Zn—Mn alloy by tailoring the morphology of long period stacking ordered phase. Materials Science and Engineering: A 2018;733:267—75.

[3] Wang J, et al. Enhanced strength and ductility of Mg—RE—Zn alloy simultaneously by trace Ag addition. Materials Science and Engineering: A 2018;728:10—19.

[4] Fan TW, et al. First-principles study of long-period stacking ordered-like multi-stacking fault structures in pure magnesium. Scripta Materialia 2011;64(10):942—5.

[5] Hagihara K, Yokotani N, Umakoshi Y. Plastic deformation behavior of $Mg_{12}YZn$ with 18R long-period stacking ordered structure. Intermetallics 2010;18(2):267—76.

[6] Liu S, et al. Ageing behavior and mechanisms of strengthening and toughening of ultrahigh-strength Mg—Gd—Y—Zn—Mn alloy. Materials Science and Engineering: A 2019;758:96—8.

[7] Luo SQ, et al. Effect of mole ratio of Y to Zn on phase constituent of Mg-Zn-Zr-Y alloys. Transactions of Nonferrous Metals Society of China 2011;21(4):795—800.

[8] Xu SW, Zheng MY, Kamado S, et al. Dynamic microstructural changes during hot extrusion and mechanical properties of a Mg—5.0Zn—0.9Y—0.16Zr (wt%) alloy. Materials Science and Engineering: A 2011;528:4055—67.

Index